68/69 Advances in Polymer Science

Fortschritte der Hochpolymeren-Forschung

Cationic Ring-Opening Polymerization

2. Synthetic Applications

By S. Penczek, P. Kubisa, and K. Matyjaszewski

Preface by J. P. Kennedy

With 40 Figures and 71 Tables

Springer-Verlag
Berlin Heidelberg New York Tokyo

Prof. Dr. S. Penczek, Dr. P. Kubisa, Dr. K. Matyjaszewski

Polish Academy of Sciences, Centre of Molecular
and Macromolecular Studies, Boczna 5, PL-90-362 Łódź, Poland

Part I is published in Vol. 37 of this series:
Cationic Ring-Opening Polymerization of Heterocyclic
Monomers: Mechanisms

ISBN-3-540-13781-5 Springer-Verlag Berlin Heidelberg New York Tokyo
ISBN-0-387-13781-5 Springer-Verlag New York Heidelberg Berlin Tokyo.

Library of Congress Catalog Card Number 61-642

Printed in GDR

Typesetting and Offsetprinting: Th. Müntzer, Bad Langensalza;
Bookbinding: Lüderitz & Bauer, Berlin
2154/3020–543210

Editors

In Memoriam

Professor Dr. Werner Kern, one of the pioneers of polymer chemistry and particularly of cationic ring-opening polymerization, died on January 18, 1985 at the age of 79.

Already in the late 1920s, while working with the Nobel Prize winner Herman Staudinger, Professor Kern contributed to the ionic polymerization of carbonyl and cyclic monomers. This contribution was a cornerstone in the development of polymer science that gave a basis for the contemporary treating of polymerization products as consisting of long chains and not merely of aggregates of low-molecular-weight compounds as was believed earlier.

Professor Kern was forced to leave academia in the 1930s but shortly after World War II started to build the Institute of Organic Chemistry at the University of Mainz with a large section devoted to polymer chemistry. This institute has become one of the world's leading centers of polymer chemistry; the personal achievements of Professor Kern in many areas of polymer chemistry, as e.g. polyelectrolytes, redox polymerization, ring-operning polymerization, polymer end groups, are examples of excellence guiding the next generation of chemists.

The close contact of Professor Kern with these authors was established in the 1960s: his visit to Poland, and particularly to Łódź, and our visits to Mainz formed a link that is still vital.

As one of the editors of "Advances in Polymer Science", originally Professor Kern invited the senior author of this volume a long time ago to write a review on cationic ring-opening polymerization. I regret that this has been accomplished only after Professor Kern's untimely death.

Stanisław Penczek

* 9. 2. 1906 † 18. 1. 1985

Editorial

With the publication of Vol. 51 the editors and the publisher would like to take this opportunity to thank authors and readers for their collaboration and their efforts to meet the scientific requirements of this series. We appreciate the concern of our authors for the progress of "Advances in Polymer Science" and we also welcome the advice and critical comments of our readers.

With the publication of Vol. 51 we would also like to refer to a editorial policy: *this series publishes invited, critical review articles of new developments in all areas of polymer science in English (authors may naturally also include workes of their own)*. The responsible editor, that means the editor who has invited the author, discusses the scope of the review with the author on the basis of a tentative outline which the author is asked to provide. The author and editor are responsible for the scientific quality of the contribution.

Manuscripts must be submitted in content, language, and form satisfactory to Springer-Verlag. Figures and formulas should be reproducible. To meet the convenience of our readers, the publisher will include "volume index" which characterizes the content of the volume.

The editors and the publisher will make all efforts to publish the manuscripts as rapidly as possible, i.e., at the maximum six months after the submission of an accepted paper. Contributions from diverse areas of polymer science must occasionally be united in one volume. In such cases a "volume index" cannot meet all expectations, but will nevertheless provide more information than a mere volume number.

Starting with Vol. 51, each volume will contain a subject index.

Editors　　　　Publisher

Preface

The scientific community will be grateful to Prof. S. Penczek and his coworkers Doz. P. Kubisa and Dr. K. Matyjaszewski for this volume on Cationic Ring-Opening Polymerization: Synthetic Applications. This volume — together with the previously published first part on Cationic Ring-Opening Polymerization of Heterocyclic Monomers (Adv. Polym. Sci., *37*, 1–156 (1980)) — represents a comprehensive review of the subject, indeed a definitive treatment of the field. This whole work is a truly comprehensive review and not a "telephone directory": in addition to detailed discussions of the greatest variety of cationic heterocyclic polymerizations it is chockfull of thoughtful criticism, analysis, and sometimes reanalysis of published information, with deep insight into explaining heretofore seemingly conflicting observations or conclusions. The authors' interpretation of "unusual" copolymerization data is particularly powerful and will be thoroughly studied and often quoted in the expert literature.

The authors combine synthesis thermodynamics and kinetics, and view their field in a holistic all-embracing manner. There is a constant interplay between polymer synthesis and mechanistic interpretation. They strive to generalize and to simplify, and helped by their deep understanding of fundamentals often succeed in deriving conclusions. When they are stymied they so state and call for additional experiments to fill the gaps in our knowledge. And they pinpoint these gaps.

The organization of the book follows a logical sequence: After a thorough presentation of basic thermodynamic principles and the Jacobson-Stockmayer cyclization theory, the authors discuss in depth all kinds of aspects of the various heterocyclic compound classes. In addition to detailed discussions of mechanisms, many other facets of heterocyclic polymerizations are treated, e.g., monomer synthesis, contemporary research trends, industrial significance. The treatise ends with an excellent up-to-date discussion of random, block and graft copolymerizations of heterocyclics.

The authors are superbly qualified to treat their subject and they have made significant contributions to the field. While the editor's assignment was not always easy, it was always a learning experience and a pleasure because he felt that in a small way he was contributing to an important scientific undertaking.

Akron, Ohio
April 1985

Joseph P. Kennedy

Table of Contents

1 Introduction

The present Volume II of Cationic Ring-Opening Polymerization has the subtitle "Synthetic Applications". In this volume we first intended to describe reliable synthetic methods which could be used in preparing polymers of desired degrees of polymerization and end-groups. The synthesis of well-defined polymers is facilitated in living systems. However, the number of heterocyclic monomers that provide systems close to living polymer is still quite limited in cationic ring-opening polymerizations (a few cyclic ethers, acetals, iminoethers, and amines). In many other systems one or more required features are not achieved and these polymerizations deviate more or less from living ones.

Thus, we decided to explain why living conditions cannot be attained for some monomers and we attempted to formulate requirements of achieving transfer- and terminationless processes. However, we soon realized that it is impossible to answer this question in a simple manner and that knowledge of thermodynamics (not treated in Vol. I — Adv. Polym. Sci. *37* (1980)) and mechanistic peculiarities of the given group of monomers is required. These considerations have led to the present form of Volume II.

Thus, we first discuss thermodynamics, paying attention to features that are important for polymer synthesis (e.g., dependence of equilibrium monomer concentration on polymerization variables); then we describe kinetics and thermodynamics of macrocyclization, trying to combine these two related problems, usually discussed separately. In particular we present the new theory of kinetic enhancement and kinetic reduction in macrocyclics. Thereafter, we describe the polymerization of several groups of monomers, namely: cyclic ethers (oxiranes, oxetanes, oxolanes, acetals, and bicyclic compounds) lactones, cyclic sulfides, cyclic amines, lactams, cyclic iminoethers, siloxanes, and cyclic phosphorus-containing compounds, in this order. We attempted to treat the chapters uniformly; we discuss practical methods of synthesis of the corresponding polymers (monomer syntheses and polymer properties are added), and conditions of reaching systems state and reasons of deviations. However, for various groups of monomers the quality of the available information differ so much, that this attempt of uniformity can not be fulfilled.

In a separate chapter we treat copolymerizations. The importance of reversibility is stressed and it is shown that a large number of what were assumed to be the valid reactivity ratios (r_1 and r_2), were determined by the oversimplified two parameter scheme (four reactions: two homo-propagations, and two cross-propagations). Reversibility of some steps requires a consideration of depropagation; the corresponding methods are described. The chemistry of alternating, block, and graft copolymers is also presented.

2 Thermodynamics of Cationic Ring-Opening Polymerization

A number of general reviews exist on the thermodynamics of vinyl and ring-opening polymerization and thus there is no need to repeat the derivations of the basic relationships. The interested reader is referred to the comprehensive and lucid reviews by Ivin [1,2], which include all the relevant information for understanding the phenomena discussed in this section.

First we describe the major symbols, then the methods for the determination of thermodynamic parameters relevant for cationic ring-opening polymerization, the factors affecting these parameters, and finally deviations of real systems from ideality. On this basis we discuss fundamental principles and peculiarities observed in the thermodynamics of cationic ring-opening polymerization, and, eventually we show how these principles are applied for preparation of some homo- and copolymers.

2.1 Terms and Symbols and Their Relationships

The free energy change due to polymerization is equal to the difference of the Gibbs free energies of the polymer segment and the monomer molecule:

$$\begin{aligned} \Delta G_p &= G_{pol} - G_{mon} = (H_{pol} - T \cdot S_{pol}) - (H_{mon} - T \cdot S_{mon}) \\ &= \Delta H_p - T \cdot \Delta S_p \end{aligned} \qquad (2\text{-}1)$$

Such a definition implies that the polymer is composed entirely of repeating units of one kind, while real linear macromolecules are terminated at both ends with structurally different units: end groups. For high-molecular weight polymers the contribution of these end groups is so small that it can be neglected. The large majority of polymers described in this chapter are also uniform in structure. Thus, from the thermodynamic point of view a polymerization is sufficiently well described by a single propagation reaction.

Both enthalpy and entropy of monomer (initial state) and polymer segment (final state) depend on the state of association; consequently, the thermodynamic parameters of polymerization have to be referred to these specific conditions denoted by subcripts. Thus, ΔG_{gg} corresponds to the free energy change during polymerization of gaseous monomer into gaseous (hypothetical) polymer. Examples of other notations are given in Table 2.1.

Table 2.1. Notations of initial and final states in polymerization

Monomer State	Polymer State	Notation
Gas	Gas (hypothetical)	gg
Gas	Condensed (liquid or amorphous)	gc
Liquid[a]	Solution in monomer	ls
Liquid	Condensed	lc
Liquid	Crystalline	lc′
Solution	Solution	ss
Solution	Condensed	sc
Solution	Crystalline	sc′

[a] i.e. bulk

When a polymerization is accompanied by phase transition, the overall thermodynamic parameters are the sum of parameters of the chemical reaction and phase transition (cf. p. 11). Thus, for instance, thermodynamics of polymerization in the crystalline state from a liquid monomer will be given by the thermodynamics of formation of the amorphous (condensed) polymer and polymer crystallization, provided, that polymerization proceeds in the solid state with monomer packing in the crystalline state simultaneously with propagation.

2.2 Relationship Between Thermodynamic Parameters and Polymerizability

Spontaneous polymerization requires $\Delta G_p < 0$. Ionic polymerizations are commonly carried out in solution, thus, ΔG_p depends on monomer and polymer concentrations (activities).

Entropies of monomer and polymer increase with dilution by $-R \ln [M]_0$ and $\frac{1}{n} R \ln \{[M]_0 - [M]_e)/n\}$, respectively ($[M]_0$ and $[M]_e$ are initial and equilibrium concentrations of monomer, n is the degree of polymerization).

The polymer concentration is expressed by the concentration of macromolecules as one species and not by the polymer segments ($R \ln \{[M]_0 - [M]_e\}$). The dilution factor of the entropy relies on translation of the whole macromolecule, the polymer segment is considered to have no translational freedom.

Thus, the total change of the free energy of conversion of 1 mole of monomer to polymer is:

$$\Delta G_p = \Delta G_p^0 - RT \cdot \ln [M]_0 + \frac{[M]_e}{[M]_0 - [M]_e} RT \cdot \ln \frac{[M]_e}{[M]_0} + \frac{1}{n} RT \cdot \frac{[M]_0 - [M]_e}{n} \tag{2-2}$$

(where ΔG_p^0 corresponds to the standard conditions, i.e. at $[M]_0 = 1.0 \text{ mol} \cdot l^{-1}$).

Usually high molecular weight polymer is formed ($n \gg 1$) and the last term of the right hand side (RHS) of Eq. (2-2) can be neglected. When the monomer equilibrium concentration is much lower than the initial concentration of monomer then Eq. (2-2) is reduced to the well-known Eq. (2-3):

$$\Delta G_p = \Delta G_p^0 - RT \ln [M]_0 \tag{2-3}$$

For initial concentrations of monomer close to the equilibrium, i.e. when the overall change of the free energy is close to zero, $\Delta G_p^0 = RT \ln [M]_0$ and the equilibrium constant of polymerization becomes:
$K_e = \exp(-\Delta G_p^0/RT) = [M]_0^{-1} = [M]_e^{-1}$. Eq. (2-3) indicates that if the initial concentration of monomer is high polymerization may still be thermodynamically feasible even if ΔH_p^0 has a small *positive* value.

The standard free energy change is given by a change of enthalpy and entropy:

$$\Delta G_p^0 = \Delta H_p^0 - T\,\Delta S_p^0 \tag{2-4}$$

Thus, four general cases are possible, depending whether ΔH_p^0 and ΔS_p^0 are positive or negative:

1. $\Delta H_p^0 < 0$ and $\Delta S_p^0 < 0$

For the large majority of heterocyclic monomers polymerization is exothermic due to the release of the ring strain. The entropy of the system usually also decreases ($\Delta S_p^0 < 0$) due to the loss of translational entropy. Thus, for a system with $\Delta H_p^0 < 0$ and $\Delta S_p^0 < 0$ polymerization is possible only below a certain temperature $T_c^0 = \Delta H_p^0/\Delta S_p^0$, called the ceiling temperature. This temperature is highest for bulk polymerization and decreases with the finitial monomer concentration:

$$T_c = \Delta H_p^0/(\Delta S_p^0 + R \ln [M]_0) \tag{2-4a}$$

(note that ΔS_p^0 and $R \ln [M]_0$ have opposite signs for $[M]_0 > 1.0$; thus, with increasing $[M]_0$ the overall value of the denominator decreases and T_c increases).

The ceiling temperature can be observed only for systems in which ΔH_p^0 has a relatively small negative value ($> -40\ kJ \cdot mol^{-1}$). For higher negative values of ΔH_p^0 (e.g. for most vinyl polymerizations) T_c is well above the decomposition temperature of monomer and polymer.

The position of the polymer-monomer equilibrium (i.e. the equilibrium monomer concentration $[M]_e$) is given by

$$\ln [M]_e = \Delta H_p^0/RT - \Delta S_p^0/R \tag{2-5}$$

Examples of the thermodynamic parameters are given in Table 2.2. Data collected in Table 2.2 indicate that the position of the equilibrium, at least at moderate temperatures, depends mostly on ΔH_p.

Table 2.2. Thermodynamic parameters of polymerization of some typical monomers. Comparison of vinyl and ring-opening polymerization

Monomer		$-\Delta H_p^0$ $kJ \cdot mol^{-1}$	$-\Delta S_p^0$ $J \cdot mol^{-1} \cdot {}^0K^{-1}$	$[M]_e^{25°}$ $mol \cdot l^{-1}$	Ref.
Ethylene	gg	93.24	142	$1 \cdot 10^{-9}$	[3]
	lc	108.78	174	$1 \cdot 10^{-10}$	[4]
Butadiene	lc	73.9	584	$3 \cdot 10^{-9}$	[5]
Isobutylene	lc	54.2	121	$7 \cdot 10^{-4}$	[5]
Styrene	gg	76	148	$3 \cdot 10^{-6}$	[3]
	lc	70.1	105	$2 \cdot 10^{-7}$	[5]
α-Methylstyrene	gg	34	147	51	[3]
	lc	35.3	104	0.18	[5]
Methyl-methacrylate	lc	55.4	117	$3 \cdot 10^{-4}$	[5]
Formaldehyde	gc	55.4	175	0.28	[4,6]
		72.2	183	$9 \cdot 10^{-4}$	[7]
Tetrahydrofuran	lc	22.3	75	1.11	[5]
	ss(CH_2Cl_2)	15.1	63	4.9	[8]
Trioxane	gc	16.1	56	1.39	[9]

2. $\Delta H_p^0 > 0$; $\Delta S_p^0 > 0$

Cationic polymerization of some larger cyclic siloxanes [10] and 6-membered cyclic phosphates [11] belong to this group. Besides, polymerization of some unstrained heterocyclics in which repulsion between substituents or lone electron pairs is stronger in the polymer than in the monomer can also be endothermic:

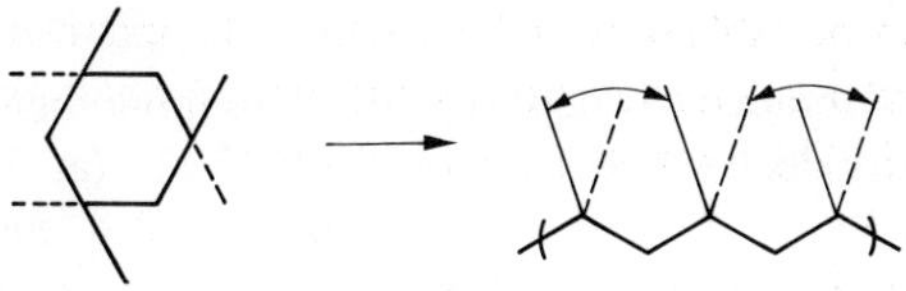

(2-6)

The positive change of entropy of polymerization is due to the fact, that the rotational entropy of the flexible polymer chain is higher than that of the rigid cyclic monomer molecule and this entropy gain may overcome, at least for some systems, the loss of entropy due to the decrease of translational entropy. Elemental sulfur (8-membered ring) is the best known example of this class of monomers. Some cyclic siloxanes and 6-membered cyclic phosphates with large exocyclic groups also belong in this class.

3. $\Delta H_p^0 < 0$; $\Delta S_p^0 > 0$

Polymerization would be possible over the entire temperature range at a standard concentration of $[M]_0 = 1 \text{ mol} \cdot l^{-1}$. At low monomer concentrations ΔG_p may become positive.

4. $\Delta H_p^0 > 0$; $\Delta S_p^0 < 0$

Polymerization is not possible ($\Delta G_p^0 > 0$). For small negative ΔS_p at high $[M]_0$, ΔG_p^0 may, however, become negative, making polymerization possible (case 2).

2.3 Kinetic Aspects of the Monomer-Polymer Equilibrium

Monomer is at equilibrium with its own growing polymer when the rates of propagation and depropagation are equal:

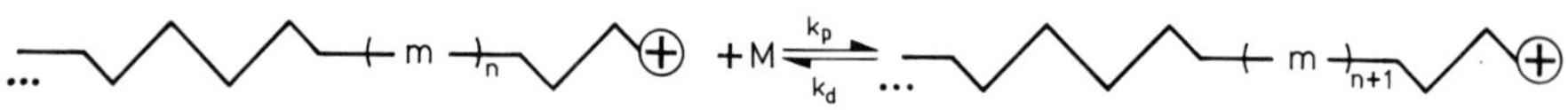

(2-7)

(where m is a monomer unit in the chain and M is a monomer molecule).

At equilibrium $k_p[M]_e = k_d$. If we denote the equilibrium constant as $K_e = k_p/k_d$, then $[M]_e = 1/K_e$. It follows that:

$$\ln K_e = -\Delta G_p^0/RT = -\Delta H_p^0/RT + \Delta S_p^0/R \tag{2-8}$$

The last equation can be rearranged using activation parameters of propagation and depropagation:

$$\ln K_e = (\Delta H_d^{\ddagger}/RT - \Delta H_p^{\ddagger}/RT) + (\Delta S_p^{\ddagger}/R - \Delta S_d^{\ddagger}/R) \tag{2-9}$$

Equation (2-9) allows the calculation of any pair of activation parameters, provided that the other pair is known from independent measurements. For example, tetrahydrotiophene cannot be polymerized to the corresponding polymer, but it, can be obtained by polycondensation:

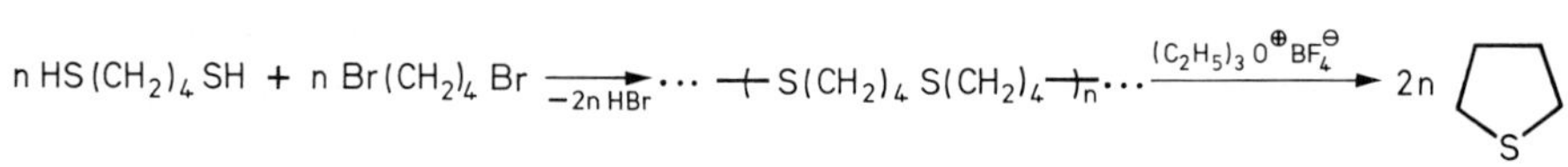

(2-10)

This polymer was cationically degraded (by depropagation) to monomer and the rate constant of depropagation was determined as $k_d = 9 \cdot 10^{-6}\ s^{-1}$ (at 35 °C); $\Delta H_d^{\ddagger} = 99\ kJ \cdot mol^{-1}$ $\Delta S_d^{\ddagger} = -24\ J \cdot mol^{-1} \cdot deg^{-1}$ [12]. Then, the free energy change for tetrahydrotiophene polymerization was estimated to be $17\ kJ \cdot mol^{-1}$ at 35 °C ($\Delta H_p^0 = -8.0\ kJ \cdot mol^{-1}$ and $\Delta S_p^0 \simeq -85\ J \cdot mol^{-1} \cdot deg^{-1}$). Thus, using these parameters and Eq. (2-9) we calculated the rate constant of propagation as $k_p = 10^{-8}\ mol^{-1} \cdot l \cdot s^{-1}$ (at 35 °C) and $[M]_e$ (35 °C) $= 800\ mol \cdot l^{-1}$ which is almost one hundred times higher than the concentration of tetrahydrotiophene in bulk. Comparison of the rate constants k_p and k_d shows that ring closure is so fast, that at any monomer concentration its rate cannot be balanced by the rate of propagation ($k_p \cdot [M]$).

2.4 Thermodynamics of Real Systems. Non-Ideality of Systems

Influence of solvent properties

The thermodynamics of conversion of a monomer (M) to a polymer segment (m) has been discussed in previous sections in terms of the ideal system, i.e. without taking into account specific interactions. The monomer-polymer, solvent-monomer, and solvent-polymer interactions may contribute to the overall change of ΔG_p. Solvation and desolvation of monomer and polymer are particularly important in ionic processes when highly polar solvents are used. Thus, the polymerization process should formally be written as:

$$M \cdot \bar{x}S \rightarrow m \cdot \bar{y}S + (\bar{x} - \bar{y})\,S \qquad (2\text{-}11)$$

This notation indicates that the monomer M, solvated by an average number $\bar{x}$ of solvent molecules S is converted into the polymeric segment m, solvated by an average number $\bar{y}$ of solvent molecules. Polymerization of tetrahydrofuran exemplifies the influence of solvent on thermodynamics. Indeed, $[M]_e$ determined for otherwise equal conditions can change by a factor of two when the solvent is changed [16].

When a sufficiently high concentration of polymer is formed at equilibrium, monomer-polymer interactions have to be taken into account.

If all above discussed interactions are simultaneously considered, the following equation for the change of the free energy of the system has to be used [13]:

$$\Delta G^0_{lc}/RT = \ln \varphi_M + 1 + \mu_{MP}(\varphi_P - \varphi_M) + \varphi_S(\mu_{MS} - \mu_{PS} \cdot V_M/V_S) \qquad (2\text{-}12)$$

where: φ_M, φ_P, and φ_S are volume fractions of monomer, polymer, and solvent, respectively, V_M and V_S are the corresponding molar volumes, and μ_{MP}, μ_{MS}, and μ_{SP} denote the interaction parameters between constituents indicated by subscripts.

When interactions involving solvent are neglected or cancel each other (bulk polymerization), Eq. (2-12) reduces to its simpler form, involving only monomer-polymer interactions [14]:

$$\Delta G^0_{lc}/RT = \Delta H^0_{lc}/RT - \Delta S^0_{lc}/R = \ln \varphi_M + 1 + \mu_{MP}(\varphi_P - \varphi_M) \qquad (2\text{-}13)$$

According to this equation, the determination of ΔH^0_{lc} and ΔS^0_{lc} requires plotting the right-hand side (RHS) as a function of 1/T. On the RHS φ_M and φ_P are known from experiment, whereas the interaction parameter μ_{MP} has to be chosen in such a way that the plot of the RHS of Eq. (2-13) against 1/T gives a straight line. It may happen that no single value of μ_{MP} will linearize Eq. (2-13); this means that μ_{MP} is a function of temperature and two or more values of μ_{MP} have to be used for different ranges of temperature. Once μ_{MP} is known, ΔG^0_{lc} can be computed for several temperatures, and from the plot of ΔG^0_{lc} ($= \Delta H^0_{lc} - T\,\Delta S^0_{lc}$) vs T both ΔH_{lc} and ΔS^0_{lc} can be determined.

Another complication comes from the dependence of ΔH^0_p and ΔS^0_p on temperature:

$$\Delta H^0_{P(T)} = \Delta H^0_{P(298)} + \int_{298}^{T} \Delta C_p \, dT \qquad (2\text{-}14)$$

and

$$\Delta S^0_{P(T)} = \Delta S^0_{P(298)} + \int_{298}^{T} \Delta C_p/T \, dT \qquad (2\text{-}15)$$

where ΔC_p is the difference between the heat capacities of monomer and polymer segment. Usually ΔC_p is less than 40 $J \cdot mol^{-1} \cdot deg^{-1}$ and due to the low precision of measurements of $[M]_e$ (or φ_P and φ_M) no reliable deviation from linearity of $\Delta G_{lc}/RT$ vs $1/T$ is observed for a limited temperature range (e.g. for $\Delta C_p = 40\ J \cdot mol^{-1} \times deg^{-1}$ $\delta\,\Delta H_p = 2\ kJ \cdot mol^{-1}$ within 50° range).

Influence of the Initial Monomer Concentration

It has been observed in the cationic polymerization of THF [15,16] and, to a lesser extent in the polymerization of 1,3-dioxolane [17], that the equilibrium monomer concentration $[M]_e$ increases linearly with the initial monomer concentration $[M]_0$. Moreover, the stronger the monomer-solvent interaction (judged from the heat of mixing) — the more pronounced the dependence of $[M]_e$ on $[M]_0$ will be. It can also be assumed, that the reverse dependence would be observed if the solvent-polymer interaction is stronger than that of monomer-solvent. This provides a way of increasing the polymer yield given by $([M]_0 - [M]_e)/[M]_0$, by the proper choice of solvent.

In the studies of dependence of $[M]_e$ on $[M]_0$ in the cationic polymerization of THF, the linear dependence of polymer volume fraction (φ_P) on monomer volume fraction (φ_M) was observed by Leonard [15]:

$$\varphi_M = \varphi_M^0 + b\varphi_P \tag{2-16}$$

To quantitatively relate φ_M^0 and b to thermodynamics, ln φ_M was first expanded into the *a* series:

$$\ln \varphi_M = \ln a + (\varphi_M - a)/a - (\varphi_M - a)^2/2a^2 + \ldots \tag{2-17}$$

Substitution of the first two terms of Eq. (2-17) into Eq. (2-12) gives:

$$\varphi_M = \frac{-\Delta G_{lc}^0/RT + \ln a + \beta}{\beta + \mu_{MP} - 1/a} + \frac{\mu_{MP} - \beta}{\beta + \mu_{MP} - 1/a} \cdot \varphi_P \tag{2-18}$$

$$\beta = \mu_{MS} - \mu_{SP}\frac{V_M}{V_S} \tag{2-19}$$

Comparing Eqs. (2-16) and (2-19) it follows that:

$$b = (\mu_{MP} - \beta)/(\beta + \mu_{MP} - 1/a) \tag{2-20}$$

while φ_M^0 is equal to:

$$\varphi_M^0 = \frac{-(\Delta G_{lc}^0/RT) + \ln a + \beta}{\beta + \mu_{MP} - 1/a} \tag{2-21}$$

In the ideal case, *a* should be equal to φ_M^0. This happens when φ_M is very close to φ_M^0. Then, the *a* series can be reduced to the first two terms of Eq. (2-17). In that case one calculates β from the known values of b, $a = \varphi_M^0$ and μ_{MP} [14]. Consequently, $\Delta G_{lc}^0/RT$ is computed from Eq. (2-21), knowing φ_M, *a*, β, and μ_{MP}. In this way Leonard [15] calculated the thermodynamic parameters of polymerization of THF in benzene. Plotting $\Delta G^0_{lc}/RT$ vs T^{-1} he obtained $\Delta H_{lc} = -15\ kJ \cdot mol^{-1}$ and $\Delta S_{lc} = -50\ J \times mol^{-1} \cdot K^{-1}$.

The interaction parameter μ_{MS} is experimentally available from the heats of mixing:

$$\mu_{MS} = \frac{\Delta H_{mix}(\varphi_S + \varphi_M)}{V_M \cdot RT \cdot \varphi_S} \tag{2-22}$$

where V_M is the monomer molar volume. Leonard [15] calculated β values from Eq. (2-19) using μ_{MS} from Eq. (2-22) and estimating $\mu_{SP} = 0.3$. These β values deviate considerably from those calculated by assumming that $a = \varphi^0_M$. Indeed, recalculation of Leonard's data shows that the a values computed from β and b values (available experimentally) are twice as low as φ^0_M values. The large difference between a and φ^0_M means that the expansion of ln φ_M into a series cannot be limited to the first two terms only. Recalculation of the thermodynamic parameters gives $\Delta H^0_{lc} = -13.5\ kJ \times mol^{-1}$ and $\Delta S^0_{lc} = -50\ J \cdot mol^{-1} \cdot K^{-1}$ i.e., values close, to those reported by Leonard.

Influence of the Degree of Polymerization

The equilibrium monomer concentration $[M]_e$ depends on $\overline{DP}_n$ for shorter chains, because the influence of the end-groups changes the vicinity of the statistical unit. It would be equivalent to say that the rate constants of propagation and depropagation, that govern the position of the monomer-polymer equilibrium (Eq. 2-1) depend on chain length, and become independent of $\overline{DP}_n$ only after a certain $\overline{DP}_n$ is reached.

Let us consider a series of equilibria including a reversible initiation step:

$$\begin{aligned} I + M &\overset{K_0}{\rightleftharpoons} P^*_1 \\ P^*_1 + M &\overset{K_1}{\rightleftharpoons} P^*_2 \\ P^*_n + M &\overset{K_n}{\rightleftharpoons} P^*_{n+1} \end{aligned} \tag{2-23}$$

When all the equilibrium constants are the same (i.e. $K_0 = K_1 = \cdots = K_n = \cdots K$), the equilibrium monomer concentration, according to Tobolsky [18], is given by the following equation:

$$\frac{[M]_0 - [M]_e}{[I]_0} = \frac{K[M]_e}{1 - K[M]_e} \tag{2-24}$$

Thus, independently of $([M]_0 - [M]_e)/[I]_0$ constant K values are obtained. If, however, the equilibrium constants for each step are different, the values of K determined from Eq. (2-24) will vary with $([M]_0 - [M]_e)/[I]_0$. The equilibrium constants can be determined by Szwarc's treatment outlined below [19].

Extrapolation to zero and infinity enables determination of K_0 and K_n:

$$\frac{[M]_0 - [M]_e}{[I]_0} \rightarrow 0 : K \rightarrow K_0 , \qquad \frac{[M]_0 - [M]_e}{[I]_0} \rightarrow \infty : K \rightarrow K_n \tag{2-25}$$

When K_0 is known, the concentration of species P_1^* may be related to K_0. Thus, for the set of equilibria, from which the first equation is already excluded, the following equation is obtained:

$$K_1 = \frac{1}{[M]_e} - \frac{K_0}{2}\left(\sqrt{1 + 4[I]_0/K_0[M]_e([M]_0 - [M]_e)} - 1\right) \tag{2-26}$$

Again, the constancy of K_1 determined from Eq. (2-26) for different $([M]_0 - [M]_e)/[I]_0$ ratios means that $K_1 = \cdots = K_n$. If K_1 is not constant (i.e. $K_1 \neq \cdots \neq K_n$), the extrapolation may give K_1 and K_n' values. In principle, such a treatment can be used for determination of K_2, K_3 and so on, however, due to the accumulation of experimental errors in determination of $[M]_e$, such a procedure becomes impractical for the determination of K_3 [19].

A different approach was used by Tobolsky [18]. It requires the knowledge of $\overline{DP}_n$ to determine K_0 and $K = K_1 \cdots = K_n$.

$$\overline{DP}_n = (1 - K[M]_e)^{-1} \tag{2-27}$$

For determination of K_0 the value of $[I]_e$ should also be known.

$$[I]_0 = [I]_e(1 + K_0[M]_e\,\overline{DP}_n) \tag{2-28}$$

or:

$$[M]_0 = [M]_e\,(1 + K_0[I]_e/\overline{DP}_n^2) \tag{2-29}$$

If in Eq. (2-23) the first equilibrium is shifted to the left and some initiator remains unreacted ($[I]_e$), then $\overline{DP}_n \neq ([M]_0 - [M]_e)/[I]_0$.

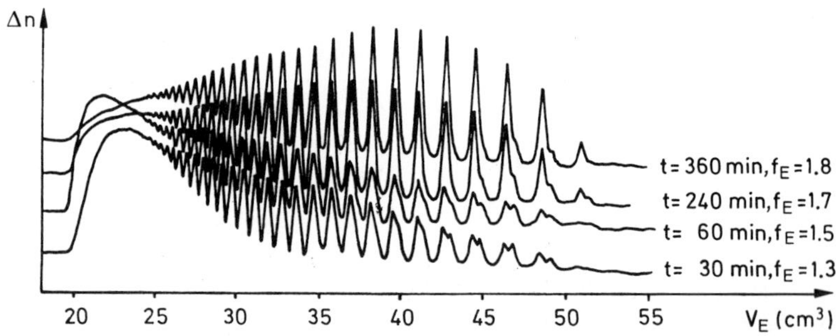

Fig. 2.1. GPC-traces of polyTHF initiated by the $HSbF_6/Ac_2O$ system (f_E-content of ester groups in macromolecules) [20]

Recently Heitz et al. [20] have used cationic oligomerization of THF in the presence of acetic anhydride to calculate K_0 and K. They have assumed that the distribution of the oligomers carrying acetoxy end groups, determined by gpc, corresponds to the distribution of the living oligomers. The GPC chromatogram is shown in Fig. 2.1.

The apparent equilibrium constants for every addition step have been calculated from the corresponding peak area with correction of intensities. At 10 °C in CH_2Cl_2 solution, $K_1 = 1.7\ mol^{-1} \cdot l$ and $K_n = 0.256\ mol^{-1} \cdot l$ were obtained. Only above $\overline{DP}_n > 15$ was the constant value $[THF]_e = 3.8\ mol \cdot l^{-1}$ observed.

Influence of Phase Separation

When liquid monomer or monomer in solution is converted into insoluble polymer, the following possibilities arise:

- solid amorphous polymer precipitates out of the saturated solution
- solid crystalline polymer precipitates out of solution
- polymerization proceeds in the solid crystalline state.

Only the polymerization in the solid crystalline state is thermodynamically different from the polymerization of the same monomer performed in homogeneous solution. In the first two instances $[M]_e$ would be the same for both homogeneous and heterogeneous conditions. However, this does not mean that the yields of polymer, i.e. $([M]_0 - [M]_e)/[M]_0$, would be the same because when polymer precipitates the volume of the mixture becomes smaller; although $[M]_e$ remains the same it is in the precipitating system in a smaller volume so that the amount of monomer at equilibrium is decreasing compared with homogenous conditions.

Polymerization in the crystalline state is, as stated above, thermodynamically different from the same polymerization in the homogeneous medium. Polymerizations, which would not be allowed under homogeneous conditions, may be accomplished in the crystalline state, i.e. monomer may add directly to the active species in the crystal and polymerization is accompanied by simultaneous crystallization. In this process each propagation step is equivalent to an increase of the crystal size.

The change of the free energy, when the polymerization proceeds in the crystalline state, is different from that in the homogeneous solution by the heat of crystallization, provided that in solution the monomer-polymer-solvent interactions can be neglected. Thus:

$$\Delta G^0_{ss} = RT \ln [M]_{e,ss} \tag{2-31}$$

$$\Delta G^0_{sc'} = RT \ln [M]_{e,sc'} \tag{2-31a}$$

and

$$[M]_{e,ss}/[M]_{e,sc'} = \exp (\Delta G_{ss} - \Delta G_{sc'})/RT \tag{2-32}$$

where

$$\Delta G_{sc'} = \Delta G_{ss} + \Delta G_{cryst} \tag{2-33}$$

thus,

$$[M]_{e,ss}/[M]_{e,sc'} = \exp (-\Delta G_{cryst})/RT\ . \tag{2-33a}$$

Thus, for larger heats of crystallization, resulting in higher overall change of the free energy, the tendency for polymerization increases.

It will be shown in Chapter 7 that the low $[M]_e$ (and therefore high yields) in 1,3,5-trioxane polymerization is due to the polymerization proceeding simultaneously with crystallization. Otherwise, only a small fraction of monomer could polymerize at moderate temperatures.

2.5 Methods for Determining ΔH_p^0 and ΔS_p^0

The simplest method to determine enthalpy and entropy of polymerization is based on application of Eq. (2-34):

$$\ln K_e = \ln(1/[M]_e) = -\Delta H_p^0/RT + \Delta S_p^0/R \tag{2-34}$$

Equation (2-34) gives the standard entropy of polymerization ΔS_p^0. The overall entropy change due to polymerization is equal to $\Delta S_p = \Delta S_p^0 + RT \ln [M]$.

In practice, one has to allow the polymerization to reach equilibrium at a series of temperatures and determine $[M]_e$ for the equilibrated systems. This can be done by gravimetry, dilatometry, gas chromatography, spectrometric methods (e.g. NMR) or any other suitable analytical method that yields accurate measures of monomer and polymer. The measured conversion should not be due to termination and/or deactivation of the active species before the true equilibrium is established. Thus, when a plateau is reached on the time-conversion curve and there is no direct proof that the active species persists in the system, more initiator should be added to check that the monomer concentration does not decline. A decrease in the monomer concentration would indicate that the plateau is of kinetic origin.

We wish to stress this point because one can find examples of improper treatment of experimental data in the literature on cationic polymerization. For instance, in a recent paper on the polymerization of α-epichlorohydrin the authors observed limited conversions of monomer [21]. The polymer yield was increasing with decreasing temperatures. The authors assumed that the monomer concentration at the plateau was the equilibrium monomer concentration, and calculated "thermodynamic parameters". The ring strain, expressed by ΔH_p^0 thus obtained for α-epichlorohydrin was unrealistically low (only $-23\ kJ \cdot mol^{-1}$) when compared with heats of polymerization of other 3-membered cyclic ethers (cf. Table 2.7). Other reports on α-epichlorohydrin polymerization have shown that nearly quantitative conversions are possible even at higher temperatures, provided that termination is suppressed [22,23].

Another way of determining whether the observed plateau does really correspond to $[M]_e$ is to perform polymerization at various ratios of $[M]_0$ to $[I]_0$. The properly determined value of $[M]_e$ should be independent of this ratio, provided that $[M]_0/[I]_0$ is not too small. Thus, for non-living systems it is necessary to carry out polymerizations with increasing initial initiator concentration until a constant ultimate monomer conversion is reached. This method of approaching the equilibrium concentration gave reliable thermodynamic parameters for the cationic polymerization of cyclic esters of phosphoric acid, in spite of termination observed in these systems [11].

2.5.1 Other Sources of Errors in Determining $[M]_e$

By applying NMR methods one has to remember that the integral is not necessarily directly proportional to the concentration of the observed nucleus; this proportionality does not hold, for instance, in ^{13}C-NMR. NMR is however a method of choice, particularly for living systems, permitting the determination of $[M]_e$ at various temperatures, and using only one sample (sealed in the NMR tube). It is advisable to determine $[M]_e$ going up and down with temperature several times. Agreement between the experimental points of various sets of experiments indicates that the reaction follows sufficiently rapidly temperature changes and sufficient time exists to reestablish a new equilibrium. A recent example of determining $[M]_e$ in this way is illustrated in Fig. 2.2.

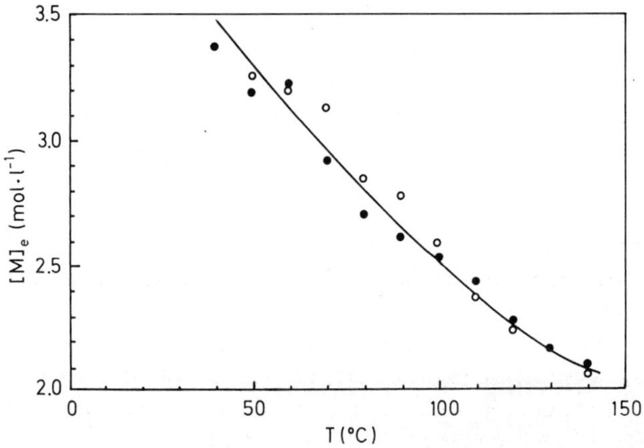

Fig. 2.2. Dependence of $[M]_e$ on T in polymerization of 2-oxo-1,3,2 λ^5 dioxaphosphorinane. (○): Temp. increasing (●): Temp. decreasing [11]

For gaseous monomers measurements of equilibrium monomer pressure permits the determination of $[M]_e$. As pointed out, however, absorption of monomer in the condensed phase may lead to serious errors [24].

Gravimetric measurements may be erroneous if the polymerization is not properly terminated. Special care has to be taken to ensure complete termination exactly at the temperature of measurement.

2.5.2 Thermochemical Measurements

Apart from methods based on measurements of $[M]_e$ there are also methods based on the direct calorimetric determination of ΔH_p or based on calculation of ΔH_p from heats of combustion of monomer and polymer. This latter approach requires, however, that two large values are substracted one from another to obtain a comparatively small value. This method may give large errors stemming from small errors in measurements (e.g. heats of combustion of liquid 7-oxabicyclo [2.2.1]heptane and its solid polymer are equal to $-\Delta H = 3566.3$ kJ · mol^{-1} and $\Delta H_c = -3522.0$ kJ · mol^{-1} yielding $\Delta H_{lc} = -44.3$ kJ · mol^{-1}, i.e. a value close to 1% of the measured values).

2.5.3 Empirical and Semiempirical Methods of Determining ΔH_p and ΔS_p

Two different approaches need to be considered. In the first one the thermodynamic parameters are calculated on the basis of spectroscopic and specific heat data of monomer and its polymer. The estimation of entropy from specific heat values is based on the assumption that at 0 °K the entropy of the perfect crystal equals zero. Since polymers are never perfectly crystalline this method is based on a rough approximation.

Empirical methods rely on the already known ΔH_p and ΔS_p, determined experimentally for a given class of monomers, and relate structural and electronic features of these monomers to known thermodynamic parameters. Thus, for the four-membered oxygen containing monomers the following equation was offered [26]:

$$\Delta H_p = 26.4 + 1.9\, n \cdot x \cdot m \pm 2 \text{ kcal} \cdot \text{mol}^{-1} \qquad (2\text{-}35)$$

where n is the number of the sp^2 hybridized C atoms in the ring, m the number of substituted C atoms in the ring, and x = 1 for one oxygen atom, otherwise x = 0.

2.6 Factors Affecting Polymerizability. Enthalpy of Polymerization

We shall follow Carothers' definition of polymerizability [27], described in thermodynamic terms, and relating the polymerizability to the free energy of polymerization (ΔG_p^0). According to this definition, the more negative ΔG_p^0 the higher the polymerizability. Since $\Delta G_p^0 = RT \ln [M]_e$, the more negative ΔG_p^0, the lower the equilibrium monomer concentration.

This definition was criticized by Goldfarb [28], giving cyclopropane as an example, known to be a highly strained compound, thus having (hypothetically) a large negative value of ΔG_p^0, but not giving high polymers. One may argue, however, that nonpolymerizability of cyclopropane is due to the limitation of our present knowledge, because the proper system (mostly initiator) has not yet been found for this monomer. More recent examples of strained but "nonpolymerizable" monomers are cyclic sulfones and sulfoxides:

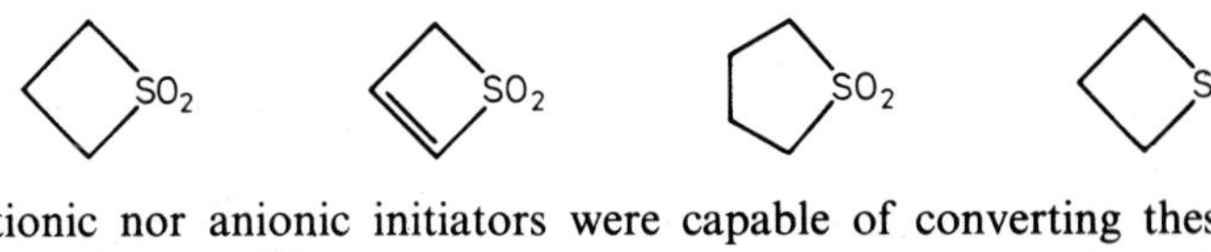

Neither cationic nor anionic initiators were capable of converting these strained monomers into polymer [29]. The reason of the low reactivity of sulfones is probably the strong 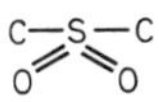 bond stabilized by two oxygen atoms. There are however, a number of monomers that were considered "nonpolymerizable" in the past and later were polymerized when the proper system was found. Thus, the only objective definition of polymerizability is the above described thermodynamic definition of Carothers.

The change of the free energy can be dissected into enthalpic and entropic contributions. The change of enthalpy is mostly due to the release of the ring strain, being the major driving force of ring-opening polymerizations. This strain is generally attributed to bond angle deformations, although bond stretching or compression as well as non-bonding interactions may also contribute to the overall strain. Let us compare alicyclic compounds with the corresponding heterocyclics polymerized by ring-opening polymerization.

Unperturbed bond angles between four orbitals of the sp^3 hybridized carbon atoms are equal to 109°28′. The differences between a bond angle in planar rings and an unperturbed valency angle (Δ) are given in Table 2.3.

Table 2.3. Differences between bond angles in planar rings and unperturbed valency angles

Number of atoms in the ring:	3	4	5	6	8
Δ, in deg (°)	49° 28′	19° 28′	0° 32′	10° 32′	25° 32′

These angular distortions are reduced because of the non-planar conformations of all cyclics with more than 3 atoms in the ring.

Preferred conformations of some cyclic ethers and cyclic acetals are shown below:

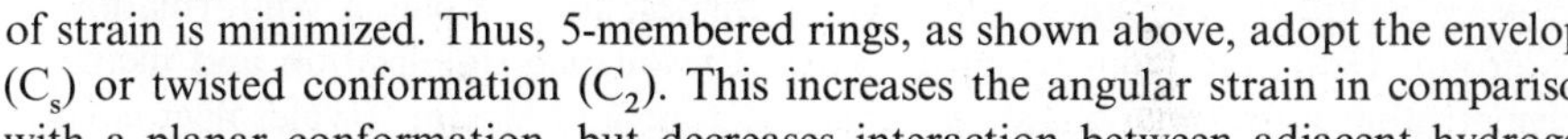

Consequently, due to the ability to attain the favorable conformation, the ring strain for the 6-membered cyclic ether (tetrahydropyrane) is lower than that for the 5-membered one (THF). The same is true for cyclic acetals.

Any ring tends to adopt a conformation in which the sum of all possible sources of strain is minimized. Thus, 5-membered rings, as shown above, adopt the envelope (C_s) or twisted conformation (C_2). This increases the angular strain in comparison with a planar conformation, but decreases interaction between adjacent hydrogen atoms (eclipsed in the planar conformation).

For larger rings transannular strain may become important. This type of strain is caused by nonbonding interactions between atoms on the opposite sides of the ring.

In still larger rings ($n > 15$) this kind of strain is reduced because these large cycles are able to form two parallel chains when fully extended.

Thus, as shown in Table 2.4 for unsubstituted cycloalkanes, ring strain is high for small rings ($n = 3, 4$), approaches zero at $n = 6$, increases again to a shallow maximum and decreases to a small value for large rings. The data for unsubstituted cycloalkanes are compared with recently determined values of ΔH_p by Yamashita [35] for

Table 2.4. Comparison of ring strains (in $kJ \cdot mol^{-1}$)

n	3	4	5	6	7	8
cycloalkanes[a] [30]	113.4 [31]	105.4 [32]	21.8 [32]	−2.9 [32]	21.4 [32]	34.9 [32]
cyclic 1,3-formals[b] [35]	—	—	21.3 [33]		14.6 [34]	16.5

n	9	10	11	12	13	14	15	16	17
cycloalkanes[a] [30]	47.0	48.3	45.4	14.3	22.3	7.1	12.2	8.0	8.4
cyclic 1,3-formals[b] [35]	—	—	—	—	—	12.2	—	5.0	—

n	22	24	28	32	33	34	40	42
cyclic formals, contd:	3.3	0	7.5	0	2.1	3.3	0	0

[a] — ΔH^0_{lc}
[b] — ΔH^0_{ss}

macrocyclic formals. For larger rings, namely for $n > 44$ (up to 98) ΔH_p was found to be zero. Some strain observed in large rings ($n \approx 30$) is tentatively explained by Yamashita as resulting from repulsion between the quasi axial hydrogen atoms caused by chain folding.

Introduction of an heteroatom into the ring does not change bond lenghts or bond angles considerably (sulfur is an exception) although differences in ring strain can be quite large (cf. Table 2.4).

Table 2.5 shows unperturbed bond lengths and bond angles for carbon-heteroatom bonds.

Similarities in the bond lengths and angles lead to similar thermodynamic polymerizabilities of cycloalkanes and heterocyclics.

Lactones and lactams behave differently, due to the contribution of $-\overset{\overset{O}{\|}}{C}-O-$ or $-\overset{\overset{O}{\|}}{C}-NH-$ group resonance, which is the highest for the planar conformation of this group. Deviation from planarity reduces resonance stabilization and increases polymerizability. In the 6-membered cyclic lactone and lactam (δ-valerolactone and

Table 2.5. Bond lengths and bond angles of carbon-heteroatom bonds

Type of bond	Length, A^0	Angle
C—C—C	1.54	109° 28′
C—N—C	1.47	109°
C—O—C	1.44	111°
C—S—C	1.82	100°

2-pyrrolidone, respectively) the ester and amide groups cannot adopt the planar conformations without introducing strain due to repulsion of the eclipsed hydrogen atoms. Moreover, up to n = 8 the amide group is forced to adopt cis-conformation [36] which is less stable by 5.9 kJ mol^{-1} than the trans-conformation [37]. Therefore, both 6-membered lactones and lactams are able to polymerize.

Heterocycles containing sulfur atoms are usually less strained than the corresponding ethers. Thus, the 5-membered cyclic sulfide-tetrahydrotiophene (THT) does not polymerize in contrast to THF (cf. additional discussion on THF on p. 23).

2.6.2 Entropy of Polymerization

The following factors affecting the entropy of polymerization have to be considered: translational, rotational, and vibrational entropy. Conversion of a cyclic monomer into a chain segment is accompanied by the decrease of translational and external rotational entropy. At the same time, however, due to the higher conformational flexibility of the polymer chain, internal rotational and vibrational entropy increases. The loss of the external rotational entropy is nearly balanced in the gas phase (ΔS_{gg}) by the gain in internal rotational and vibrational entropy. Thus, the overall entropy of polymerization is determined mainly by the change of translational entropy, being close to 120 J · mol^{-1} · deg^{-1}. The translational entropy is fairly insensitive to the molecular weight and structure of monomer.

Larger rings may exhibit a positive change of entropy. For instance, the entropy of polymerization of elemental sulfur (8-membered ring) is $\Delta S_p^0 = 20$ J · mol^{-1} × deg^{-1} [38] due to the gain in the vibrational and internal rotational entropy.

Table 2.6. Thermodynamic parameters of polymerization of cycloalkanes [39,40]

n	$-\Delta H^0_{lc}$, kJ · mol^{-1}	$-\Delta S^0_{lc}$, J · mol^{-1} · K^{-1}	$-\Delta G^0_{lc}$, kJ · mol^{-1}
	$\lfloor(CH_2)_n\rfloor$		
3	113.4	69.3	92.8
4	105.4	55.4	89.0
5	21.9	42.8	9.2
6	−2.9	10.5	−5.9
	$CH_3-\lfloor CH(CH_2)_{n-1}\rfloor$		
3	105.4	84.8	80.2
4	106.4	72.2	79.0
5	17.2	64.3	−2.1
6	−9.2	31.9	−18.9
	$(CH_3)_2-\lfloor C(CH_2)_{n-1}\rfloor$		
3	97.9	93.7	69.7
4	93.7	75.6	67.2
5	13.4	65.9	−6.3
6	−7.6	35.7	−18.1

The entropy of polymerization of higher unstrained cycles increases and, for e.g., the 42-membered cyclic hexamer of ε-caprolactone, exceeds $40\ J \cdot mol^{-1} \cdot deg^{-1}$. Similar values were reported for the polymerization of cyclic oligomers ($n = 3$–7) of 1,3,6-trioxacyclooctane [35]. This phenomenon has its origin in the increase of vibrational and rotational freedom in the large cycle.

Effect of Substitution

Interaction between substituents are generally more pronounced in linear than in cyclic molecules. Thus, substitution, mostly affecting the enthalpy of polymerization, usually decreases polymerizability. Polymerization of monomers with increasing substitution, becomes less exothermic and, eventually, their free energy may become positive, as shown in Table 2.6. The effect of substitution is also demonstrated in the conformational strain, caused by the interaction of groups located either in the immediate neighbourhood or across the ring (transannular strain). The conformational strain in the macromolecule formed may, however, be higher than the conformational strain in the monomer molecules. A typical case where the substituents are kept apart in the monomer but may interact in the macromolecule may be shown schematically as follows:

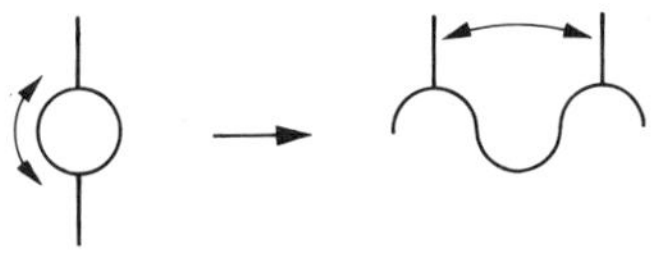

(2-38)

In small rings the angular strain is sufficiently high to compensate the conformational strain in the macromolecule. Thus, even fully substituted oxiranes can polymerize. Cationic polymerization of 1,1,2,2-tetramethyloxirane is a good example of this countereffect:

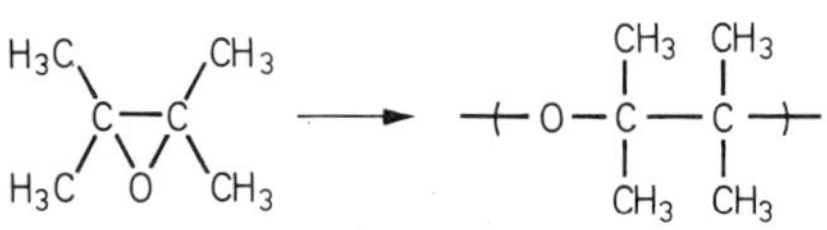

(2-39)

(the heat of polymerization has not been measured) [41].

Table 2.7 shows the free energies of polymerization of monosubstituted oxiranes. There are two conclusions coming from this Table. No direct correlation exists with

Table 2.7. ΔG_p^0 in the polymerization of substituted oxiranes R–CH–CH$_2$ (–O–) at 25° < (Ref. [42])

R	C_2H_5	CH_3	$C_6H_5CH_2$	H	CH_3OCH_2	C_6H_5	$C_6H_5OCH_2$	CH_2Cl
$-\Delta G_p$ $kJ \cdot mol^{-1}$ (25 °C)	72	75	84	93	94	97	106	113
pK_B of monomer	6.84	7.02	7.70	7.91	7.97	8.1	8.57	8.94

the size of substituent and polymerizability. $|\Delta G_p^0|$ increases with increasing pK_B, thus, with decreasing basicity which may be affected by substituents. The lone electron pairs on the oxygen are too far from each other in the macromolecule for a strong repulsion but can interact with substituents. Another possibility is that monomer-solvent and polymer-solvent interactions are stronger for the more basic monomers. However, to discuss this point in detail we would need not merely the pK_B of monomer but the pK_B of monomer and of the corresponding polymer unit. It is possible that a relatively strong monomer-solvent interaction of the more basic monomer renders polymerization energetically less favorable, on account of the larger energy demand to break the solvent-monomer interaction. How much of this energy is released because of the emerging polymer-solvent interaction is not known for cationic ring opening polymerizations studied to date.

Substituted oxetanes (4-membered rings) polymerize as readily as oxiranes, particularly those disubstituted at the 3-position. Thus, even oxetanes with large substitutents like benzoxymethyl or iodomethyl:

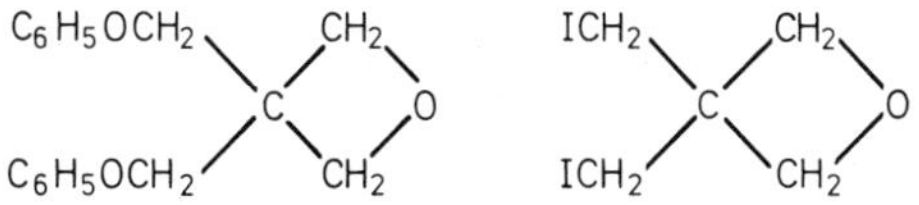

(2-40)

readily polymerize and give high-molecular weight polymers. There are contradicting reports on the polymerizability of substituted oxolanes. According to Garrido et al. [45] for the polymerization of 2-methyltetrahydrofuran $\Delta H_p = -23\ \mathrm{kJ \cdot mol^{-1}}$, $\Delta S_p^0 = -101\ \mathrm{J \cdot mol^{-1} \cdot deg^{-1}}$, and $[M]_e$ (bulk, 0°) = $8.8\ \mathrm{mol \cdot l^{-1}}$. In contrast, there is at least one report stating that this monomer does not polymerize [46]. One has to remember that at $[M]_e = 8.8\ \mathrm{mol \cdot l^{-1}}$ it is easy to miss the polymerization, particularly if slightly higher temperatures and lower monomer concentrations are used.

Polymerizability of 5-membered cyclic acetals (1,3-dioxolanes) have been studied in much greater detail. 1,3-Dioxolanes adopt the stable envelope conformation. One of the carbon atoms in the ethylene group is located at the tip of the flap, giving a dihedral angle between the cis- neighbouring hydrogens of 35° [47].

Polydioxolanes exist predominantly in gauche conformation, because in the anti- form the polymer would be of higher energy due to the repulsion of the non-bonded electrons of the oxygen atoms ("rabbit ears" effect):

(2-41)

The difference in energies of these interactions for the anti- and gauche-forms is approx. 1.0 kcal $\mathrm{mol^{-1}}$.

Substitution in dioxolanes decreases their polymerizability. Although the strain in the monomer may increase with substitution, the conformational strain of the macromolecule apparently prevails. The increase of the strain that destabilizes the monomer comes from the replacement of the cis-geminal C_4—H and C_5—H bond oppositions with greater C_4—H and C_5-substitutent interactions (for mono-substitution)[48]. In the open chain the conformational strain comes from gauche-interactions of substitutents, e.g. for disubstituted rings:

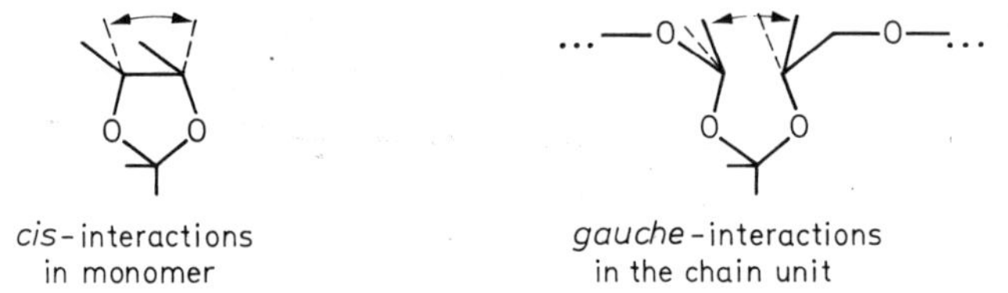

(2-42)

Okada calculated the approximate ΔH_{lc} values for various methyl-substituted 1,3-dioxolanes, taking 1,3-dioxolane ($\Delta H_{lc} = -17.6 \text{ kJ} \cdot \text{mol}^{-1}$) as the parent monomer[49]. The calculated values are given in Table 2.8. Only in one case Okada could compare these values with the experimentally determined ones (indicated below in brackets):

Table 2.8. ΔH_p in the polymerization of substituted 1,3-dioxolanes[49]

Substituent	$-\Delta H_{lc}$ kJ · mol^{-1}
4-methyl	13.5 (13.5)
4,4-dimethyl	9.3
cis-4,5-dimethyl	12.2
trans-4,5-dimethyl	7.6
4-ethyl	12.6 (13.0)
4-i-propyl	12.2

For 4-methyl-1,3-dioxolane there was good agreement between the calculated (−13.5) and measured (−13.5) ΔH_{lc}. According to Plesch, however, $\Delta H_{lc} = -19$ kJ $\times \text{mol}^{-1}$[50]. Thus, in this series the general rule is confirmed: the higher the substitution the lower the polymerizability. In the light of these data the lack of polymerizability of higher substituted 1,3-dioxolanes, recently observed by Jedliński[51], agrees well with expectations.

Polymerizability of substituted 7-membered cyclic acetals: 1,3-dioxepanes (6-membered 1,3-dioxanes do not polymerize) was also studied by Okada[52,53]. Major information is gathered in Table 2.9.

Table 2.9. Thermodynamic parameters in the polymerization of substituted 1,3-dioxepanes[52,53]

Substituent	$-\Delta H_p^0$ kJ · mol^{-1}	$-\Delta S_p^0$ kJ · mol^{-1} · deg^{-1}	T_c, °C (for $[M]_0 = 1 \text{ mol} \cdot l^{-1}$)
none	15.1	48	300
2-methyl	8.8	37	236
4-methyl	9.3	39	240

The bulk polymerization of 4-methyl-1,3-dioxepane at 0 °C yields 60% polymer. Further substitution decreases polymerizability as with 1,3-dioxolanes. 2,4-Dimethyl-1,3-dioxepane gives oligomers in limited yield, the 2,2-dimethyl-derivative dimerizes and the 4,4-dimethyl derivative does not polymerize at all [53].

The size of substituents may also influence the polymerizability of cyclic monomers. This was shown for the cationic (and anionic) polymerization of cyclic esters of phosphoric acid [11]:

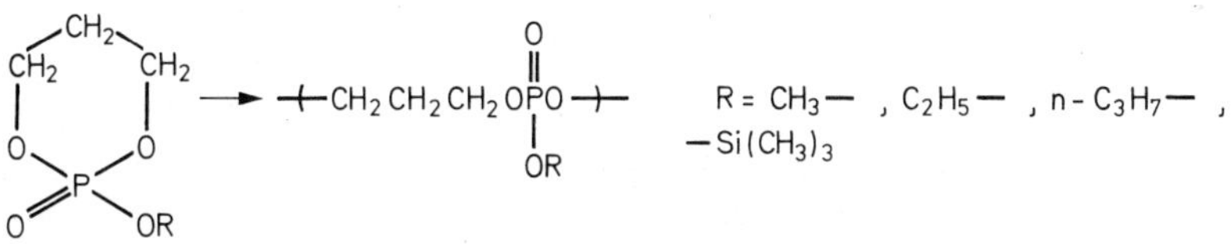

(2-43)

In these systems a linear correlation has been observed between enthalpy and entropy of polymerization (compensation plot) (Fig. 2.3). For a small exocyclic group ($R{=}CH_3$) polymerization is exothermic and exoentropic, thus, the driving force for the polymerization comes from the gain of energy due to release of ring strain.

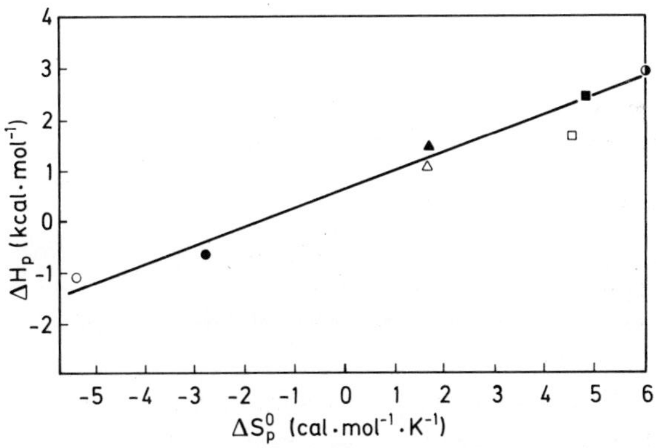

Fig. 2.3. Isoequilibrium dependence for 2-substituted 2-oxo-1,3,2-λ⁵-dioxaphosphorinanes (○): Methyl (cationic process), (●): Methyl (anionic process), (△): Ethyl, (▲): Hydrogen, (□): Propyl, (◑): Trimethylsilyl [11]

With an increase in the size of substituents, the enthalpy of polymerization becomes positive because of the strong interactions between the large substituents in the polymer chain. At the same time, however, the entropy of polymerization increases due to the gain in the rotational entropy when passing from monomer to polymer. Rotation of the $—Si(CH_3)_3$ group, for instance, is strongly restricted in the monomer because of interactions of axial H-atoms of this substituent with axial H-atoms of CH_2 groups in the ring:

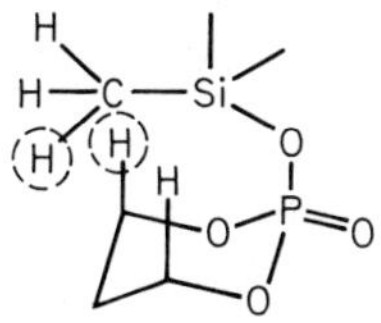

(2-44)

These interactions are much less pronounced in the polymer and the resulting gain in internal rotational entropy provides a driving force for the polymerization. Thus, by increasing the size of substituents one may pass from a system for which ΔH_p^0 and ΔS_p^0 are both negative and which exhibits a ceiling temperature, to a system with positive ΔH_p^0 and ΔS_p^0 characterized by a floor temperature.

2.7 Bicyclic Monomers

Polymerizability of bicyclic monomers is generally higher than the polymerizability of disubstituted cyclic monomers. This is because closure of an additional ring usually increases the angular strain; moreover, conformational strain may also be increased due to the forced unfavourable conformation in the monomer molecule. In 1,4-epoxycyclohexane (7-oxabicyclo|2,2,1|heptane) the 6-membered ring has to adopt the boat-like conformation while in the chain segment a more stable chair conformation may be attained (cf. Sect. 2.6.1):

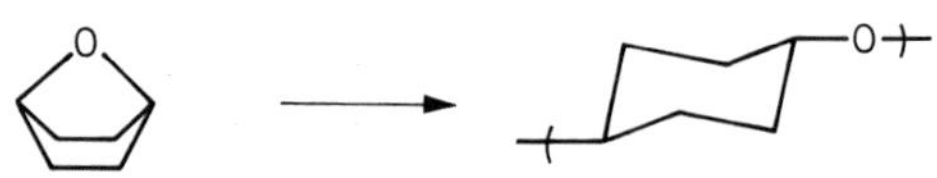

(2-45)

Thus, according to Plesch et al. [25,54] enthalpies of polymerization determined from the heats of combustion of 7-oxabicyclo|2,2,1|heptane and its 2-methyl derivatives are considerably higher than those for unsubstituted and substituted 5-membered rings (Table 2.10).

Table 2.10. ΔH_{lc}^0, ΔS_{lc}^0, and T_c in the polymerization of 7-oxabicyclo|2,2,1|heptanes [25]

Monomer	$-\Delta H_{lc}^0$ kJ · mol^{-1}	$-\Delta S_{lc}^0$ J · mol^{-1} · K^{-1}	T_c [a] °C
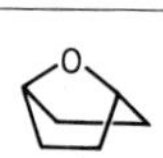	44.3	75.3	320
exo 	49.7	96.2	240
endo 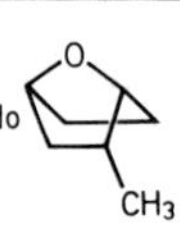	45.4	96.2	200

[a] for $[M]_0 = 1$ mol · l^{-1}

The enthalpy of polymerization is markedly different for the exo- and endo-isomers. The dependence of the polymerizability on configuration has been observed also by other authors [55,56]. Thus, endo-exo-2,6-dimethyl-7-oxabicyclo|2,2,1|heptane gave polymers:

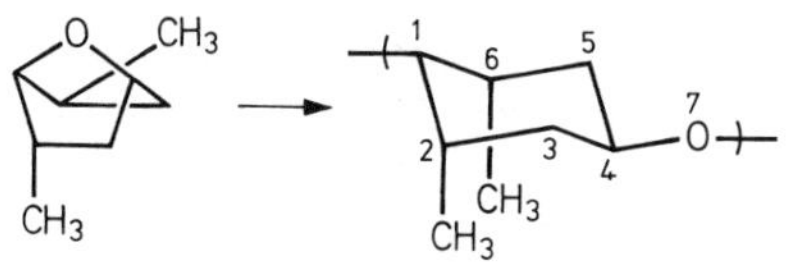

(2-46)

while the exo-exo-2,6-dimethyl isomer failed to polymerize.

Exo-exo-, and endo-exo-2,3-dimethyl derivatives were polymerized at 0 °C with 60% yield to polymer having $\overline{M}_n = 1000$, whereas under the same conditions exo-exo- and endo-endo-2,5-dimethyl monomers gave only 15% polymer, and for exo-endo-derivatives the yield was 25% [55]. Polymerization of these monomers with Lewis acid initiators is acompanied by termination. Thus, it is not clear whether the reported limited conversions reflect genuine monomer-polymer equilibria or are due to termination (cf. p. 12).

Another example of the relationship between configuration and polymerizability concerns the polymerization of 8-oxabicyclo|4,3,0|nonane [57]:

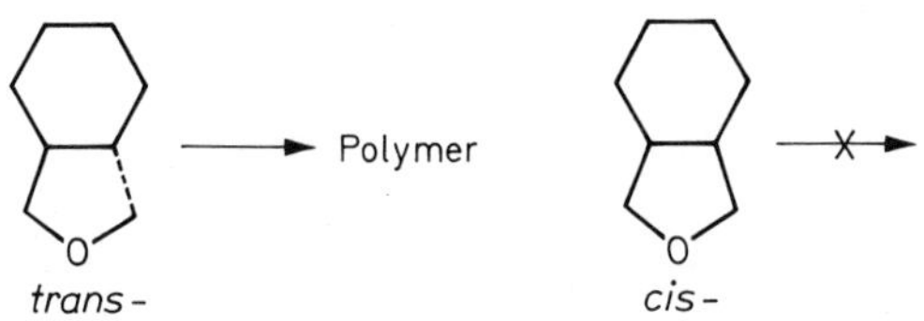

(2-47)

Trans-8-oxabicyclo|4,3,0|nonane was polymerized to high molecular weight polymer, whereas the cis-isomer did not polymerize at all (although it could undergo copolymerization).

Bicyclic acetals, containing a 5-membered dioxolane ring, show similar behaviour [58]. Trans-7,9-dioxabicyclo|4,3,0|nonane gives high molecular weight polymer while the cis-monomer yields a cyclic dimer:

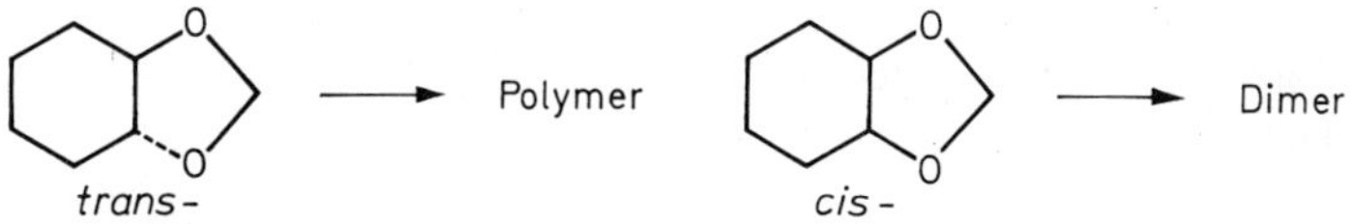

(2-48)

Recently, the thermodynamics of polymerization of bicyclic monomers has been extensively reviewed by Hall [59] and for more details one should refer to this work; this is the only comprehensive review of the polymerization of this class of compounds.

2.8 Thermodynamics of Polymerization of Selected Monomers

THF is a good model for thermodynamic studies, because under properly chosen conditions, it gives a living cationic polymerization. The ceiling temperature T_c

= 80° (polymerization in bulk) [60] and it changes with pressure. At 2.5 kbar it increases to 129° [61].

In Table 2.11 we collected available information on the thermodynamics of polymerization of THF. Direct comparison of these data is difficult, because, according to the indexes given in the second column, the state of monomer and polymer varied. Moreover, due to the different methods used, errors peculiar to these methods occurred. Thus, we shall refrain from discussing the ΔH and ΔS obtained for gaseous states and those calculated from heats of formation.

Table 2.11. Thermodynamics of THF polymerization

No.	Phase transition	Solvent	T, °C	$-\Delta H^0$, $kJ \cdot mol^{-1}$	$-\Delta S^0$, $J \cdot mol^{-1} \cdot K^{-1}$	Method	Ref.
1	gg		20	21	—	A	62)
2	gg		25	12	—	A	63)
3	gc		25	39.8	112	B	64)
4	gc		25	11.2	34	B[a]	24)
5	gc		25	—	139	D	65)
6	gc′		25	—	177	D	65)
7	lc′		25	—	100	D	65)
8	lc		25	25.5	62	D, C	65)
9	lc′		25	7.1	20	A, D	65)
10	lc		25	4.4	11	A, D	66)
11	lc′		25	38	—	C	63)
12	lc		25	7.5	16	B	64)
13	lc		50	12.5	41	B	14)
14	lc		30	15	49	B	15)
15	ls		25	19.1	74	B	67)
16	ls		25	18	71	B	68)
17	ls			23	67		60)
18	ss	ether	40	23	87	B	69)
19	ss	CH_3NO_2	25	9.2	46	B	70)
20	ss	CCl_4	25	18	71		8)
21	ss	CH_2Cl_2	25	15	63		8)
22	ss	CH_3NO_2	25	14	63		8)

A — semiempirical calculations (from heats of formation);
B — thermodynamic approach;
C — combustion;
D — 3rd law or statistical calculations;
[a] — corrected for the absorbed monomer

It is rather remarkable that the data obtained by the majority of authors for bulk and solution polymerization are very similar and oscillate within the range from 15 to 23 $kJ \cdot mol^{-1}$ for ΔH_{ls} (and ΔH_{ss}) and from 50 to 90 $J \cdot mol^{-1} \cdot deg^{-1}$ for ΔS_{ls} (and ΔS_{ss}). Within this range some additional dependences can be observed, i.e., for CCl_4, CH_2Cl_2, and CH_3NO_2 solvents both ΔH_{ss} and ΔS_{ss} increase slightly in this order. Clearly, monomer-solvent and polymer-solvent interactions are responsible for this trend, as indicated in Sect. 2.4.

Ivin and Leonard [14] have discussed the general approach to evaluate the thermodynamic parameters related strictly to the monomer-polymer unit change and free from the monomer-polymer or monomer(polymer)-solvent interaction. This approach has been used in analysis of the THF data and No. 13 in Table 2.11 gives ΔH_{lc} and ΔS_{lc} calculated in this manner by using Eq. 2-13 and taking $\mu_{MP} = 0.3$.

For bulk polymerization $[M]_e$ can have only one unique value at a given temperature. When solvent is added and the corresponding parameters include the interaction with solvent, $[M]_e$ depends on the nature and volume fraction of solvent, thus on $[M]_0$ as discussed on p. 6. This effect has been neglected in earlier work which led to serious mistakes in the kinetic treatment of THF polymerization (cf. Part I, and Ref. [80]). Leonard has attempted to account for the interactions between monomer, polymer and solvent (Eq. 2-21) in evaluating ΔH_{lc} and ΔS_{lc}. These values (No. 14) have already been discussed in Sect. 2.4.

It has also been observed that $[M]_e$ depends differently on $[M]_0$ in various solvents [80]. The stronger the solvent-monomer interaction, the higher $[M]_e$ for a given $[M]_0$. The explanation given for this phenomenon involves the higher nucleophilicity of THF than that of the polymer segments [80]. Thus, the stronger interaction between THF and solvent is not fully compensated by the comparatively weaker solvent-polymer interaction. The stronger the interaction, the higher $[M]_e$ for a given $[M]_0$. Thus, $[M]_e$ increases in the following order $CCl_4 < C_6H_6 < CH_2Cl_2 < CH_3NO_2$. The same explanation may apply to the observed change of ΔH_{ss} and ΔS_{ss} measured for these solvents.

Table 2.12. Thermodynamics of 1,3-dioxolane polymerization

No.	Phase transition	Solvent	T, °C	$-\Delta H^0$, $J \cdot mol^{-1}$	$-\Delta S^0$, $J \cdot mol^{-1} \cdot K^{-1}$	Method	Ref.
1	gg		20	26	—	A	[72]
2	gc		25	50	133	B	[64]
3	gc		25	54	155	B[a]	[24]
4	lc		100	24	76	A	[71]
5	lc		25	15	37	B	[64]
6	lc		40	23	63	B	[17]
7	ls		55	6.5	—	B	[74]
8	ss	CH_2Cl_2	0	22	78	E, B	[17]
9	ss	CH_3CH_2Cl	20	27	94	B	[76]
10	ss	C_6H_6	30	15	59	B	[77]
11	ss	CH_2Cl_2	70	17.5	—	—	[78]
12	gc′		25	—	205	D	[73]
13	gc		25	—	167	D	[73]
14	lc′		25	—	100	D	[73]
15	lc		25	—	67	D	[73]
16	lc		25	17.5	48	B	[79]

A — semiempirical calculations (from heats of formation);
B — thermodynamic approach;
D — 3rd law or statistical calculations;
E — calorimetry;
[a] — corrected for the absorbed monomer

1,3-Dioxolane (DXL)

This monomer has been less thoroughly studied than THF, but the general findings are similar to those described in the previous section. DXL is slightly more strained than THF as confirmed by the higher $T_c = 165\ ^\circ C$ [71]. Values of ΔH_p^0 and ΔS_p^0 as determined by various authors are listed in Table 2.12. $[M]_e$ has been found to be almost independent of the solvent used. This is probably due to the low basicity of DXL and therefore weak interaction of this monomer even with strong acceptors. The difference in basicities of DXL and the corresponding polymer unit is also much smaller than the same difference in the polymerization of THF.

1,3,5-Trioxane (TXN)

Polyoxymethylene (POM) can be synthesized cationically not only from TXN but also from formaldehyde (FA) and 1,3,5,7-tetraoxocane (TTXN). In fact, when TXN polymerizes all these products coexist in multiple equilibria. The polymer is crystalline and precipitates out from the reaction mixture. Thus, the concentration of the components of the system depends on the thermodynamics of the equilibria:

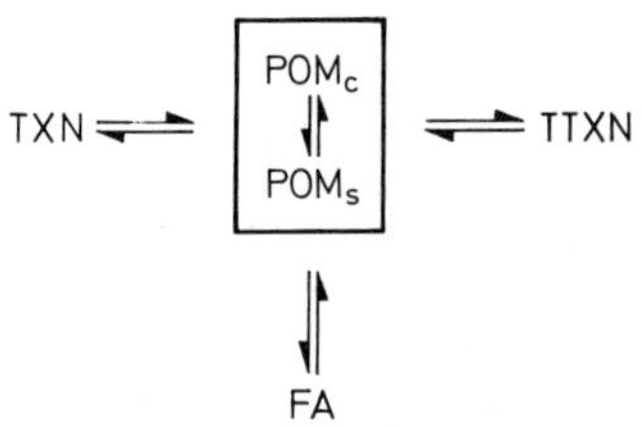

(2-49)

Application of ^{1}H-NMR spectroscopy to the polymerization conducted in $C_6H_5NO_2$ solvent yielded the thermodynamic parameters shown in Table 2.13.

Different states of monomer and polymer are involved and, therefore, ΔH_p^0 and ΔS_p^0 of each monomer depend on the final and initial states. Table 2.12 shows the corresponding values. Some of them have been determined directly from polymerization data (plotting $[M]_e$ vs 1/T) others have been calculated from heats of sublimation, fusion, and evaporation.

Table 2.13. Thermodynamic parameters of polyoxymethylene synthesis [81]

No.	System	Solvent	$-\Delta H$, $kJ \cdot mol^{-1}$	$-\Delta S^0$, $J \cdot K^{-1} \cdot mol^{-1}$	$-\Delta G^0_{298}$, $kJ \cdot mol^{-1}$
1	TXN_s-POM_s	$C_6H_5NO_2$	8.4 ± 0.2	3.1 ± 0.8	—0.8 ± 0.4
2	TXN_l-POM_s	—	9.2 ± 0.6	9.2 ± 0.8	
3	TXN_l-POM_c	—	21.6 ± 0.8	49 ± 8	7 ± 2
4	TXN_c-POM_c	—	6.3 ± 0.6	3 ± 8	5.5 ± 2.5
5	$TTXN_s$-POM_s	$C_6H_5NO_2$	9.2 ± 0.4	12 ± 1.7	5.5 ± 0.8
6	$TTXN_l$-POM_s	—	10.9 ± 0.8	—7 ± 2	
7	$TTXN_l$-POM_c	—	26.4 ± 1.2	49 ± 13	12 ± 4
8	$TTXN_c$-POM_c	—	3.8 ± 0.6	—12 ± 12	7.5 ± 4
9	FA_l-POM_c	—	46.5 ± 2.0	87 ± 9	
10	FA_c-POM_c	—	39 ± 3	38 ± 16	

Table 2.14. Thermodynamic parameters of 1,3,5-trioxane polymerization in C_6H_6 solvent [82)]

System	$-\Delta H$, $kJ \cdot mol^{-1}$	$-\Delta S$, $J \cdot K^{-1} \cdot mol^{-1}$	$[TXN]_e^{22°}$, $mol \cdot l^{-1}$
TXN_s-POM_s	8 ± 0.9	27 ± 3	0.88
TXN_s-POM_c	28 ± 0.6	69 ± 2	0.04

The most negative ΔH_p^0 and ΔS_p^0 values were obtained for the polymerization of liquid TXN to crystalline polymer, since the fusion of both monomer and polymer is exothermic with ΔH (fusion) $= -15\ kJ \cdot mol^{-1}$ (monomer), ΔH (fusion) $= -20\ kJ \times mol^{-1}$ (polymer) and endoentropic with ΔS (fusion) $= 44\ J \cdot mol^{-1} \cdot K^{-1}$ (monomer), ΔS (fusion) $= 49\ J \cdot mol^{-1} \cdot K^{-1}$ (polymer) [81)].

Recently a model system involving TXN, TTXN, and low molecular weight polyformaldehyde of the formula $CH_3OCH_2O(CH_2O)_4CH_2OCH_3$ was studied using C_6H_6 solvent by ^1H- and ^{13}C-NMR spectroscopy to determine the thermodynamic parameters of polymerization. Results are shown in Table 2.14.

The equilibrium concentration of TXN depends on the state of polymer (crystalline or in solution). $[TXN]_e$ is also solvent dependent and correspondingly $[TXN]_{e,sc} = 0.14$, 0.10, and $0.003\ mol \cdot l^{-1}$ in CH_2Cl_2, $C_6H_5NO_2$, and $C_{10}H_{22}$, respectively [81)]. These data have been interpreted in terms of the influence of the dielectric constant of the solvent, assuming that the strongest solvent-monomer interactions occur for compounds of similar polarity. We think, however, that these data should rather be correlated with ability of solvent to form complexes with TXN (i.e. with solvent acidity expressed by pK_a). CH_2Cl_2 as the most acidic among the studied solvents may form complexes with monomers whereas other solvents give weaker interactions. n-Decane is the weakest acid and it gives the lowest $[TXN]_e$. Similar behaviour is observed for TTXN which is more strained and thus gives lower values of equilibrium concentration. In the range of 20–25° $[TTXN]_{e,sc'} = 7.5 \cdot 10^{-3}$, $6.5 \cdot 10^{-3}$, and $5 \cdot 10^{-4}\ mol \cdot l^{-1}$ in CH_2Cl_2, $C_6H_5NO_2$, and $C_{10}H_{22}$, respectively.

The thermodynamics of polymerization and crystallization of TXN is discussed in more detail in Chapter 7.

2.9 Thermodynamics of Copolymerization

General Considerations

The scheme of fully reversible copolymerization of monomers M_1 and M_2, proceeding with the active species m_1^* and m_2^*, is shown below:

$$\begin{aligned}
m_1^* + M_1 &\overset{K_{11}}{\rightleftharpoons} m_1m_1^* \\
m_1^* + M_2 &\overset{K_{12}}{\rightleftharpoons} m_1m_2^* \\
m_2^* + M_1 &\overset{K_{21}}{\rightleftharpoons} m_2m_1^* \\
m_2^* + M_2 &\overset{K_{22}}{\rightleftharpoons} m_2m_2^*
\end{aligned} \tag{2-50}$$

Let us denote the proportions of various chain ends as:

$$\alpha = \frac{[m_1m_1^*]}{[m_1^*]}, \qquad \beta = \frac{[m_2m_2^*]}{[m_2^*]} \tag{2-51}$$

where $0 < \alpha, \beta < 1$.

The equilibrium concentrations in copolymerization according to Eq. 2-50 are expressed through the proportions of the chain ends and equilibrium constants:

$$\begin{aligned}[M_1]_e &= \alpha/K_{11} \\ [M_2]_e &= \beta/K_{22}\end{aligned} \tag{2-52}$$

The equilibrium constants K_{11} and K_{22} are reciprocals of the monomer equilibrium concentrations in homopropagations. Therefore, as α and β are less than unity, the equilibrium concentration of monomers, M_1 and M_2 is reduced in copolymerization ($[M]_e < [M]_e$ (homo)):

$$\begin{aligned}[M_1]_e &= \alpha[M_1]_e \quad \text{(homo)} \\ [M_2]_e &= \beta[M_2]_e \quad \text{(homo)}\end{aligned} \tag{2-53}$$

Monomer equilibrium concentrations in copolymerization will be discussed in more detail in the next section, which is devoted to the polymerization of 1,3,5-trioxepane.

According to Scheme (2-50) copolymerization consists of four different reactions characterized by specific enthalpy and entropy changes. If the comonomer M_1 content in the copolymer equals x, the total enthalpy and entropy change will be:

$$\begin{aligned}\Delta H &= x \cdot \Delta H_{11} + (1 - x) \cdot \Delta H_{22} + x \cdot (1 - x) \\ &\quad \times \Psi(\Delta H_{12} + \Delta H_{21} - \Delta H_{22} - \Delta H_{11}) \\ \Delta S &= x \cdot \Delta S_{11} + (1 - x) \cdot \Delta S_{22} + x \cdot (1 - x) \\ &\quad \times \Psi(\Delta S_{12} + \Delta S_{21} - \Delta S_{22} - \Delta S_{11}) + \Delta S_{mix}\end{aligned} \tag{2-54}$$

where Ψ is the randomness parameter, i.e. the ratio of the probability of finding M_2 next to M_1 (P_{21}) to the proportion of M_1 molecules in copolymer x. Thus $\Psi = P_{21}/x$. $\Delta S_{mix} = R \cdot (1 - \Psi \cdot x) \cdot \ln(1 - \Psi \cdot x) + \Psi \cdot x \ln(\Psi \cdot x)$; for exactly alternating ($P_{21} = 1$) and block ($P_{21} = 0$) copolymers $\Delta S_{mix} = 0$.

If the enthalpies of the cross-propagation steps are more negative than those of the homo-propagation steps, then copolymerization will be enhanced (the overall $-\Delta H$, is higher). When the opposite is true, copolymerization will be hampered. For identical enthalpies and entropies of homo- and cross-propagations, there still remains the term ΔS_{mix}, which for ideally random copolymers ($\Delta S_{mix} \simeq 8\ J \cdot mol^{-1} \times deg^{-1}$) reduces the equilibrium concentration of each comonomer by a factor of two. In such cases $\alpha = \beta = 1/2$.

If, however, repulsion between unlike units is larger than in the homopolymers, block copolymer will be formed ($\Psi = 0$).

A small amount of comonomer is, however, very easily introduced into the copolymer due to the marked increase of randomization.

Equilibrium and Stationary Concentrations in Copolymerization

The case of 1,3,5-trioxepane

Certain cyclic monomers are build up from two or more other simpler compounds which can serve as monomers by themselves. Thus, during the polymerization of more complex cyclic monomers simpler structures may emerge during the polymerization process. Szwarc [83] for the first time quantitatively treated the thermodynamics of polymerization of such a composite monomer, reevaluating the experimental data of Schulz et al. [84]. Below we follow Szwarc's reasoning applied for the cationic polymerization of 1,3,5-trioxepane (TXP), a monomer comprising formaldehyde (FA) and 1,3-dioxolane (DXL). Under equilibrium conditions besides the unreacted monomer both FA and DXL are also present:

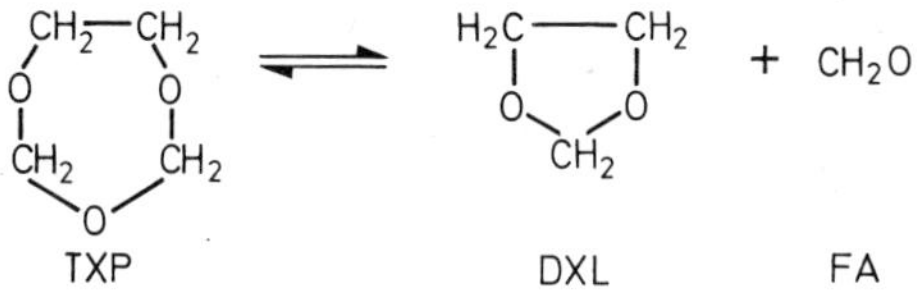

(2-55)

Hence, this system can also be treated as a copolymerization (or terpolymerization) of three monomers. The rate of propagation of each monomer is proportional to the total concentration of the active centers and monomer concentration. Depropagation can proceed for a given monomer only within the fraction of active species. Thus, assuming, for the sake of simplicity, the carbenium nature of the active species, the following scheme may be considered:

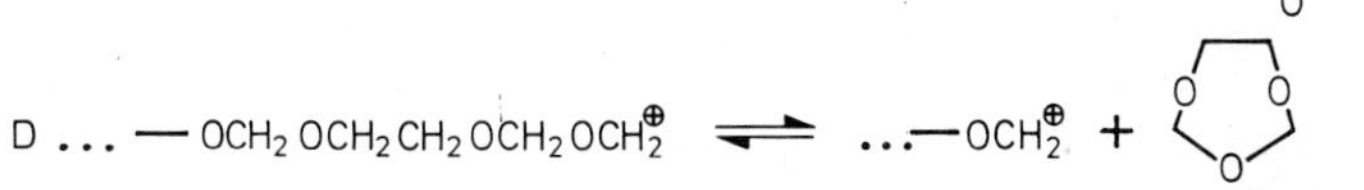

(2-56)

(We have indicated in Vol. I (Adv. Pol. Sci. *37* (1980)) that the predominant structures are oxonium ions; the use of carbonium ions, shown in the above scheme, leads to the same cruclusions).

By 'equilibrium concentration' Szwarc meant a concentration of a given monomer in equilibrium with its own growing species. If there is a multiplicity of growing structures, this monomer is involved in a number of equilibria and the measured value is the copolymerization equilibrium concentration ('stationary concentration'), differing from the equilibrium concentration in homopolymerization.

Szwarc concluded that the observed stationary concentrations for each monomer are equal to the product of the equilibrium concentrations measured in homopolymerization and fraction of the corresponding species (e.g., for DXL: A (f_A)). Of course $f_A > f_C$, $f_B > f_D$, and $f_A + f_B = 1$, using the abbreviations from Eq. (2-56).

Szwarc assumed that the dyad distribution in the copolymer corresponds to the distribution of the related active species. It follows that the ratio of concentrations of segments $[OCH_2CH_2O]/[OCH_2O] = \varrho$ was equal to fraction f_A, $f_B = (1 - \varrho)$ and $f_C = f_D = \varrho \cdot (1 - \varrho)$. Thus, according to Szwarc, the stationary monomer concentrations in contact with living polymers are as follows:

$$
\begin{aligned}
[DXL]_{ST} &= [DXL]_e \cdot \varrho \\
[FA]_{ST} &= [FA]_e \cdot (1 - \varrho) \\
[TXP]_{ST} &= [TXP]_e \cdot \varrho \cdot (1 - \varrho)
\end{aligned}
\qquad (2\text{-}57)
$$

There is a formal analogy between Eq. (2-57) and Eq. (2-53) that describes the simple two comonomer system.

The major assumption concerning the similar distribution of dyads in the chain and active species is not necessary to be fulfilled. The concentrations of the heterodyads always have to be [12] = [21], however, the concentrations of active species do not ($[P^*_{12}] \neq [P^*_{21}]$) although it may happen that $[P^*_{12}] = [P^*_{21}]$.

In the original paper by Schulz [84] the equilibrium concentrations of TXP and DXL ($[M_1]_e$), as well as the copolymer composition were given at different temperatures, allowing calculations of K and the thermodynamic parameters of TXP polymerization. These values, calculated directly from dependence of $[TXP]_e$ on temperature by Schulz, and Szwarc (Eq. 2-57), are given in Table 2.15.

Table 2.15. Thermodynamic parameters of 1,3,5-trioxepane polymerization at 25 °C

	Schulz [84]	Szwarc (2-57) [83]
ΔH^0_{ss}, $kJ \cdot mol^{-1}$	−6.6	−6.9
ΔS^0_{ss}, $J \cdot mol^{-1} \cdot K^{-1}$	−18.9	−31.5

The thermodynamic parameters of polymerization of 1,3-dioxepane, which is the closest model of TXP, are as follows: $\Delta H_{ss} = -13.4\ kJ \cdot mol^{-1}$, $\Delta S^0_{ss} = 39\ J \cdot mol^{-1} \cdot K^{-1}$.

Thus, using the above treatment of Szwarc it has been possible to estimate the thermodynamic parameters of conversion of TXP into its regular polymer, which is not otherwise available.

Distinctive Features of the Cationic Ring-Opening Copolymerizations

The active species in the large majority of cationic polymerization of heterocyclics are onium ions. Therefore, the depropagation step is always accompanied by the formation of a strained onium ion from the penultimate unit. Thus, the presence of a penultimate unit that arose of a strained monomer, will retard depropagation although depropagation occurs easily in the homopolymer. Copolymerization of BCMO with tetrahydropyrane is a good example of such a system [85].

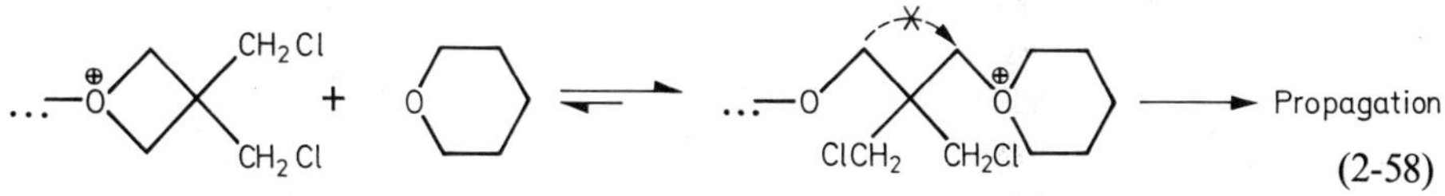

(2-58)

Once the tetrahydropyranium cation is formed it can not depropagate if the penultimate unit is BCMO. This is because the formation of the four-membered ring of BCMO from its linear segment must overcome an energy barrier of more than 100 kJ $\times \mathrm{mol}^{-1}$ in order to form the corresponding oxetanium cation and to replace the tetrahydropyranium cation ($\sim$THP$^+$) at the chain end.

Therefore, $\sim$THP$^+$ (as shown above on the left hand side of Eq. 2-65) will only propagate (with BCMO or THP). Indeed up to 50% THP units have been found in the corresponding copolymer (cf. Table 2.16).

When the penultimate unit effect has to be taken into account the set of four equations in Scheme (2-50) has to be replaced by eight equations. For example, instead of the first equation from Scheme 2-50 two new equilibria must be considered:

$$m_1 m_1^* + M_1 \overset{K_{11}}{\rightleftharpoons} m_1 m_1 m_1^*$$
$$m_2 m_1^* + M_1 \overset{K_{21}}{\rightleftharpoons} m_2 m_1 m_1^* \qquad (2\text{-}59)$$

Table 2.16. Examples of copolymerization of non-homopolymerizable monomers

No.	Monomer	Comonomer	Conditions	Products	Ref.
1	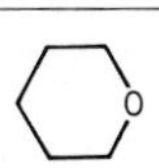	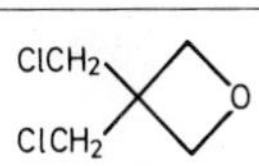	$AlEt_3$—H_2O (2:1), 5%, bulk, [THP]:[BCMO] = 6:4, 50°, 24 days	80% yield, 50% THP content, η_{sp}/c = 0.3 dl/g	85)
2	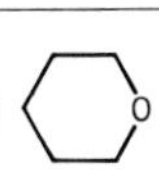	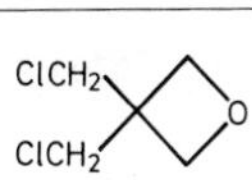	$BF_3 \cdot OEt_2$ 5%, bulk, 1:1 ratio, 0°, 7 days,	82% yield, 33% THP content, soluble wax	86)
3	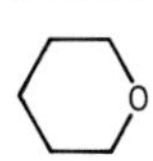	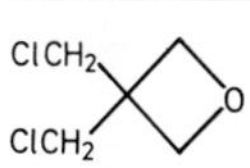	$BF_3 \cdot OEt_2$, bulk, 0°, 7 days	72% yield, 25% THP content, resinous solid soluble in $CHCl_3$	87)
4	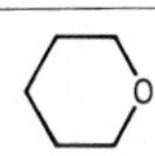	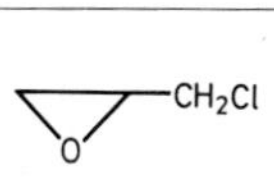	$BF_3 \cdot OEt_2$, 0.5%, bulk, 0°, 7 days	12% yield, 52% THP content, viscous liquid	87)
5	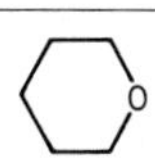	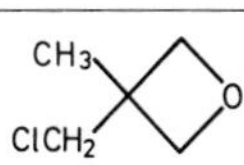	$BF_3 \cdot OEt$, $AlEt_3$-H_2O (1:1)	low content of THP	88)
6	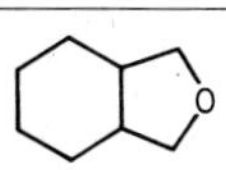 *cis*	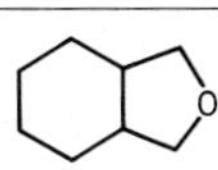 *trans*	PF_5, 0°, bulk	up to 20% of cis izomer content, $\overline{M}_n = 3 \cdot 10^4$, $\overline{M}_w = 2 \cdot 10^5$	89)

In this system the thermodynamics of any of the four steps from Eq. 2-50 depends on the structure of the penultimate unit in propagation and the antepenultimate unit in depropagation. Such systems will be described in more detail in Chapter 15.

Copolymerization of Monomers Unable to Homopolymerize

A number of cyclic monomers are known to be unable to homopolymerize mostly because of low ring strain. Some of the medium strained rings polymerize only at a sufficiently low temperature or at a sufficiently high monomer concentration. The latter, however, is limited by the bulk concentration, and the former by the kinetics.

Copolymerization with a properly chosen comonomer, however, may be a way to introduce into the polymer chain monomers that are otherwise unable to polymerize, or to enhance the conversion of monomers showing limited polymerizability. The previous section describes one of the possibilities related to retardation of depropagation. An analogous system is the radical copolymerization of SO_2 with olefins.

Some examples of the cationic copolymerization of monomers unable to homopolymerize are given in Table 2.16.

The kinetic treatment of the copolymerization involving non-homopolymerizable monomer is discussed in detail in Chapter 15.

2.10 References

1. Ivin, K. J.: In: Encyclopedia of Polymer Science and Technology, Suppl. Vol. 2, p. 700, Wiley, New York 1977
2. Ivin, K. J.: In: Reactivity, Mechanism and Structure in Polymer Chemistry, Jenkins, A. D., Ledwith, A. (Eds.) Wiley Interscience, London 1974
3. Joshi, R. M., Zwolinski, B. J.: Macromolecules *1*, 25 (1968)
4. Parks, G. S., Mosher, H. P.: J. Polym. Sci. *A1*, 1979 (1963)
5. Joshi, R. M., Zwolinski, B. J.: In: Polymerization of Vinyl Monomers, Ham, G. (Ed.), Marcel Dekker, New York 1969
6. Dainton, F. S., Evans, D. M., Hoare, F. E., Melia, T. P.: Polymer *3*, 263 (1962)
7. Melia, T. P.: Polymer *7*, 640 (1966)
8. Matyjaszewski, K.: unpublished results
9. Melia, T. P., Bailey, D., Tyson, A.: J. Appl. Chem. *17*, 15 (1967)
10. Carmichael, J. B., Winger, R.: J. Polym. Sci. *A3*, 971 (1965)
11. Kałużynski, K., Penczek, S.: Makromol. Chem. *180*, 2289 (1979)
12. Van Ooteghem, D., Deveux, R., Goethals, E. J.: J. Polym. Sci. Polym. Symp. *56*, 459 (1976)
13. Ivin, K. J., Leonard, J.: Eur. Polym. J. *6*, 331 (1970)
14. Ivin, K. J., Leonard, J.: Polymer *6*, 621 (1965)
15. Leonard, J., Maheux, D.: J. Macromol. Sci. *A7*, 1421 (1973)
16. Penczek, S., Matyjaszewski, K.: J. Polym. Sci., Polym. Symp. *56*, 255 (1976)
17. Kuzub, L. I., Markevitch, M. A., Berlin, Al. Al., Enikolopyan, N. S.: Vysokomol. Soedin. *A10*, 2007 (1968)
18. Tobolsky, A. V., Eisenberg, A.: J. Amer. Chem. Soc. *82*, 289 (1960)
19. Szwarc, M.: Carbanions Living Polymers and Electron Transfer Process, Interscience, New York 1968
20. Heitz, W., Stix, W., Kress, J.: Preprints, Makro Mainz 1979 vol. 1, p. 200
21. Mateva, R., Kabaiwanov, W.: Makromol. Chem. *178*, 2609 (1977)
22. Ivanchev, S. S., Pavlyuchenko, V. N., Khaikin, S. Ya., Vladimirova, L. G., Balaev, G. A., Schevchuk, V. I.: J. Polym. Sci., Polym. Chem. Ed. *18*, 2051 (1980)
23. Penczek, S.: Makromol. Chem., Rapid Commun. *2*, 401 (1981)

24. Favier, J. C., Leonard, J.: Macromolecules *11*, 424 (1978)
25. Andruzzi, F., Pilcher, G., Virmani, Y., Plesch, P. H.: Makromol. Chem. *178*, 2367 (1977)
26. Joshi, R. M.: J. Macromol. Sci. Chem. *9*, 1309 (1975)
27. Carothers, W. H., Dorough, G. L., Van Natta, F. J.: J. Amer. Chem. Soc. *54*, 761 (1962)
28. Goldfarb, J. L., Belenkij, L. I.: Usp. Chim., (USSR) *29*, 470 (1960)
29. Kingston, S. B., Plesch, P. H.: Makromol. Chem. Rapid Commun. *1*, 749 (1980)
30. Van Kamp, H., Coops, J., Lambregts, W. A., Visser, B. J., Dekker, H.: Rec. Trav. Chim. *79*, 1226 (1960)
31. Kaarsemaker, S., Coops, J.: Rec. Trav. Chim. *71*, 261 (1952)
32. Dainton, F. S., Delvin, T. R. E., Small, P. A.: Trans. Faraday Soc. *51*, 1710 (1955)
33. Plesch, P. H., Westermann, P. H.: J. Polym. Sci. *C16*, 3837 (1968)
34. Plesch, P. H., Westermann, P. H.: Polymer *10*, 105 (1969)
35. Yamashita, Y., Mayumi, J., Kawakami, Y., Ito, K.: Macromolecules *13*, 1075 (1980)
36. Huisgen, R., Brade, H., Waltz, H., Glogger, I.: Chem. Ber. *90*, 1437 (1957)
37. Gutowsky, H. S., Holm, C. H.: J. Chem. Phys. *25*, 1228 (1956)
38. Tobolsky, A. V.: J. Polym. Sci. *C12*, 71 (1966)
39. Sawada, H.: "Thermodynamics of Polymerization", New York, Marcel Dekker, 1976
40. Dainton, F. S., Ivin, K. J.: Quart. Rev. *12*, 61 (1958)
41. Cairns, T. L., Joyce, R. M.: US patent 2455 912 (12.14.1948)
42. Kazanskij, K. S.: in: Soviet Encyclopedia of Polymers, Vol. I, p. 413, Moscow 1974
43. Kambara, S., Hatano, M.: J. Chem. Soc. Jap., Ind. Chem. Sect. *61*, 904 (1958)
44. Farthing, A. C.: J. Chem. Soc. *1955*, 3648
45. Garrido, L., Guzman, J., Riande, E.: Macromolecules *14*, 1132 (1981)
46. Kawai, W.: J. Polym. Sci. *A1*, *6*, 137 (1968)
47. Pihlaja, K.: Acta Chim. Scand. *25*, 451 (1971)
48. Penczek, S., Kubisa, P.: ACS Symp. Ser. *59*, 60 (1977)
49. Okada, M., Mita, K., Sumitomo, H.: Makromol. Chem. *177*, 2055 (1976)
50. Firat, Y., Plesch, P. H.: Makromol. Chem. *176*, 179 (1975)
51. Jedliński, Z., Łukaszczyk, J., Dudek, J., Gibas, M.: Macromolecules *9*, 622 (1976)
52. Okada, M., Yagi, K., Sumitomo, H.: Makromol. Chem. *163*, 225 (1973)
53. Okada, M., Sada, H., Sumitomo, H.: Makromol. Chem. *179*, 959 (1978)
54. Andruzzi, F., Barnes, D. S., Plesch, P. H.: Makromol. Chem. *176*, 2053 (1975)
55. Saegusa, T., Motoi, M., Suda, M., Kasai, K., Ueyama, A.: Polymer J. *11*, 113 (1979)
56. Kops, J., Spanggaard, H.: Makromol. Chem. *151*, 21 (1972)
57. Kops, J., Spanggaard, H.: Makromol. Chem. *175*, 3077 (1974)
58. Kops, J., Spanggaard, H.: Makromol. Chem. *176*, 299 (1975)
59. Yokoyama, Y., Hall, H. K.: Adv. Pol. Sci. *42*, 107 (1982)
60. Bawn, C. E. H., Bell, R. M., Ledwith, A.: Polymer *6*, 95 (1965)
61. Rahman, M., Weale, K. E.: Polymer *11*, 122 (1970)
62. Skuratov, S. M., Strepikheev, A. A., Kozina, M. P.: Dokl. Akad. Nauk SSSR *117*, 452 (1957)
63. Cass, R. C., Fletcher, S. E., Mortimer, C. T., Springall, H. D., White, T. R.: J. Chem. Soc. *1958*, 1406
64. Busfield, W. K., Lee, R. M., Merigold, D.: Makromol. Chem. *156*, 183 (1972)
65. Clegg, C. A., Gee, D. R., Melia, T. P., Tyson, A.: Polymer *9*, 501 (1968)
66. Lebedev, B. V., Litiagov, B. Ya.: Vysokomol. Soed. *A19*, 2283 (1977)
67. Dreyfuss, M. P., Dreyfuss, P.: J. Pol. Sci. *A3*, 3261 (1965)
68. Sims, D.: J. Chem. Soc. *1964*, 864
69. Rozenberg, B. A., Chekhuta, O. M., Lyudvig, E. B., Gantmakher, A. R., Medvedev, S. S.: Vysokomol. Soed. *6*, 2090 (1964)
70. Pruckmayr, G., Wu, T. K.: ACS Symp. Ser. *103*, 237 (1979)
71. Kharitonova, L. A., Rahova, G. V., Shaginyan, A. A., Enikolopyan, N. S.: Vysokomol. Soed. *A9*, 2586 (1967)
72. Skuratov, S. M., Strepikheev, A. A., Shtekher, S. M., Volokhina, A. V.: Dokl. Akad. Nauk SSSR *117*, 263 (1957)
73. Clegg, G. A., Melia, T. P.: Polymer *10*, 912 (1969)
74. Kučera, M., Pichler, Y.: Vysokomol. Soed. *7* 10 (1965)
75. Plesch, P. H., Westermann: J. Polym. Sci. *C16*, 3837 (1965)

76. Berman, E. L., Lyudvig, E. B., Ponomarenko, V. A., Medvedev, S. S.: Vysokomol. Soed. *A11*, 200 (1969)
77. Yamashita, Y., Okada, M., Suyama, K., Kashara, M.: Makromol. Chem. *114*, 1461 (1968)
78. Gevorgyan, G. M., Bonetskaya, A. K., Shuratov, S. M., Enikolopyan, N. S.: Vysokomol. Soed. *A9*, 1363 (1967)
79. Binet, R., Leonard, J.: Polymer *14*, 355 (1973)
80. Penczek, S., Matyjaszewski, K.: J. Polym. Sci. Polym. Symp. *56*, 256 (1976)
81. Berlin, Al. Al., Bogdanova, K. A., Rakova, G. V., Enikolopyan, N. S.: Vysokomol. Soed. *A17*, 643 (1975)
82. Oreshnikova, T. F., Gruznov, A. G., Bulay, A. H., Slonim, I. Y., Romanov, L. M.: Vysokomol. Soed. *A19*, 24 (1977)
83. Szwarc, M.: Makromol. Chem., Suppl. *3*, 327 (1979)
84. Schulz, R. C., Albrecht, K., Rentsch, C., Tran Thi, A. V.: ACS Symp. Ser. *59*, 77 (1977)
85. Tsuda, T., Yamashita, Y.: Makromol. Chem. *99*, 297 (1966)
86. Furukawa, J.: Polymer *3*, 487 (1962)
87. Ishigaki, A., Shono, T., Hachihama, Y.: Makromol. Chem. *79*, 170 (1965)
88. Geller, N. M., Kropachev, V. A., Dolgoplosk, B. A.: Vysokomol. Soed. *A9*, 575 (1967)
89. Kops, J., Spanggaard, H.: Makromol. Chem. *175*, 3077 (1974)

3 Cyclic Oligomers in the Cationic Ring Opening Polymerization

3.1 Introduction

Macrocyclics (starting with dimers) have been observed in every cationic ring-opening system, provided the authors specifically searched for cyclic entities. In the polymerization of THF the first report indicated that macrocyclics are not formed [1] but shortly thereafter Pruckmayr corrected this view and described a series of cyclic oligomers identified by gas chromatography coupled with mass spectroscopy [2,3].

The content and distribution of macrocyclics depend very much on monomer structure. Ethylene oxide exclusively gives a dimer (1,4-dioxane), while in other systems (e.g. 1,2-propylene oxide, epichlorohydrin, tert-butyl ethylene oxide) cyclic tetramers are predominant products [4]. 1,3-Dioxolane or hexamethyltrisiloxane yield a complete sequence of cyclic oligomers, without any preference of a specific product [5].

It is even more striking that different cyclic oligomers can be formed depending on initiator used. Cyclic tetramer is formed by polymerization of epichlorohydrin initiated with the $BF_3 \cdot$ ether complex [6], whereas with protonic acid as initiator, cyclization predominantly gives cyclic dimer [6]. Synthetic aspects of macrocyclization in the cationic ring-opening polymerization have been reviewed by Goethals [4]. Semlyen discussed the application of the Jacobson-Stockmayer theory for these systems [5].

This Section gives a uniform treatment of the mechanism of macrocyclizations which may be applied to all systems mentioned above and listed in Table 3.1.

Dimers, trimers and tetramers form either separately or as members of a continuous spectrum of cyclic oligomers. These products are formed either by end-to-end cyclization (end-biting) or by back-biting reactions involving tertiary onium ions as discussed further in this section:

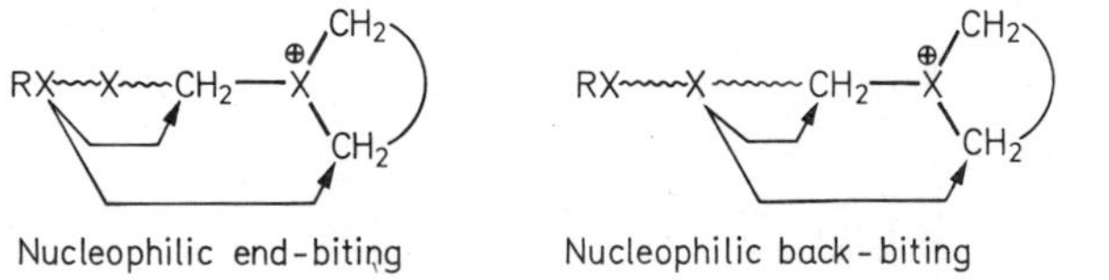

(3-1)

To form a given macrocycle, the chain must assume a conformation enabling ring closure and then the chemical reaction may occur. Thus, the probability of cyclization depends on the conformational probability which is governed by such factors

Table 3.1. Cyclic oligomers in the cationic polymerization of heterocyclic monomers

Ring-size	Cyclic ethers	Cyclic acetals	Cyclic sulfides	Cyclic amines
3	*Oxiranes* (cf. 4.2.2.) Extensive, fast cyclization (up to 100%) in the presence of typical cationic initiators (BF_3, $Et_3O^+A^-$). Strong preference for cyclic dimer, sometimes tetramer, less for trimer, pentamer. Cyclization hampered in the presence of -OH containing compounds. Tendency for cyclization considerably reduced when polymerization initiated with R_3Al/H_2O.	Not known	*Thiiranes* [4] Significant (up to 50%), slow cyclization (usually following polymerization) in polymerizations initiated by onium ions. Strong preference for cyclic dimer and/or tetramer; cyclization often accompanied by side-reactions leading to cycles containing disulfide bonds.	*Aziridines* [4] Extensive (up to 100%) slow cyclization (usually following polymerization) in the presence of typical cationic initiators; strong preference for cyclic dimer, less for cyclic tetramer.
4	*Oxetanes* (cf. Sect. 5.3.5.) Significant, fast cyclization (up to 20%) in the presence of BF_3 and $Et_3O^+A^-$. Strong preference for cyclic tetramer, less for trimer. Cyclization considerably reduced in $R_3Al—H_2O$ initiated polymerization	Not known	*Thietanes* [4] No data	*Azetidines* [4] No data
5	Tetrahydrofuran [2,3] Limited, slow cyclization (up to ~3%) in the presence of $HOSO_2CF_3$. Tendency for cyclization reduced in $ROSO_2CF_3$ initiated polymerization. The whole spectrum of oligomers formed (n = 2 to n = 8 detected) with slight preference for cyclic tetramer.	1,3-Dioxolane [14] Limited, fast cyclization (up to a few % in bulk) in the presence of BF_3. Distribution of oligomers as predicted by the J—S theory (with exception of n = 2 to n = 5) (cf. Sect. 3.2.2).	No data	No data

6	Non-polymerizable	1,3-dioxane: non-polymerizable Formation of cyclic dimer and tetramer was reported [31]. 1,3,5-trioxane: formation [32,33] of cyclic oligo- and polymers reported; no quantitative data.	Non-polymerizable Siloxanes [5] Hexamethyltrisiloxane Considerable cyclization. Distribution of cyclic oligomers as predicted by the J—S theory	Non-polymerizable
6	No data	Formals of di-, tri-, tetraethylene glycols Extensive, fast cyclization in the presence of $HOSO_2CF_3$ [7,8]. Kinetic enhancement observed (cf. Sect. 3.3). Concentration of lower cyclic oligomers (n = 2, 3, 4) high at the early stage of reaction, decreases in the course of the process.		

as ring size, substitution and chain flexibility, as well as on the relative nucleophilicity of the heteroatom in the chain which will react to form a particular ring.

Usually, the macrocyclics are observed and isolated at thermodynamic equilibrium. It may, however, happen, as it has been shown in the pioneering work of Schulz et al. [7,8], that the concentration of some macrocyclics goes through a maximum as a function of monomer conversion. This indicates end-to-end cyclization.

These and related phenomena can be explained in terms of the thermodynamic theory of macrocyclics distribution, formulated by Jacobson and Stockmayer [9] and its kinetic extension [10]. The Jacobson-Stockmayer theory, relating the distribution of cyclic oligomers to the conformational probability of ring closure, does not take into account kinetic limitations and has mostly been used as a convenient tool for studying the conformation of macromolecules in solution [5]. A number of papers appeared in which distribution of cyclic oligomers was studied with this aim and which ignored mechanistic and kinetic aspects of the cyclization processes.

Various authors, who studied the mechanism of cationic ring-opening polymerization, observed the formation of some specific cyclic oligomers; there were, however, no attempts to rationalize observed distribution on theoretical grounds, i.e. the preferential formation of certain oligomers or the apparent absence of cyclic oligomers in some systems. Several reviews have been published on macrocyclization but usually they concentrate either on the thermodynamic [5] or synthetic aspects [4]. Indeed, both these approaches, "thermodynamic" and "synthetic", were developed simultaneously but have not correlated with each other. In the most papers concerned with synthesis the Jacobson-Stockmayer theory was not even quoted.

This apparent duality has led to certain misunderstanding and confusion and for the reader less acquainted with this field the relation between theoretical distribution predicted by Jacobson-Stockmayer theory and the distributions observed in real systems (frequently different from those predicted by the theory) could be not clear. Thus, in this section we describe the theories and then explain major experimental results in terms of the thermodynamics and kinetics of cyclization.

The Jacobson-Stockmayer theory is based on several assumptions and describes only systems in true thermodynamic equilibrium. Some systems fulfill these assumptions. In most real systems, however, either some of these assumptions do not apply or the state of thermodynamic equilibrium cannot practically be reached. For such systems the distribution of cyclic oligomers cannot be described exclusively by the theory based on the idealized thermodynamic approach and the influence of kinetic factors has to be taken into account.

We first present the thermodynamic theory of cyclization and then discuss possible deviations which prevent establishing thermodynamically controlled distributions and lead to kinetic control of cyclization.

3.2 The Jacobson-Stockmayer Cyclization Theory

The theory of cyclization developed by Jacobson and Stockmayer describes the following equilibrium:

$$-\mathrm{M}_y- \rightleftharpoons \llcorner \mathrm{M} \lrcorner_x + -\mathrm{M}_{\overline{y-x}}$$

where $-M_y-$ denotes the linear macromolecule with polymerization degree equal to y and $\lfloor\overline{M_x}\rfloor$ denotes the cyclic macromolecule with polymerization degree equal to x. Theory includes the following assumptions [9,11–13]:

1. chains in solution obey Gaussian statistics;
2. all rings formed are strainless;
3. reactivity of all reaction sites along the chain is the same;
4. thermodynamic distribution is attained.

Based on these assumptions the concentration of the macroring of a given size is governed solely by the probability of attaining a conformation which enables the closure of this particular macrocycle.

The final equation derived from the mathematical treatment [5,9], is

$$K_n = [\lfloor\overline{M_n}\rfloor] = A \cdot n^{-5/2} \tag{3-2}$$

where K_n is the equilibrium constant for the formation of the cyclic n-mer, $[\lfloor\overline{M_n}\rfloor]$ denotes an equilibrium concentration of the cyclic n-mer, A is a proportionality coefficient and n is the degree of polymerization.

Thus, for each individual macrocycle a certain thermodynamic equilibrium concentration exists having the same physical meaning as the equilibrium monomer concentration. Increasing the ring size n, the equilibrium oligomer concentration decreases according to Eq. (3-2).

The theory requires that cyclization proceeds with zero enthalpy change (all rings are strainless), i.e., cyclization is exclusively entropy driven and equilibrium is attained by maximization of the entropy. Thus, the equilibrium oligomer concentration, contrary to the equilibrium monomer concentration, does not depend on temperature.

As follows from Eq. (3-2), plotting log $[\lfloor\overline{M_n}\rfloor]$ vs. log n should give a straight line with a slope equal to -2.5. The linearity of such a plot for a given system indicates that the Jacobson-Stockmayer theory can be applied.

3.2.1 Application of the Jacobson-Stockmayer Theory to Cationic Ring-opening Polymerizations

The equilibrium distribution of cyclic oligomers in the cationic polymerization of heterocycles was studied systematically for two systems, the polymerization of siloxanes initiated with protonic acids and the polymerization of 1,3-dioxolane initiated with the $BF_3 \cdot Et_2O$ complex [5]. Siloxanes are especially well suited for these studies as they are thermally stable and volatile. Thus, cyclic oligomers with n as high as 100 can be isolated and analysed by, e.g. GC-MS.

Both systems gave similar linear plots of log $[\lfloor\overline{M_n}\rfloor]$ vs. log n. Fig. 3.1 shows the plot for the 1,3-dioxolane – $BF_3 \cdot Et_2O$ system [14]. According to these results the distribution of cyclic oligomers conforms to the Jacobson-Stockmayer theory only for oligomers with $n > 5$. This is understandable since this theory is based on an assumption that the macrorings are strainless, which is apparently not true for the smaller rings. The linearity of the plot for $n > 5$ indicates that with this exception cyclization in the polymerization of 1,3-dioxolane initiated with $BF_3 \cdot OEt_2$ can be described by the theory.

Agreement with theory is also observed for the polymerization of cyclic siloxanes [5].

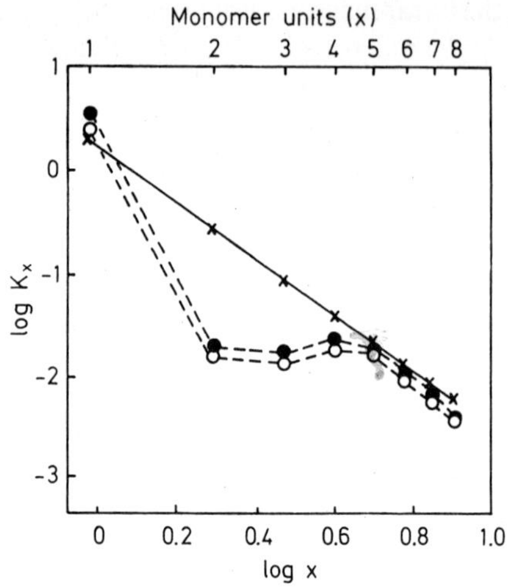

Fig. 3.1. Experimental molar cyclization equilibrium constant ($K_x = [M_x]$ mol $\cdot$ l^{-1}) in bulk (●) and solution (○) polymerization of 1,3-dioxolane compared with values (x) calculated by the Jacobson-Stockmayer theory
[1,3-dioxolane]$_0$ = bulk or 4.8 mol $\cdot$ l^{-1}, $[BF_3 \cdot OEt_2] \sim 10^{-2}$ mol $\cdot$ l^{-1}, $C_2H_4Cl_2$, 60 °C

3.2.2 Equilibrium Cyclic Oligomer Concentration

In the systems conforming to the Jacobson-Stockmayer theory (e.g. 1,3-dioxolane-BF_3) each cyclic oligomer has its own equilibrium concentration. Thus, after attaining the equilibrium, the concentration of each cyclic oligomer should be equal to this equilibrium concentration which depends neither on the initial monomer concentration nor on temperature.

If the equilibrium concentrations of the individual oligomers are taken together, the equilibrium concentration of the cyclic fraction is obtained. On the basis of the

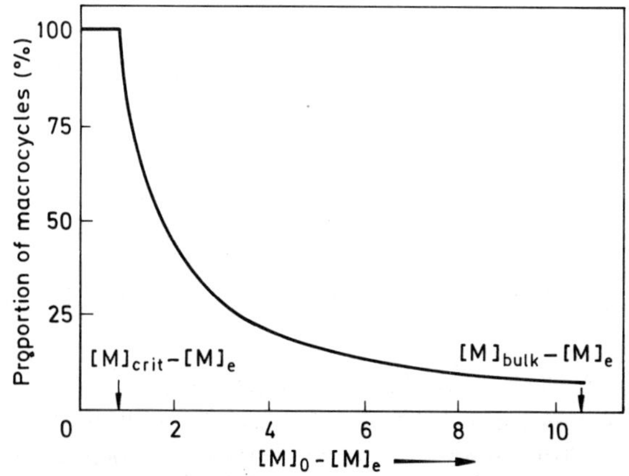

Fig. 3.2. Dependence of the proportion of macrocycles on the starting monomer concentration calculated on the basis of data reported in Ref. [14]

data given in Ref. [14] this equilibrium concentration of macrocyclics in 1,3-dioxolane polymerization is 0.85 mol $\cdot$ l^{-1} (sum from n = 2 to n = infinity). This value will be further denoted as $[M]_{crit.}$.

In the equilibrated system the polymer concentration is $[M]_0 - [M]_e$. Thus, if this value is lower than $[M]_{crit.}$ practically all polymer should exist in the form of cyclic macromolecules. This was pointed out by Jacobson and Stockmayer in their original paper [9] and discussed in detail by Enikolopyan et al. as early as 1967 [15], but has not received enough attention.

Recently we stressed that the proportion of the cyclic fraction is given by the ratio $[M]_{crit.}/([M]_0 - [M]_e - [M]_{crit.})$, thus it is different for different starting monomer concentrations, as shown in Fig. 3.2 for 1,3-dioxolane polymerization [18]. Figure 3.2 is based on the data of Ref. [14], (note that only the proportions change, the absolute value of $[M]_{crit.}$ does not depend on $[M]_0$).

Apparently, this largely forgotten phenomenon was, at least partially, responsible for the long-lasting controversy concerning the proportion of the cyclic fraction in cyclic acetal polymers [16].

3.2.3 Deviations due to Nonequivalency of Reaction Sites. End-to-end Cyclization

Equal reactivity of the reaction sites is a prerequisite to obtain the distribution of macrocycles predicted by the Jacobson-Stockmayer theory. This condition is usually fulfilled for reaction sites located along the chain as the macromolecule is constructed of the same repeating units. There is, however, at least one reaction site which is structurally different, i.e. the end-group:

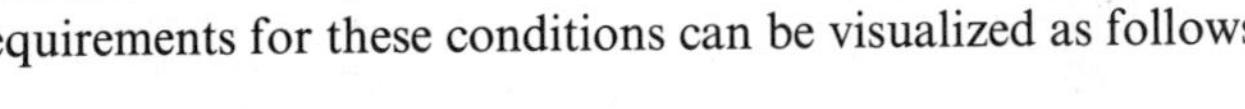

(3-4)

The reactivity of this reaction site (heteroatom as part of the end-group) toward the cationic active center can either be lower or higher than the reactivity of the heteroatom within the chain. In the former case the distribution predicted by theory is only slightly or not at all affected, whereas in the latter the distribution is altered considerably. The requirements for these conditions can be visualized as follows:

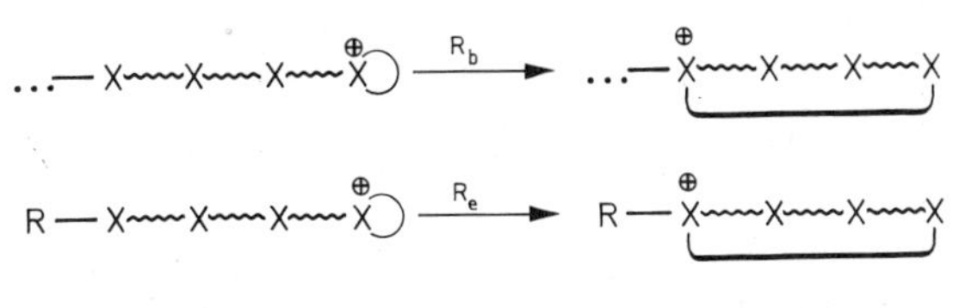

(3-5)

(where $R_e \gg R_b$ — for closure of the ring of the same size).

The mathematical treatment of the end-to-end cyclization scheme was given by us elsewhere [10]; in this review we will discuss the practical consequences of the cyclization mechanism in which end-to-end closure dominates over random back-biting.

Let us assume that for the linear n-mer, the end-to-end cyclization is favoured over both back-biting and addition of the next monomer molecule:

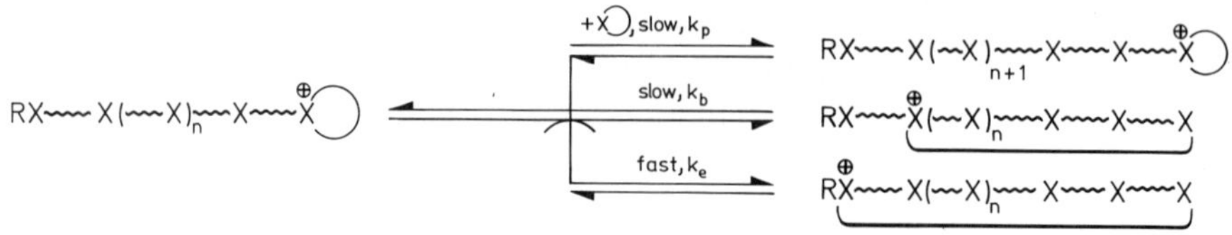

(3-6)

The rate constant of cyclization decreases with increasing n according to Eq. 3-7 [5]:

$$k_n = A' \cdot n^{-3/2} \tag{3-7}$$

(note —3/2 exponent in contrast to —5/2 for the thermodynamic equilibrium constant K_n).

In the idealized system all macromolecules grow simultaneously, i.e. at the early stages first dimers, then trimers, tetramers and so on are formed from all growing species. At the dimer stage the tendency for cylization, according to Eq. 3-7, is highest, and if the rate of cyclization is much higher than the rate of propagation then nearly all macromolecules will exist in the form of cyclic dimers. The same is true for the trimer stage, etc. The tendency for cyclization is lower, according to Eq. 3-7, for the trimer than for the dimer and thus the ratio of cyclic to linear trimer will be lower than that for a dimer. For higher oligomers the tendency for end-to-end cyclization decreases still further and the proportion of linear oligomers increases. This is illustrated in Fig. 3.3, giving the computer simulated kinetics of polymerization dominated by the end-to-end cyclization (all rings are assumed to be strainless).

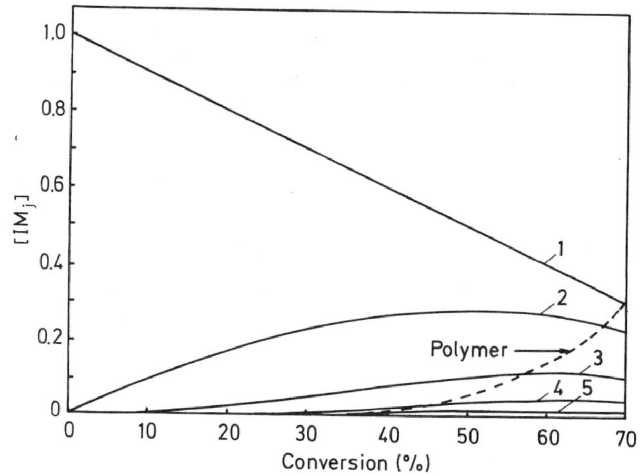

Fig. 3.3. Computer simulation of the dependence of the concentration of macrocycles $|M_j|$ (in moles of monomeric units per liter) on monomer conversion [10]

The characteristic features of the system dominated by end-to-end cyclization are[10]:

— end-to-end cylization dominates at the early stages of polymerization when the growing chains are short and it gradually decreases in importance with increasing chain length;
— the concentration of each individual cyclic oligomer passes through a maximum reflecting the changes in concentration of the corresponding linear oligomer, provided that propagation is not much faster than cyclization (cf. Sect. 3.2.5).

3.2.3.1 End-to-end Cyclization in Real Systems

The mechanism described in the previous section requires that the reactivity of the end-group towards growing species is considerably higher than the reactivity of any other reaction site in the polymer molecule. Thus, one may expect that zwitterionic polymerizations would belong to this class because end-to-end cyclization would involve reaction between two oppositely charged ions. Recently Goethals has shown that significant amounts of cyclic oligomers, mainly 13- and 16-membered rings, are formed in the spontaneous zwitter-ionic copolymerization of 1-alkylaziridines with β-propiolactone [17], a feature neglected in earlier work.

The polymerization of cyclic acetals initiated with protonic acids can also be described in terms of the end-to-end cyclization scheme [7,8,18,19].

The reactivity (nucleophilicity) of the oxygen atom in the hydroxyl end-group towards cationic growing species is so much higher than that of the backbone acetal oxygen that end-to-end cyclization is favoured over random back-biting. Consequently, the distribution of cyclic oligomers is governed mainly by the former process, at least for shorter chains. For longer chains (high $\overline{DP}_n$) the probability of both ends to meet becomes negligible and random back-biting may become the only process leading to the formation of cyclic oligomers [18].

In such a system concentration of each oligomer should pass through a maximum (provided that cyclization is fast enough to compete with chain growth) and then decreases approaching the equilibrium concentration predicted by the Jacobson-Stockmayer theory [10]. On the other hand, when random back-biting is the only

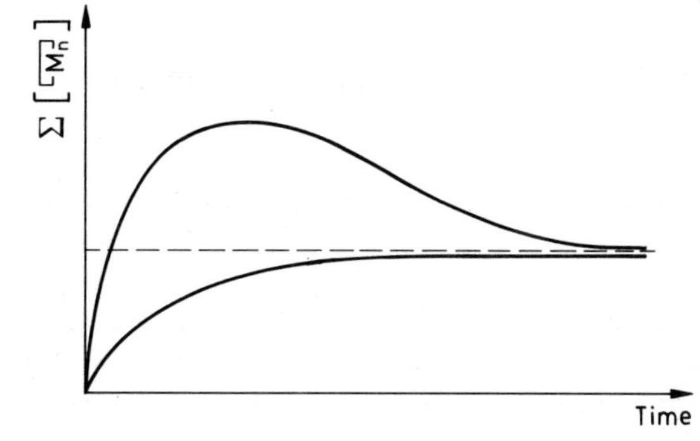

Fig. 3.4. Schematic representation of a change of concentration of a given cyclic oligomer for back-biting (lower curve) and end-biting (upper curve) schemes during polymerization. Concentration and time in arbitrary units

cyclization process possible, the concentration of a given oligomer will never exceed its final equilibrium concentration, as shown schematically in Fig. 3.4. Experimental evidence for these conditions has amply been demonstrated by Schulz [19] and Yamashita [20].

In the polymerization of 1,3,6,9-tetraoxacycloundecane (triethylene glycol formal) monomer is converted predominantly to low molecular weight cyclic oligomers during the early stages but later the concentration of macrocyclics decreases and the high molecular weight polymer is predominantly linear, as evidenced from end-group determination. Fig. 3.5, taken from a paper by Schulz and Rentsch [8] shows the changes of the cyclic dimer concentration, as determined by GPC during the polymerization.

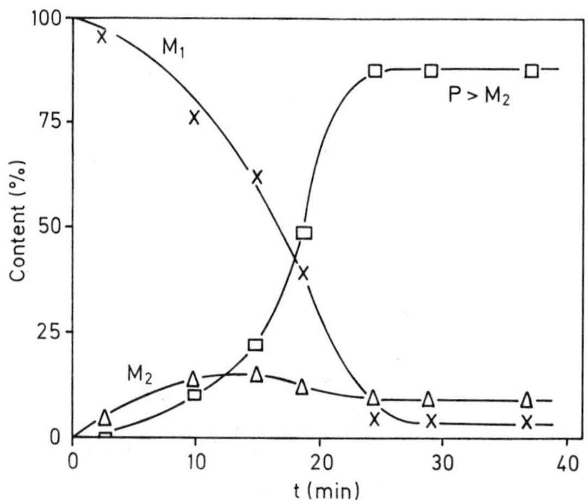

Fig. 3.5. Time-conversion curve for the consumption of monomer (M_1), formation of polymer (P) and formation of cyclic dimer (M_2) during the polymerization of triethylene glycol formal (0.5 mol · l^{-1} in CH_2Cl_2)
$|CF_3SO_3H|_0 = 5 \cdot 10^{-4}$ mol · l^{-1}, 0 °C; [8]

3.2.4 Deviations from the Theory due to Conformational Restrictions or Preferences

The Jacobson-Stockmayer cyclization theory is based on the assumption that all rings are strainless and that the conformational probability of ring closure is given by Eq. 3-2, i.e. conformational restrictions or preferences are absent. This assumption apparently cannot be fulfilled in real systems for small rings (as shown in Sect. 3.2.2). In some systems this leads to a lowering of the concentration of cyclics when compared with their equilibrium concentration. This is because the probability of small ring closure is reduced due to the strain caused not only by bond angle deformation but also by bond opposition and transannular interactions. In such a case, as in the discussed earlier 1,3-dioxolane-BF_3 system, the concentration of small cycles (up to 25–30 bonds) is lower than the calculated one [14].

In contrast, in several systems the formation of certain cyclic oligomers (for example dimers, trimers or tetramers) was observed at concentrations considerably exceeding theoretically predicted values [4]. This again shows the limitations of the theory dealing with idealized systems in which the probability of attaining a certain conformation is governed solely by the distance between the two reaction sites, without considering any specific effects caused by, e.g., substituent interactions.

Thus, it may seem that the probability of ring closure should not be higher than calculated on a theoretical basis, already including the assumption of a lack of ring strain. The theory requires, however, that all possible conformations of a macromolecule be equally probable. If we assume that, due to the interaction between substituents or electron pairs or to the presence of certain bonds in eclipsed position, some conformations are excluded, and if these factors do not operate in the conformations leading to ring closure, then the formation of this particular ring will be favoured.

This phenomenon may be illustrated by the example of polymerization of t-butyloxirane [21]. The cyclic tetramer of t-butyloxirane (formed at a yield up to 60%) is exclusively in the gauche$^+$-gauche$^+$-trans (G^+G^+T) conformation:

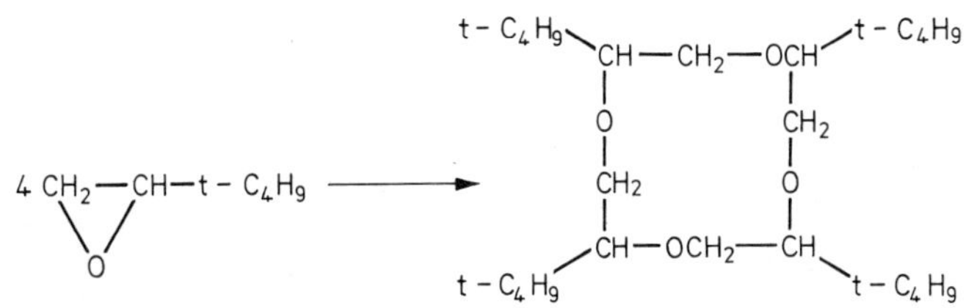

(3-8)

In contrast, the linear polymer, assumes the G^+G^-T conformation. Evidently the ability to form a cyclic tetramer depends on the conformation of the terminal fragment of the macromolecule. When it is in a right position to form the G^+G^+T conformer then cyclization occurs. When the conformation corresponds to the G^+G^-T arrangement, cyclization is less probable and the linear fragment is fixed by addition of the next monomer molecule.

The minimization of nonbonding interactions apparently also leads to the preferential formation of cyclic tetramers in the polymerization of 4-membered cyclic

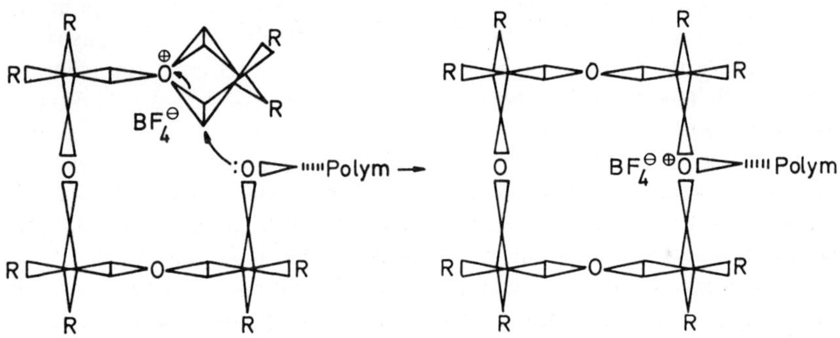

Fig. 3.6. Preferred conformation of growing polyoxetane chain leading to closure of 16 membered ring (cyclic tetramer) [22]

ethers (oxetanes). The oxetane tetramer contains 16 atoms in the ring, which allows a regular structure to be formed. This is shown in Fig. 3.6 in which all bonds have the gauche conformation [22,23]. Accordingly, there is a strong preference for the formation of cyclic tetramers in the polymerization of monomers containing 4-membered rings (oxetanes) but not 5-membered rings (1,3-dioxolane, THF) where the corresponding tetramers cannot adopt such a favourable conformation (in fact, growing chains cannot preferably attain a conformation leading to the formation of tetramer).

The conformational preference, leading to the formation of rings of a certain size, has ingeniously been used by Dale for the synthesis of crown ethers of a desired ring size by the cationic polymerization of ethylene oxide [24]. The principle of this method is based on the folding of the chain-end around a purposely introduced cation solvated specifically by a well-defined number of ethylene oxide units (cf. also Sect. 4.2.2.1):

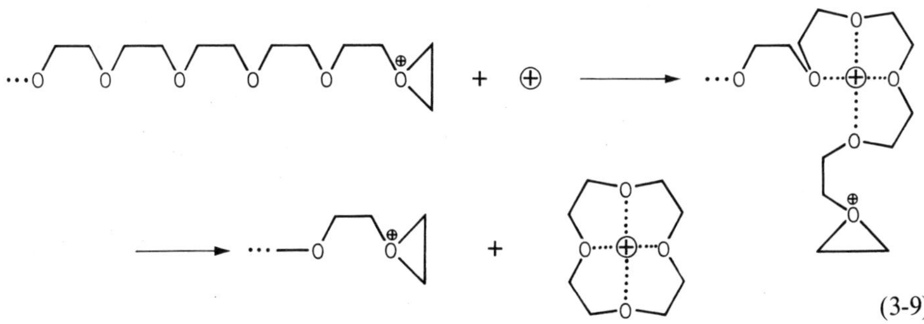

(3-9)

3.2.5 Kinetic Limitations

In the previous discussion we have neglected the kinetic factor, relating exclusively the concentration of cyclic oligomers to the conformational probability of their formation. In the majority of systems however, the kinetic factor cannot, be neglected, because usually no evidence is presented to show that the distribution already refers to equilibrium.

Until recently it was claimed that in the cationic polymerization of THF the final THF-polyTHF mixture contains no cyclic oligomers [1]. Further examination reveals that cyclic oligomers are formed, albeit slowly, when the living system is kept at the living polymer-monomer equilibrium for longer reaction times [2,3]. This is illustrated in Fig. 3.7, showing the increase of concentration of cyclic tetramer (in arbitrary units) with time [3]. Under these conditions the monomer-polymer equilibrium is reached in about 1 hr.

In polymerization studies the moment at which the final monomer conversion is reached is usually treated as the end of the reaction. Thus, the absence of cyclic oligomers observed earlier in THF polymerizations meant only that the reaction was terminated before the equilibrium oligomer distribution has been established.

The same phenomenon was observed in the polymerization of cyclic sulfides [4]. Goethals has shown that in the polymerization of propylene sulfide linear polymer is formed first which then degrades to a mixture of cyclic oligomers, mostly to the cyclic tetramer [25]. Thus, in several systems it was shown conclusively that cyclic

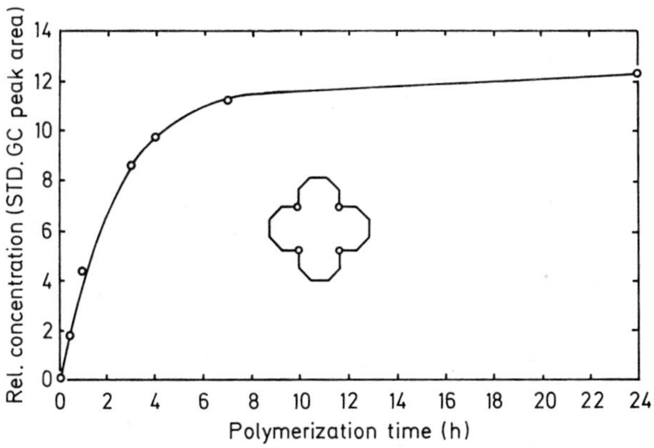

Fig. 3.7. Increase in relative concentration of cyclic tetramer of THF with polymerization time. [THF] $\sim$ 6 mol · l^{-1}, [CF_3SO_3H] $\sim$ 0.3 mol · l^{-1}, CH_3NO_2, 0 °C [3)]

biting) and propagation proceed through the same type of reaction, as shown schematically for cyclic ethers:
cally for cyclic ethers:

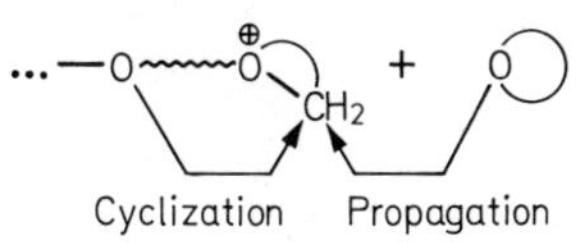

(3-10)

Depending on such factors as the relative nucleophilicity of heteroatoms in the monomer and in the polymer, the flexibility of the chain (leading to the larger or smaller conformational assistance-neighbouring group participation) the relative rates of both processes may be different. If the rate of cyclization is comparable to the rate of propagation (like in oxetane polymerization) considerable amounts of cyclic oligomers (compared to final equilibrium concentration) are formed within the time needed to attain complete monomer conversion (cf. Sect. 5.3.5). If, however, the rate of cyclization is low (e.g. THF), macrocycles will still form slowly after the monomer-polymer equilibrium has been established.

The fact, that the difference between these two systems is quantitative rather than qualitative (i.e. the same mechanism but different kinetics) is not always appreciated. For example in Ref. [26)] it is claimed, that cyclization in oxetane and THF polymerization proceed by different mechanisms. This is based on the observation that in oxetane polymerization cyclization is concurrent with propagation but in THF polymerization macrocycles appear after the monomer-polymer equilibrium has already been reached. This assumption of two different mechanisms is not necessary, however; as we have shown the difference comes only from the differences in k_p/k_b in Scheme 3.6, conforming to the polymerization of both monomers.

As shown in Sect. 3.2.1, the rate of cyclic oligomer formation (expressed by the probability of attaining the corresponding conformation) is proportional to $n^{-3/2}$. Thus, for systems conforming to the Jacobson-Stockmayer theory, the distribution of cyclic oligomers during the early stages, i.e., when the reverse reaction of "propagation" of macrocyclics can still be neglected, is governed by the $n^{-3/2}$ law (kinetic distribution). After reaching the equilibrium, this distribution changes and becomes proportional to $n^{-5/2}$ (thermodynamic distribution). This was amply demonstrated by Chojnowski for polymerization of hexamethylcyclotrisiloxane initiated with triflic acid [27]. The passage from the kinetic to thermodynamic distribution is shown in Fig. 3.8 where line 1 gives the distribution observed by GC for the early stage of polymerization (30% conversion, slope$^{-3/2}$) and line 2 the distribution determined for the later stages (100% conversion, equilibrium, slope$^{-5/2}$).

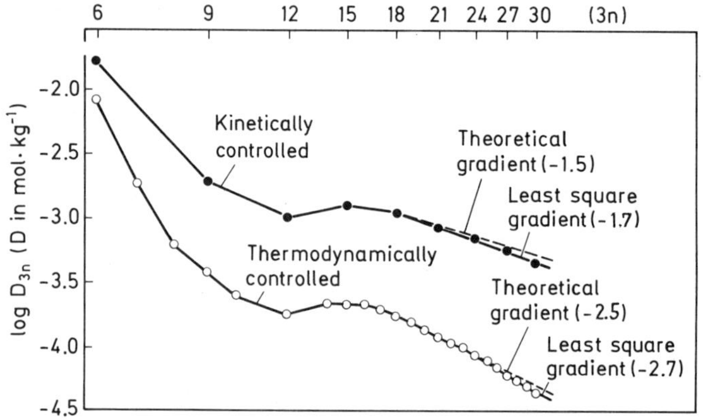

Fig. 3.8. Comparison of kinetically and thermodynamically controlled distribution of hexamethylcyclotrisiloxane (D_3). cyclic oligomers. $|D|_3 \sim 2$ mol · kg^{-1}, $|CF_3SO_3H|_0 \sim 7 \cdot 10^{-4}$ mol · kg^{-1}, °C heptane, 30 °C; ● — at 30% conversion of D_3; ○ — at equilibrium [27]

3.2.6 Distribution of Cyclic Oligomes in the Polymerization of Heterocyclic Monomers

In the majority of systems studied the influence of the conformational preferences is so important that the distribution of cyclic oligomers can no longer be described in terms of the Jacobson-Stockmayer theory which neglects these factors.

Both conformational preferences and kinetics are different for different monomers (depending on ring size, substitution, nucleophilicity of the heteroatom in monomer and polymer, flexibility of the chain) and even for a given monomer may change with changing solvent, temperature or counterion.

Thus, it is not surprising that the observed distributions could not be explained quantitatively in a number of systems. Moreover, usually it is not known to what extent cyclization was advanced when the polymerization was terminated.

The knowledge of the extent of cyclization as a function of the monomer conversion is essential for synthetic purposes. Usually it cannot be easily predicted and it is advisable to determine the concentration of cyclic oligomers (e.g. by GPC) in the products for various monomer conversions.

Table 3.1, based on Ref. [4], shows data on the formation of cyclic oligomers in the cationic polymerization of heterocyclic monomers.

3.2.7 Avoiding or Minimizing Macrocyclization in Cationic Ring-Opening Polymerization

We have shown in the previous sections that in certain systems (e.g. polymerization of cyclic formals studied by Schulz [7,8] and Yamashita [20]) one can use kinetic enhancement to obtain higher proportions of macrocyclics. This is mostly due to the enhanced contribution of the end-to-end closure. There are systems (THF[2,3], cyclic sulfides [25]) in which kinetic ring-depression was observed due to a slow rate of cyclization and thus the equilibrium concentration of the rings was attained only slowly. These two extreme cases are depicted in Fig. 3.9.

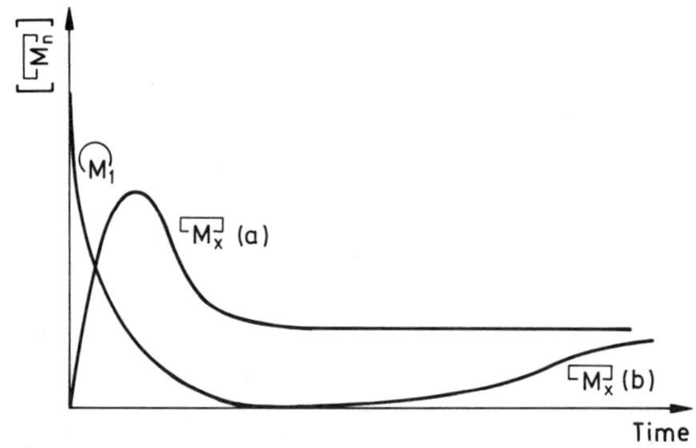

Fig. 3.9. Schematic representation of a change of concentration of a given cyclic oligomer for fast end-biting (a) and slow back-biting schemes (b). Concentrations and time in arbitrary units

Thus, the synthetic chemist interested either in having low or high concentrations of macrocyclics must first determine the relative concentrations of the macrocyclics and linear polymers as a function of monomer concentration, and then start to modify the initiating system to achieve his goal. The increase of the proportion of macrocycles may be achieved by varying the end-group structure as discussed in previous sections. The diminution in macrocyclic concentration may be the highest if the mechanism of polymerization is changed.

The competition between propagation and cyclization may be discussed in terms of the following scheme:

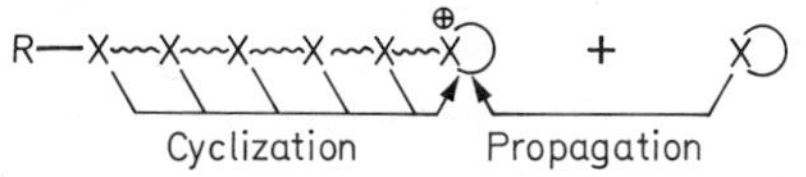

(3-11)

In polymerization the nature of the reactive sites in the monomer and along the polymer chain is determined by the monomer structure. The ratio of their reactivity therefore is invariant. Thus, one may influence the proportion of macrocyclization either by changing the structure of the end-groups RX or by changing the structure of the active species.

The importance of the end-group nature is easier to define and already has been discussed. The more reactivis the end-group toward the growing species the higher is the rate of end-to-end cyclization and the higher is the proportion of the macrocyclics formed by this mechanism (i.e. from the relatively short polymer chains). The end-groups can be chosen in such a way, that the end-to-end closure and formation of the charged macrocycle can either allow rapid formation of the uncharged macrocycle or retard its formation. In the polymerization of oxygen-containing heterocycles initiation with protonic acids gives OH end-groups. This group is more nucleophilic than the oxygens in the polymer and it also provides, as a result of cyclization, the protonated macrocycle which is able to transfer protons easily. Thus, the proportion of macrocyclics in protonic acid initiated polymerizations will be higher than in polymerizations initiated with alkylating agents (e.g. trialkyloxonium salts or triflic esters). The latters give rise to less reactive alkoxy end-groups and less easily migrating alkyl groups:

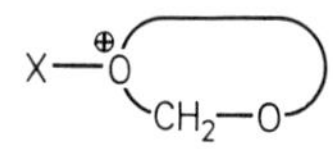

X = H, fast migration of "$\overset{\oplus}{H}$"

; X = alkyl, aryl, acyl : slow migration of $R^{\oplus}$ or $RCO^{\oplus}$ (3-12)

These effects have been experimentally observed in the polymerization of some cyclic acetals [7,8] and THF [2,3]. In the polymerization of n-tert-butylaziridine, the presence of the secondary amino groups also favours end-to-end cyclization [28].

The influence of the change of the mechanism of polymerization on the extent of macrocyclization is most dramatic when the anionic polymerization of heterocyclics is compared with the cationic process. For instance, in the polymerization of oxiranes, extensive cyclization occurs during tertiary oxonium ion propagation (ethylene oxide, propylene oxide, styrene oxide) (cf. Sect. 4), whereas macrocyclics do not form from these monomers in alcoholate anion propagated polymerizations [29]. Coordinate ionic polymerizations initiated with Al- und Zn-alkyls or modified derivatives (e.g. Vanderberg's catalysts) also proceed in the absence of cyclization (Sect. 4). This difference of behaviour stems from the fact that although the thermodynamically controlled concentrations should be the same for all mechanisms involved, the rates of equilibration may differ, and in some systems complete monomer conversion into linear polymer may occur long before the macrocyclics are formed (cf. Sect. 3.2.5).

This situation may also occur in cationic polymerizations. As described in Sect. 4, polymerization of ECH initiated with, e.g. tertiary oxonium salts, leads to extensive cyclization but initiation in the presence of high concentrations of alcohols eliminates macrocyclization. Thus, ECH gives up to 43% of a cyclic tetramer with the $BF_3 \cdot THF$ complex ($C_2H_4Cl_2$ solvent, 25°) [6] but tetramer is not observed when the polymerization is conducted in the presence of a glycol, 1:10 to ECH [30]. The rationale was put forward (cf Sect. 4) that in the former system propagation by the tertiary oxonium ions prevails, whereas in the latter the protonated monomer adds to the polymer being a macrodiol at every stage of chain growth. Therefore, in the former system macrocyclization (back-biting) competes with linear growth whereas in the latter mainly linear growth, occurs:

(3-13)

Successful competition of macrocyclization, macrocyclization enhanced

Linear growth predominates, cyclization depressed if not eliminated

These extreme cases should not be taken categorically; in certain systems both mechanisms may proceed and compete. However, this discussion shows that the situation of the synthetic chemist is not hopeless when he wishes to avoid macrocyclization and has to work with monomers that polymerize only by cationic mechanism.

3.3 References

1. Andrews, J. M., Semlyen, J. A.: Polymer *12*, 642 (1971)
2. McKenna, J. M., Wu, T. K., Pruckmayr, G.: Macromolecules *10*, 877 (1977)
3. Pruckmayr, G., Wu, T. K.: Macromolecules *11*, 265 (1978)
4. Goethals, E. J.: Adv. Polym. Sci. *23*, 103 (1977)
5. Semlyen, J. A.: Adv. Polym. Sci. *21*, 43 (1976)
6. Ito, K., Usami, N., Yamashita, Y.: Polymer J. *2*, 171 (1979)
7. Schulz, R. C. at al.: ACS Symposium Series, *59*, 77 (1977)
8. Rentsch, C., Schulz, R. C.: Makromol. Chem. *178*, 2535 (1977)
9. Jacobson, W., Stockmayer, W. H.: J. Chem. Phys. *18*, 1600 (1950)
10. Matyjaszewski, K., et al.: Makromol. Chem. *181*, 1469 (1980)
11. Kuhn, W.: Kolloid.-Z. *68*, 2 (1934)
12. Flory, P. J.: Macromolecules *7*, 381 (1974)
13. Flory, P. J., Semlyen, J. A.: J. Am. Chem. Soc. *88*, 3209 (1968)
14. Andrews, J. M., Semlyen, J. A.: Polymer, *13*, 141 (1972)
15. Berlin, Al. Al., Iwanov, W. W., Enikolopyan, N. S.: Vysokomol. Soed. *9*, 61 (1967)
16. Szymanski, R., Kubisa, P., Penczek, S.: Macromolecules *16*, 1000 (1983)
17. Goethals, E. J., Bouquet, G., Simonds, R.: 6th Intern. Symposium on Cationic Polymerization, Ghent, 1983, p. 131
18. Kubisa, P., Penczek, S.: Makromol. Chem. *180*, 1821 (1979)
19. Schulz, R. C., et al.: Pure Appl. Chem. *53*, 1763 (1981)
20. Yamashita, Y., Kawakami, Y.: ACS Symposium Series *59*, 77 (1977)
21. Sato, A.: Polymer J., *9*, 209 (1977)
22. Bucquoye, M. R., Ph. D. Thesis, Ghent, 1978
23. Bucquoye, M. R., Goethals, E. J.: Makromol. Chem. *179*, 1681 (1978)
24. Dale, J., Daasvatn, K.: J. Chem. Soc., Chem. Commun. *1976*, 295
25. Simmonds, R., Goethals, E. J.: Makromol. Chem. *179*, 1689 (1978)
26. Dreyfuss, P., Dreyfuss, M. P.: Polymer J. *8*, 81 (1975)
27. Chojnowski, J., Ścibiorek, M., Kowalski, J.: Makromol. Chem. *178*, 1351 (1977)
28. Goethals, E. J. et al.: Makromol. Chem. Rapid Comm. *3*, 515 (1982)
29. Cyclic Monomers, Frisch, K. C. Ed., Wiley Interscience, New York, 1972 p. 37
30. Kuzaev, A. N., Olkhova, O. M.: Vysokomol. Soed. *24*, 2197 (1982)
31. Plesch, P. H., Westermann, P. H.: Polymer, *10*, 105 (1969)
32. Weissermel, K. et. al.: Angew. Chem. Internat. Ed. *6*, 526 (1967)
33. Jaacks, V., Makromol. Chem. *99*, 300 (1966)

4 Oxiranes

4.1 Introduction

Oxiranes (α-oxides, α-epoxides) are known to polymerize by cationic and anionic mechanisms as well as by ionic-coordinative reactions [1-3]. Their ability of ring opening by one of these processes depends mainly on the structure and number of substituents. Some examples are given below:

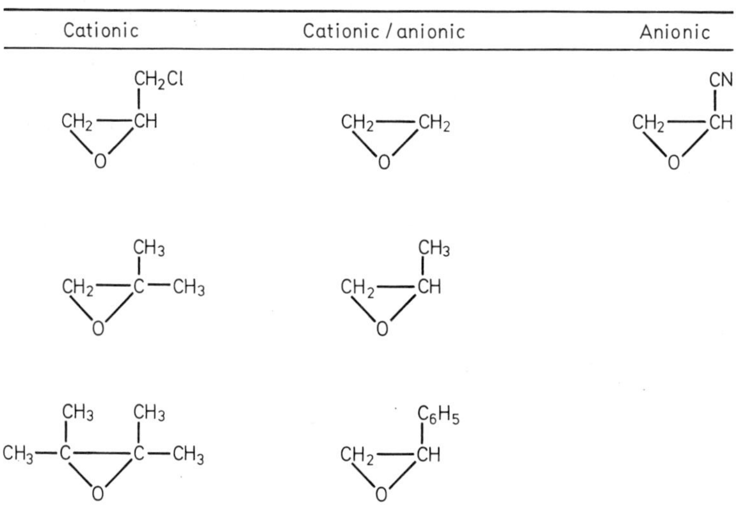

(4-1)

Some of these and related monomers give high molecular weight polymers only with initiators inducing ionic-coordinative polymerization (e.g. epichlorohydrin)[4-5]. The coordinative mechanism is sometimes further specified as cationic or anionic, although rather on the basis of the monomer structure than as of a result of studies of the polymerization mechanism.

The high thermodynamic polymerizability of oxiranes (due to the relatively large ring strain) and the availability of monomers such as ethylene oxide (EO), propylene oxide (PO) or epichlorohydrin (ECH) led to considerable efforts directed toward the preparation of high molecular weight polymers. Both cationic and anionic polymerizations were explored and it soon became clear that for several monomers only anionic polymerization gives high polymers.

Cationic processes can be advantageously used for the preparation of linear telechelics and are unique in preparing cyclic oligomers.

Cationic polymerization of oxiranes leads to products whose structures depend on the initiator used. Three groups of polyoxiranes result from polymerizations induced by various initiators, namely:

— linear high polymer:	cationic coordinate (modified alkyl metal initiators), active species and propagation mechanism are not known with certainty
— high converion to cyclic oligomers:	oxonium ions, carbenium ions or cryptoionic initiators, e.g. $R_3O^+A^-$, $ROSO_2CF_3$, BF_3 and its complexes; mostly tertiary oxonium ion propagation
— linear oligomers (telechelics) with —OH end-groups:	polymerization in the presence of proton donating compounds like H_2O and alcohols, propagation mostly by addition of the activated monomer to neutral, polymeric alcohol.

The boundaries between these three groups are not very sharp and it may happen that two chemically different active species operate at the same time in a given system leading to a mixture of products which exclusively would be formed by one kind of growing species. This may particularly be the case for protonic acids with both complex and non-complex anions as well as initiators belonging to the second of the above discussed groups. Addition of a sufficiently high concentration of a proton source may shift these intiators from the second to the third group, although in some systems protonic acids alone may lead to the linear oligomer (group three). Some other initiators, like $SnCl_4$, can be classified as intermediate between initiators leading to the first and second group of products.

Schematically the mechanisms related to the second and third group can be shown in the following way:

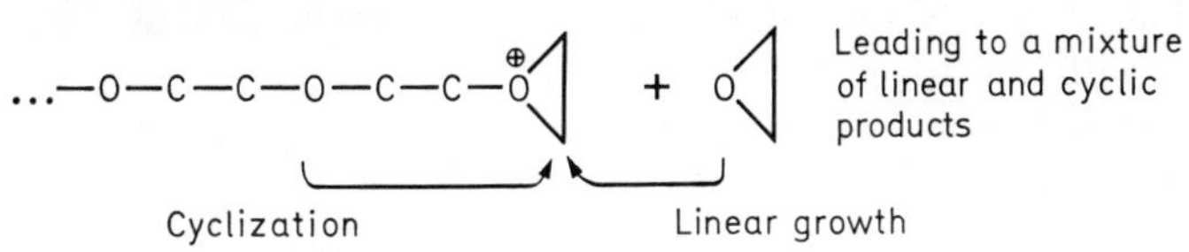

(4-2.a)

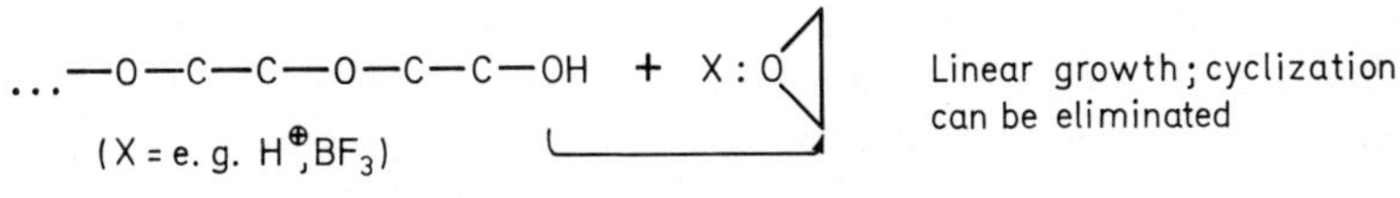

(4-2.b)

4.2 Synthetic Applications of Cationic Polymerization of Oxiranes

The three groups of products classified above have been prepared from the same monomer, depending on the structure of the initiator. Below we first discuss a few examples of high polymers, followed by cyclic oligomers and then we describe synthesis of linear oligomers (telechelics) with —OH end-groups.

4.2.1 High Polymers of Oxiranes

4.2.1.1 Oxirane (Ethylene Oxide, EO)

Linear polyEO of relatively high molecular weight ($\overline{M}_n$ up to $2 \cdot 10^4$) was obtained by cationic polymerization only with $SnCl_4$ as initiator [6]. The following mechanism, based on the well-known ability of tetravalent tin compounds to form hexacoordinate complexes, was proposed:

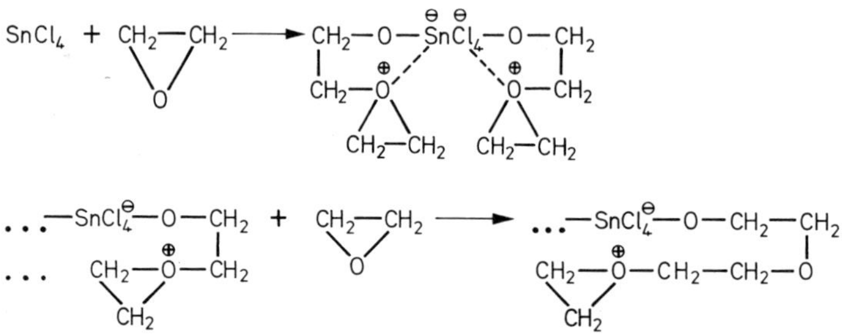
(4-3)

Thus, the growing species was regarded to be a "bifunctional internal oxonium ion" [1]. A bifunctional character of the $SnCl_4$ initiator has been recently confirmed for 1-tert-butylperoxy-2,3-epoxypropane [7]:

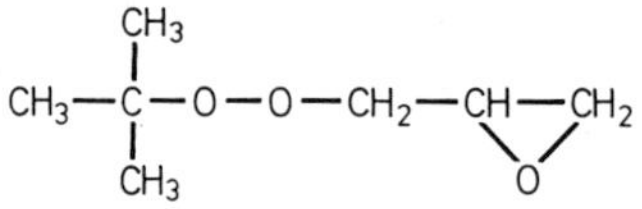
(4-4)

Balaev determined the concentration of active species for this monomer (measured by Saegusa's end-capping method [8] as $\simeq 2[SnCl_4]_0$ and being largely independent of monomer conversion. Polymerization was conducted in benzene solvent ($[M]_0 = 1\ mol \cdot l^{-1}$) at 22 °C with $[SnCl_4]_0 = 5 \cdot 10^{-2}\ mol \cdot l^{-1}$.

The mechanism shown in Scheme 4.3 does not explain why in the $SnCl_4$ initiated polymerization high molecular weight polymer ($\overline{M}_n$ up to $2 \cdot 10^4$) is obtained whereas with BF_3 initiated polymerization, also assumed to proceed by tertiary oxonium ions, the upper $\overline{M}_n$ limit of the linear fraction is only about 700.

In $SnCl_4$-initiated polymerizations the product distribution is intermediate between a typical cationic (e.g. BF_3-initiated) and coordinative polymerization.

Thus, a contribution by the coordinative mechanism cannot be exluded. It seems that such a possibility was taken into account by Eastham, as indicated by the dotted lines in Scheme 4.3. The proposed structure would then require a coordination number higher than six for the tin atom.

4.2.1.2 Methyloxirane (Propylene Oxide, PO)

High molecular weight polyPO was prepared in the polymerization initiated with the $(CH_3)_3Al—H_2O$ system (at a 2:1 molar ratio) at 80 °C in methylene dichloride solvent [9]. The polymer formed in 60–80% yield was divided into a low molecular weight fraction (~80% of the total yield) with $\overline{M}_n = 250$, composed of cyclic oligomers and a high molecular weight fraction (~20% of total yield) with $\overline{M}_n$ up to 10^6.

A similar system was studied by Furukawa and Saegusa [10]. These authors found that modification of the initiating system with an amine considerably increases the yield of the high molecular weight polymer and concluded therefore that a coordinative anionic mechanism operates in the system.

Propylene oxide-based elastomers became commercially available in 1972 under the trademark Parel® [12]. Parel elastomers are sulfur curable copolymers of PO and alkyl glycidyl ether prepared with an aluminumalkyl-water-acetylacetone catalyst. Thus, the polymerization mechanism should be classified as coordinative anionic.

4.2.1.3 Chloromethyloxirane (Epichlorohydrin, ECH)

High molecular weight polyECH as well as its copolymers with ethylene oxide (EO) are commercial products. Their production is based on the discovery made by Vandenberg [4,5] who first observed that some modified alkyl aluminiums are efficient initiators, giving high molecular weight products in the polymerization of various oxiranes.

Although sometimes it is not clear whether the actual propagation should be classified as anionic or cationic (for instance, for propylene oxide) [9] and sometimes it is assumed that both mechanisms simultaneously operate, it is quite apparent that ECH polymerizes with these initiators exclusively by a cationic mechanism.

The detailed procedure for the preparation of atactic and isotactic high molecular weight polyECHs was given by Vandenberg [11]. According to this procedure the initiator solution is prepared by adding 186 g (0.4 mol) of a 25% solution of $(C_2H_5)_3Al$ in hexane to 385 g diethyl ether, followed by addition of 4 ml water. All these operations are carried out under purified nitrogen between 20 °C and 36–38 °C (boiling point of ether). The resulting solution has a 0.58 mol $\cdot$ l^{-1} concentration with respect to aluminium and the molar ratio of $(C_2H_5)_3Al$ to H_2O is 1:0.6.

To carry out the polymerization, 125 g ECH is dissolved in 550 ml diethyl ether and the initiator solution is added at room temperature in four portions of 22 ml each (at 0, 0.5, 1.0 and 1.5 hrs). The reaction is continued for 13 hrs, and then stopped by addition of 54 ml ethanol. After washing with ether and methanol and removing the catalyst residues by slurrying with methanolic HCl followed by filtration, 97 g (78% conversion) white powder which is a mixture of atactic and isotactic polyECH is obtained. The product is separated into acetone soluble and acetone insoluble fractions. The soluble fraction gives 29% of a tough rubbery polymer having $\eta_{sp}/c = .19$ (0.1% solution in α-chloronaphthalene at 100 °C) which corresponds to a M$\overline{M}_n = 6 \cdot 10^5$. The acetone insoluble fraction gives crystalline isotactic polyECH (27%) with $\eta_{sp} = 5.8$, $\overline{M}_w = 2 \cdot 10^6$, m.p. = 117 °C, insoluble in common solvents at room temperature.

This basic procedure may be modified to obtain either rubbery atactic or crystalline isotactic polyECH as the main product.

The industrial process for making ECH elastomers (Herclor® of Hercules Co. and Hydrin® of B. F. Goodrich Chemical Co.) were developed [12] by the use of alkylaluminum initiators.

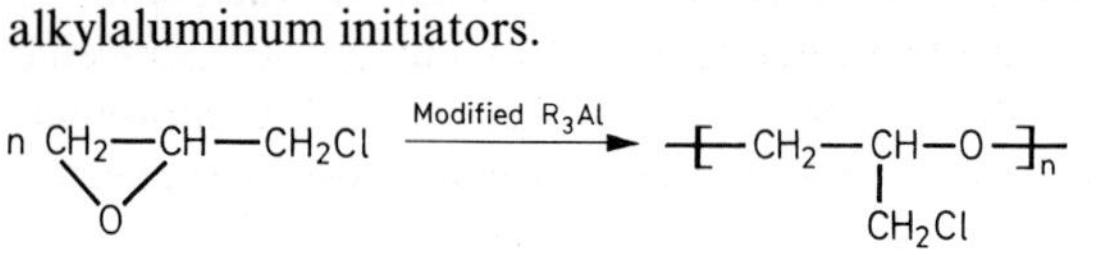

n in the range of $5 \cdot 10^3$ which corresponds to $\overline{M}_n \sim 5 \cdot 10^5$

(4-5)

The information concerning these processes is mostly from the patent literature [13]. In our description we mainly relied on the review article by Vandenberg [12].

Polymerization of ECH and its copolymerization with EO is carried out in the presence of alkylaluminium-water and alkylaluminium-water-acetylacetone catalysts in aromatic hydrocarbons or chlorinated solvents. Under these conditions the content of amorphous fraction is high and separation of crystalline material is not required. Commercial amorphous polyECH has more than 97% of head-to-tail sequences [14].

ECH-EO copolymers exhibit the best combination of properties at a 1:1 molar ratio (68% wt of ECH). Because of the much higher reactivity of EO in copolymerization to obtain the desired product, the composition of the feed should correspond to 94% ECH and 6% EO. Typical properties of these elastomers are shown in Table 4.1.

Table 4.1. Properties of typical epichlorohydrin elastomers [12]

	Homopolymer	Copolymer with EO
	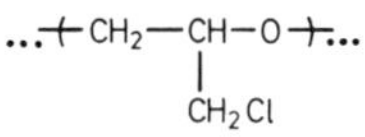	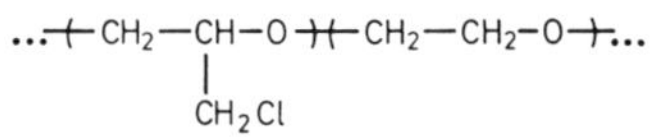
$\overline{M}_w$	450000	1400000
% ECH	100	68
% chlorine	38.4	26.0
d, g/cm^3	1.36	1.27
solvents	Toluene, acetone, methylene chloride	
η_{sp}/c, dl · g^{-1} (0.1% in α-chloronaphtalene at 100 °C)	1.4–2.0	3.0–6.0

In addition to the excellent oil resistance and heat aging resistance, the ECH homopolymer has particularly good air retention properties, by a factor of two better than butyl rubber.

ECH elastomers are easily processed and fabricated and may be vulcanised by a variety of difunctional compounds that react with chloromethyl groups, example diamines [12].

4.2.1.4 1,2-Dimethyloxirane (2,3-Epoxybutane)

Vandenberg polymerized both cis- and trans-2,3-epoxybutane to linear, high molecular weight polymers with $R_3Al—H_2O$ initiator at —78 °C [5]. The cis- isomer gave amorphous polymer while a crystalline polymer with m.p. = 100 °C was obtained from the trans- monomer. The same monomers with triflic acid or its esters as initiators gave various proportions (up to 60%) of low molecular weight isomerized products, including methylethyl ketone, substituted dioxolanes and substituted dioxanes together with an oligomer fraction (30–80%) from which cyclic dimers and tetramers were isolated. This fraction was assumed to contain cyclic oligomers of various ring size [15].

Polymerization initiated with $BF_3 \cdot O(C_2H_5)_2$ or $(C_2H_5)_3O^{\oplus}A^{\ominus}$ ($A = SbCl_6^{\ominus}$ or $BF_4^{\ominus}$) gave only the oligomer fraction with 60—90% yield.

Linear polymers with fairly high molecular weights ($\eta_{inh} = 0.6$ for 0.5% solution in benzene at 30 °C) were also obtained in the polymerization of 1,1,2-trimethyloxirane (2,3-epoxy-2-methylbutane) with R_3Al-H_2O initiator [16].

4.2.1.5 1,2-bis(Chloromethyl)Oxirane (1,4-Dichloro-2,3-Epoxybutane)

The other two highly substituted α-epoxides which polymerize to high polymers are cis- and trans-1,4-dichloroepoxybutanes, (cD and tD, respectively) [12]:

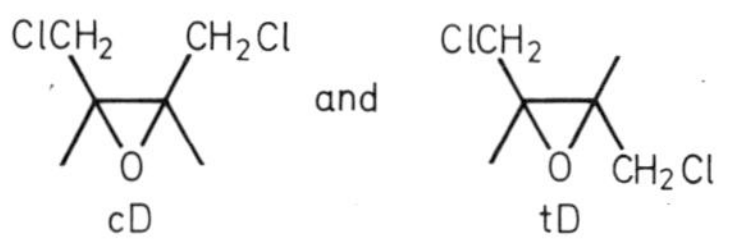

(4-6)

These polymers, developed by Vandenberg, are the natural extension of the successful development of two other "chlorinated polyethers", namely poly-3,3-bis-(chloromethyl)oxetane (polyBCMO, Penton® of Hercules Co) and the previously mentioned epichlorohydrin elastomers. The presence of two bulky chloromethyl groups hampers cyclization and increases the stability of the cationic active centers.

These monomers (a mixture of cis- and trans-isomers) may be obtained according to the following scheme:

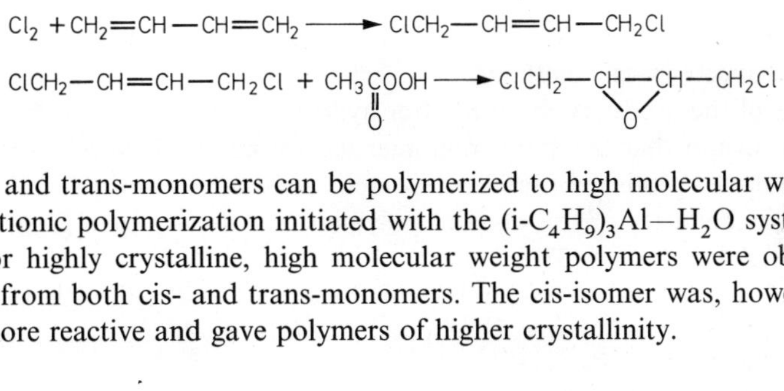

(4-7)

Both cis- and trans-monomers can be polymerized to high molecular weight products by cationic polymerization initiated with the $(i\text{-}C_4H_9)_3Al—H_2O$ system. With this initiator highly crystalline, high molecular weight polymers were obtained in high yields from both cis- and trans-monomers. The cis-isomer was, however, considerably more reactive and gave polymers of higher crystallinity.

Table 4.2. Polymerization of cis- (cD) and trans- (tD) 1,4-dichloro-2,3-epoxybutanes, $[monomer]_0 = 0.7\ mol \cdot l^{-1}$ in CH_2Cl_2, —78 °C [12]

Initiator	Monomer	Time h	CH_2Cl_2 insoluble fraction Conversion %	m.p. °C		CH_2Cl_2 soluble Conversion %
$(i\text{-}C_4H_9)_3Al$-H_2O $4 \cdot 10^{-2}\ mol \cdot l^{-1}$	cD	3	88	235	1.8	2
	tD	21	66	145	0.58	3
BF_3 $2 \cdot 10^{-2}\ mol \cdot l^{-1}$	cD	20	50	229	0.22	32 ($\eta_{inh} = 0.05$)
	tD	20	49	149	0.12	6

[a] 0.1% soln. in DMF at 50 °C

Both cis- and trans-monomers also polymerized with BF_3 initiator but as with other oxiranes the molecular weights were much lower, as shown in Table 4.2.

With the $R_3Al—H_2O$ system polymer molecular weight and crystallinity depend strongly on temperature, keeping the other parameters constant. Increasing the temperature from —78 °C to 0 °C leads to a decrease in the molecular weight from $5.6 \cdot 10^5$ to 10^4 and in the degree of crystallinity from 30% to 0.

Highly crystalline polymers prepared at —78 °C were shown to be stereoregular. As each polymer unit contains two assymetric carbon atoms:

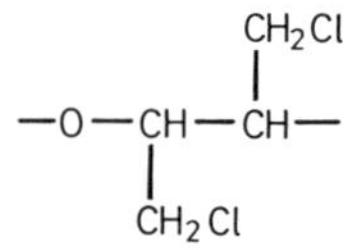

(4-8)

different steric arrangements are possible depending on the mechanism of ring opening and direction of monomer addition. In terms of diads the following sequences may exist:

meso-diisotactic	-RS-RS-
racemic-diisotactic	-RR-RR- or -SS-SS-
$meso_1$-disyndiotactic	-RS-SR-
$meso_2$-disyndiotactic	-RR-SS-

Each diad corresponds to one and only one mode of monomer addition and ring opening (e.g. with inversion or retention, head-to-head or head-to-tail and with stereoselection or stereoalternation). The various possible microstructures can be detected by spectroscopic methods.

Analysis of the products obtained after hydrolysis under controlled conditions led to the conclusion that the trans-monomer gave a meso-diisotactic structure, while the cis-monomer produced racemic-diisotactic polymer. A more detailed discussion of the stereochemistry of this class of polymers has already been given in Part I of this review (Adv. Pol. Sci. **37**).

Cis- and trans-monomers can be copolymerized and although the cis-isomer homopolymerizes much faster, in copolymerization both monomers enter the copolymer with comparable rates.

4.2.2 Cationic Cyclooligomerization of Oxiranes

4.2.2.1 Oxirane (Ethylene Oxide)

Ethylene oxide (EO), the three-membered cyclic ether, gives mostly the six-membered cyclic dimer: 1,4-dioxane. This has been observed in a series of the classical papers by Eastham, who used BF_3 as initiator [17,18] and was confirmed later by a number of authors, including recently work by Saegusa [19].

These systems, as well as the cationic polymerization of epichlorohydrin (ECH) (next section) represent the best examples of the cyclooligomerization and competition between linear growth and cyclization:

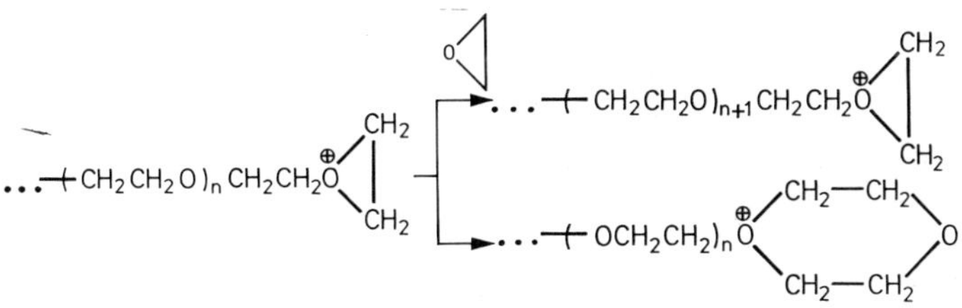 (4-9)

In the presence of hydroxyl-containing compounds, in addition to propagation by tertiary oxonium ion, another kind of linear growth may also occur with activated monomer (cf. following Sect.).

Eastham found that when EO polymerization was initiated in ethyl chloride with BF_3 in the presence of water, a relatively fast polymerization proceeded until a molecular weight of about 700 was reached [17]. At this point the polymer yield remained constant and only the formation of 1,4-dioxane was observed. Thus, at $[EO] = 2.0\ \text{mol} \cdot l^{-1}$ and $[BF_3] = 2 \cdot 10^{-2}\ \text{mol} \cdot l^{-1}$ (in C_2H_5Cl at room temperature) approximately 80% 1,4-dioxane was formed at $[H_2O] = 2 \cdot 10^{-2}\ \text{mol} \cdot l^{-1}$, whereas at $|H_2O| = 7.4 \cdot 10^{-2}\ \text{mol} \cdot l^{-1}$ the yield of 1,4-dioxane was reduced to approximately 40%.

By the use of deuterated polymer it was shown that 1,4-dioxane is also formed from the already existing polymer, thus the reaction path was: monomer-polymer unit-cyclic dimer [18].

These results indicate that the rate of polymerization is higher at lower conversions than the rate of 1,4-dioxane formation from the polymer. At certain conversions, i.e. at a certain polymerization degree, these rates become equal to each other and only the formation of 1,4-dioxane is observed. This is illustrated in Scheme (4-9) as a competition between bimolecular growth and unimolecular cyclization. Depletion of monomer decreases the rate of propagation, however, the rate of cyclization remains invariant.

More recently oligomerization of EO was studied by Saegusa with superacids and their derivatives as initiators [19]. The following conclusions were reached:

- Alkyl derivatives of superacids effectively consumed EO (~100%) and selectively gave 1,4-dioxane (~90%).
- Superacids showed high activity for EO consumption (100%) but produced 1,4-dioxane less selectively (20–80%). The remaining product was polyEO.
- Oxonium salts, e.g. $(C_2H_5)_3O^+BF_4^-$ effectively consumed EO (~100%) and gave 1,4-dioxane at a 90% yield.

It was thus concluded that "the larger the amount of protons, the less the extent of 1,4-dioxane production". Evidently, propagation by the activated monomer mechanism increased the overall extent of linear growth in the presence of acids.

An ingenious method of altering the distribution of cyclic oligomers in the EO-BF_3 system was developed by Dale [20]. This method involves formation of insoluble complexes of macrocylics with metal ions shifting the equilibrium to the direction of the desired macrocyclic products. Under typical conditions a mixture of 5 g EO (71% wt) and 2 g 1,4-dioxane (29% wt) was converted into products containing 8% wt EO, 85% wt 1,4-dioxane, 1% wt cyclic trimer, 4% wt cyclic tetramer and 1% wt of each cyclic pentamer and hexamer. This distribution may be considerably altered if the polymerization is carried out in the presence of salts containing alkaline or alkaline earth metal (template effect).

Typically, when EO is slowly added to a cold suspension of a salt in CH_2Cl_2 saturated with BF_3, the salt dissolves and a complex of cyclic oligomers with the introduced cation starts to precipitate. The complex is separated and cyclic oligomers are liberated by pyrolysis. By this method cyclic oligomers with n = 4, 5 and 6 could preferentially be formed, and their proportion depends on the nature of the salt used. Representative examples are shown in Table 4.3.

Table 4.3 Composition of cyclic oligomer mixtures in the polymerization of ethylene oxide in the presence of various complexing salts [20]

Salt	Tetramer %	Pentamer %	Hexamer %
$LiBF_4$	30	70	—
$NaBF_4$	25	25	50
KBF_4	—	50	50
$RbBF_4$	—	—	100
$CsBF_4$	—	—	100
$Ca(BF_4)_2$	50	50	—
$Cu(BF_4)_2$	5	90	5

4.2.2.2 Methyloxirane (Propylene Oxide)

The polymerization of propylene oxide (PO) in the absence of hydroxyl groups containing compounds leads to cyclic oligomers in high yield. Polymerization in bulk with $(C_2H_5)_3O^{\oplus}BF_4^{\ominus}$ (1 mol%) or BF_3 gives at room temperature and after 1 hr the distribution of products described in Table 4.4 [21].

Katnik and Schaefer have shown, using gas chromatography for isolation and NMR and mass-spectrometry for identification, that the distribution of isomers, resulting from both structural (head-to-head and head-to-tail) and stereochemical

Table 4.4. Distribution of cyclic oligomers in the polymerization of PO initiated with $(C_2H_5)_3O^{\oplus}BF_4^{\ominus}$ (1 mol%) at room temperature after 1 hr [21]

Fraction boiling range °C	Amount % wt	Molecular weight	Principal species
117–130/720 mm Hg	5.3	116	Dimer
61–65/0.07 mm Hg	51.5	232	Tetramer m.p. = 92–93 °C
77–83/0.07 mm Hg	11.5	306	Pentamer
90–150/0.07 mm Hg	4.6	369	Hexamer
Residue	27.5	609	Polymer

(cis-, trans-)isomerism, is close to random. Thus, 19 of cyclic tetramer isomers were identified in the reaction products [22].

4.2.2.3 1,1-Dimethyloxirane (Isobutylene Oxide)

According to Eastham isobutylene oxide in the presence of BF_3 gives only cyclic dimer [1]. Polymer with $\overline{M}_n$ in the range of 4500–5500 was prepared with $Ph_3C^+PF_6^-$ in 1,2-dichloroethane at 0 °C, but significant amounts of cyclic oligomers were formed in the later stages of the reaction [23].

Similar results were obtained by Vandenberg [24]. Thus, the polymerization of isobutylene oxide with BF_3 (5 mol%) at 0 °C after 1 hr. gave 3% of ether insoluble protic acids as initiators [$SnCl_4$, $BF \cdot O(C_2H_5)_2$, $HOSO_2CF_3$] and found a 100% conversion to cyclic oligomers with n = 2–5 [28].

4.2.2.4 Phenyloxirane (Styrene Oxide)

The formation of cyclic dimer (2–10% yield) in the cationic polymerization of styrene oxide was reported by different authors [25–27]. More recently, Yamashita studied the catonic oligomerization of styrene oxide in the presence of various Lewis and protic acids as initiators ($SnCl_4$, $BF_3 \cdot O(C_2H_5)_2$, $HOSO_2CF_3$) and found a 100 ₀ conversion to cyclic oligomers with n = 2–5 [28].

4.2.2.5 Chloromethyloxirane (Epichlorohydrin, ECH)

Yamashita observed by HPLC the formation of cyclic oligomers from dimer to heptamer in the polymerization of ECH initiated with various Lewis and protic acids [29]. The results are given in Table 4.5.

During distillation of the oligomer mixture needle-like crystals sublimed at 60 °C at 0.15 mm Hg. After recrystallization from benzene a product with m.p. = 109 to 110.5 °C was identified as a cyclic dimer on the basis of its mass-spectrum. The fraction distilling at 185–195 °C/0.5 mm Hg, gave another crystalline product which after recrystallization from toluene had m.p. = 135–136.5 °C. The product was identified as a cyclic tetramer which was earlier already isolated by R. J. Kern [21].

On the basis of ^{13}C-NMR spectra (only three signals at 43.45, 72.02 and 80.56 ppm δ) it was concluded that the tetramer had regular all-cis or all-trans configuration.

The same fraction boiling at 185–195 °C/0.5 mm Hg also gave another crystalline product melting at 151–152 °C. The mass spectrum indicated that this compound is also cyclic tetramer. Because six signals were present in the ^{13}C-NMR spectrum (at 42.98, 43.17, 70.24, 71.75, 79.96 and 80.28 ppm) it was assumed that this cyclic tetramer is an irregular stereochemical isomer, such as cis-cis-cis-trans.

Results shown in Table 4.5 indicate that the yield and composition of the oligomeric fraction can be varied within broad limits. It is difficult to rationalize the effect of solvent, there is, however, a straightforward pattern when the nature of initiator is considered. BF_3 etherate gives high yields of medium size rings (tetramer, pentamer), $SnCl_4$, as in EO polymerization, favours the formation of higher molecular weight products ($\overline{DP}_n$ > 7), while triflic acid gives low over-all yields and favours the formation of cyclic dimer.

Table 4.5. Distribution of cyclic oligomers in the polymerization of α-epichlorohydrin [29]

Initiator	Solvent	$[ECH]_0$	$[Initiator]_0$	Time	Yield	Product distribution %					
		$mol \cdot l^{-1}$	$10^2\ mol \cdot l^{-1}$	h	%	n = 2	3	4	5	6	≧7
$BF_3 \cdot OEt_2$	1,2-dichloroethane	2.2	5.0	4	98	2	0	43	14	6	35
	dioxane	2.1	5.0	4	11	18	1	4	22	20	35
	toluene	1.8	5.0	4	88	1	0	26	12	7	54
$SnCl_4$	1,2-dichloroethane	2.7	5.0	3	75	5	0	9	11	7	68
	dioxane	2.5	6.3	2.5	50	15	0	5	6	6	68
	toluene	1.9	4.3	23.5	73	9	0	12	12	11	56
CF_3SO_3H	1,2-dichloroethane	2.5	5.7	7.0	2	45	0	8	21	16	10
	dioxane	2.9	7.7	23.0	12	32	0	13	17	14	24

It is difficult at present to recommend a general method for the synthesis of a given cyclic oligomer from an oxirane. The template technique of Dale seems to provide background for a general approach [20]. It follows also from the data analysed in this section that the presence of hydroxyl containing compounds has to be avoided.

4.2.3 Cationic Oligomerization of Oxiranes Leading to Linear Oligomers. Activated Monomer Mechanism

Scattered information in the literature indicates [17,30,31,32] that in the presence of water or hydroxyl group-containing organic compounds linear polyoxiranes with hydroxyl end-groups are formed. It has been taken for granted that this phenomenon is caused by chain transfer.

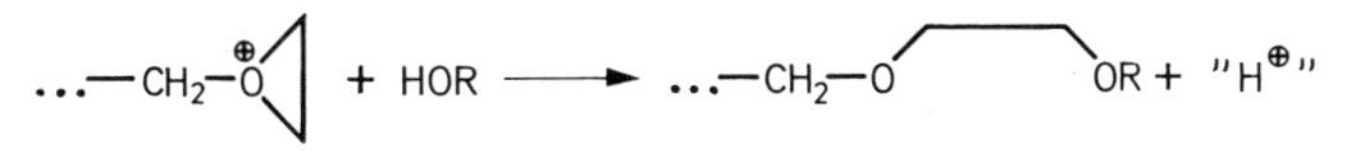 (4-10)

In the next step, protonation of the monomer produces the secondary oxonium ion which in turn starts the polymerization:

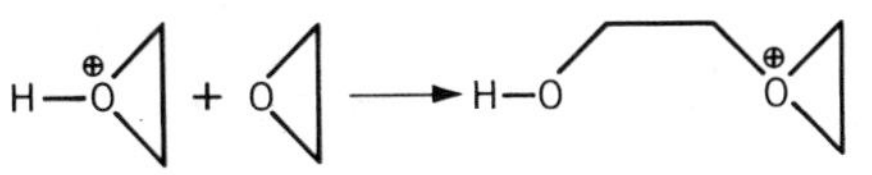 (4-11)

and forms the hydroxyl group chain-end. If a glycol or water is used as an additive macromolecular glycol results. If this mechanism were operational in the presence of glycol, the residue of glycol should be incorporated into the polymer gradually during polymerization; also the dependence of $\overline{DP}_n$ on time as well as the final distribution should be typical for a polymerization with chain transfer.

However, if the addition of the hydroxyl group-containing compounds does not change the mechanism, it would be difficult to understand why macrocyclics are completely eliminated in their presence, although growing tertiary oxonium ions persist and form macrocyclics by backbiting just as readily as in the absence of glycols.

Indeed, as indicated above, polymerization of e.g. epichlorohydrin with BF_3 alone yields mostly cyclic tetramer, whereas polymerization in the presence of BF_3 and ethylene glycol gives linear hydroxyl-ended oligomers whose $\overline{M}_n$ is governed by the ratio [epichlorohydrin (ECH)]$_0$/[ethylene glycol (EG)]$_0$ and cyclic tetramer is absent [30,31].

Typically, the reaction was carried out in 1,2-dichloromethane solution ([ECH]$_0$ $= 6\ \text{mol} \cdot l^{-1}$ [EG]$_0$ $= 0.6\ \text{mol} \cdot l^{-1}$) at 20 °C with $0.1\ \text{mol} \cdot l^{-1}$ of $BF_3 \cdot$ ether complex. At 81% conversion oligomer having $\overline{M}_n = 820$ was isolated. This value is very close to the one calculated on the assumption that each ethylene glycol molecule participates in the formation of one chain. $\overline{M}_n$ calcd. $= 800$.

According to GPC studies the molecular weight increases with conversion. This is illustrated by GPC curves obtained at various conversions of ECH polymerized in the presence of ethylene glycol (Fig. 4.1). It is remarkable that the addition product of ethylene glycol to ECH is formed first (22.4 counts) and then disappears completely

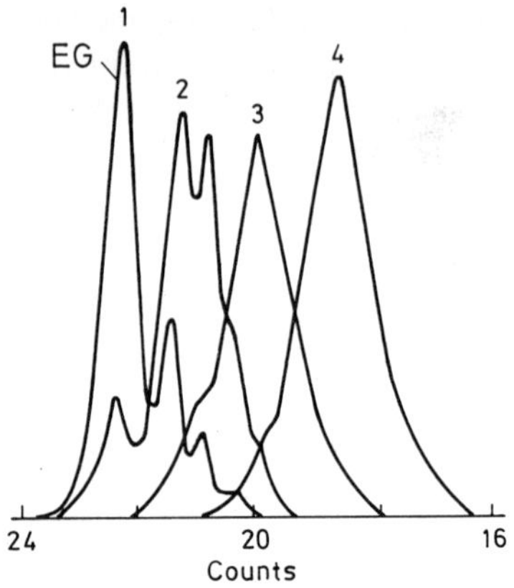

Fig. 4.1. GPC traces of products of oligomerization of epichlorohydrin (ECH) in the presence of ethylene glycol (EG) at different conversions. $[ECH]_0 = 6.0$ mol · l^{-1}, $[EG]_0 = 0.6$ mol · l^{-1}, $[BF_3 \cdot THF]_0 = 0.1$ mol · l^{-1}, $C_2H_4Cl_2$, 20 °C. ECH conversion = 1 – 8%; 2 – 28%; 3 – 50%; 4 – 81%

at higher ECH conversions. Thus, the reaction proceeds by a step by step polymerization and surprisingly the final MWD is quite narrow (1.08 at 81% conversion), lower than that calculated for the Poisson distribution.

Using similar conditions the synthesis of block copolymers was attempted [33,34]. Thus, ECH was polymerized in the presence of a macromolecular diol based on polyTHF and the products were analysed by liquid chromatography.

The product with the following structure resulted [35].

$$\text{polyECH-polyTHF-polyECH} \quad (4\text{-}12)$$

The nature of the end-groups in the polyECH (primary or secondary alcohol) has not been determined.

Another important synthetic application of this concept is in preparation of block copolymers of THF and EO (cf. Sect. 15); e.g., addition of EO to hydroxyl terminated polyTHF in the presence of strong acid.

4.3 References

1. Eastham, A.: in "The Chemistry of Cationic Polymerization" P. H. Plesch Ed. Macmillian, New York, 1963
2. Ledwith, A., Fitzimmons, C.: in "Polymer Chemistry of Synthetic Elastomers", Kennedy, J. P., Tornquist, E.G.M.: Eds. Interscience, New York, 1968, Part I, p. 337
3. Dreyfuss, P., Dreyfuss, M. P.: in "Chemical Kinetics", Bamford, C. H., Tipper, C. F. H., Eds., Elsevier, Amsterdam, Oxford, New York, 1976, Vol. 15, p. 259.
4. Vandenberg, E. J.: J. Polymer Sci. *47*, 486 (1960)
5. Vandenberg, E. J., J. Polymer Sci. *A17*, 525 (1969)
6. Worsfold, D., Eastham, A. M.: J. Am. Chem. Soc. *79*, 897 (1957)

7. Iwanchev, S. S.: et al.: J. Polymer Sci. *18*, 2051 (1980)
8. Saegusa, T., Matsumoto, S.: J. Polymer Sci. *A6*, 1559 (1968)
9. Colclough, R. O., Wilkinson, K.: J. Polymer Sci. *C4*, 311 (1963)
10. Imai, H., Saegusa, T., Furukawa, J.: Makromol. Chem. *82*, 25 (1965)
11. Vandenberg, E. J., Macromol. Synthesis *4*, 49 (1972)
12. Vandenberg, E. J., Pure & Appl. Chem. *48*, 295 (1976)
13. Vandenberg, E. J.: (to Hercules Incorporated) USP 3158580 and 3158581 (1964)
14. Vandenberg, E. J.: in Kirk-Othmer Encyclopedia of Chemical Technology, Third Edition, 1979, Vol. 8, p. 568
15. Kawakami, Y., Ogawa, A., Yamashita, Y.: J. Polymer Sci. *17*, 3785 (1979)
16. Field, N. D., Kieras, J. A., Borchert, A. E.: J. Polymer Sci. *5*, 2179 (1967)
17. Worsfold, D. J., Eastham, A. M.: J. Am. Chem. Soc. *79*, 900 (1957)
18. Latremouille, G. A., Merrall, G. T., Eastham, A. M.: J. Am. Chem. Soc. *82*, 120 (1960)
19. Kobayashi, S., Morikawa, K., Saegusa, T.: Polymer J. *11*, 405 (1979)
20. Dale, J., Daasvatn, K.: J. Chem. Soc., Chem. Commun. *1976*, 295
21. Kern, R. J.: J. Org. Chem. *33*, 388 (1968)
22. Katnik, R. J., Schaefer, J.: J. Org. Chem. *33*, 384 (1968)
23. Hornof, V., Gabra, G., Blanchard, L. P.: J. Polymer Sci. *11*, 1825 (1973)
24. Vandenberg, E. J.: J. Polymer Sci. *10*, 329 (1972)
25. Summerbell, R. K., Kland-English, M. J.: J. Am. Chem. Soc. *77*, 5095 (1955)
26. Colclough, R. O., et al.: J. Polymer Sci. *34*, 171 (1959)
27. Pasika, W. M.: J. Polymer Sci. *A3*, 4287 (1965)
28. Yamashita, Y., Iwao, K., Ito, K.: J. Polym. Sci., Pol. Lett. Ed. *17*, 1 (1979)
29. Ito, K., Usami, N., Yamashita, Y.: Polymer J. *11*, 171 (1979)
30. Entelis, S. C., Korovina, G. W., Kuzaev, A. I.: Vysokomol. Soed. *13*, 1438 (1971)
31. Kuzaev, A. I., Olkhova, O. M.: Vysokomol. Soed. *24*, 2197 (1982)
32. Blanchard, L. P., Singh, J., Baijal, M. D.: Canadian J. Chem. *44*, 2679 (1966)
33. Hammond, J. M., Hooper, J. F., Robertson, W. G. P.: J. Polymer Sci. *9*, 295 (1971)
34. Kuzaev, A. I., et al.: Vysokomol. Soed. *A12*, 1033 (1970)
35. Kuzaev, A. I., Olkhova, O. M., Entelis, S. G.: Vyskomol. Soed. *17*, 2120 (1975)

5 Oxetanes

5.1 Introduction

Oxetanes, 4-membered cyclic ethers, polymerize exclusively by cationic mechanism [1-3], although coordinative anionic homopolymerization and copolymerization with CO_2 was claimed [4,5] for the unsubstituted oxetane.

The thermodynamic polymerizability of oxetanes, in regard of their ring strain, is high: for unsubstituted oxetane $\Delta H_p \sim -80$ kJ/mol (~ -20 kcal/mol) [6]. Thus oxetanes like oxiranes may be polymerized to complete conversion ($[M]_e$ is very low) and in this respect differ considerably from the next group in the homologous series: oxolanes (5-membered cyclic ethers), for which an equilibrium character of polymerization is clearly noticeable.

As indicated in the previous section, cationic polymerization of 3-membered cyclic ethers i.e. oxiranes, is dominated by cyclization; thus, high molecular weight polymers of oxiranes cannot be obtained in the presence of, e.g. BF_3. In the polymerization of oxetanes, this tendency is considerably reduced and high molecular weight polymers may be prepared by conventional cationic polymerization although, depending on polymerization conditions, varying amounts of cyclic oligomers are formed.

Thus, the cationic polymerization of oxetanes, not having the disadvantages of polymerization of 3-membered (cyclization) and 5-membered (high $[M]_e$) cyclic ethers, is important both for providing valuable models to study the polymerization mechanism as well as for preparing high molecular weight polymers.

The scope of applications of cationic oxetane polymerization is rather limited, with one exception [3,3-bis(chloromethyl)oxetane, BCMO] polyoxetanes have not found any practical application. BCMO, is not as easily available as some of the 3- or 5-membered cyclic ethers (ethylene oxide, propylene oxide, epichlorohydrin, tetrahydrofuran) which are made from simple petrochemical products.

5.2 Monomer Synthesis

Oxetanes are generally prepared by cyclization of the corresponding α,γ-disubstituted compounds [7]:

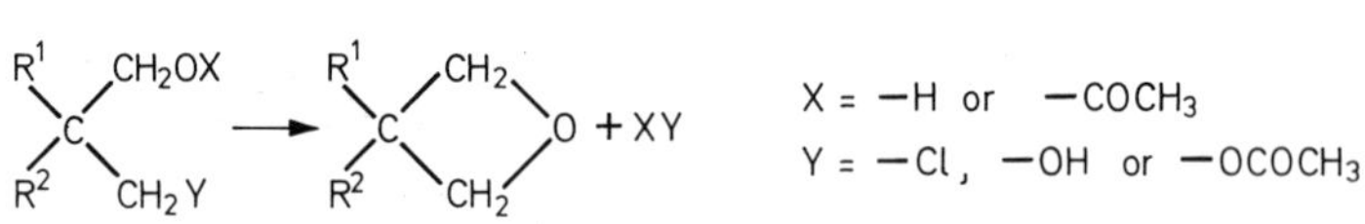

(5-1)

If difunctional substrates with suitable R groups are unvailable, the corresponding 3,3-disubstituted oxetanes may be prepared by an exchange reaction by the use of cation solvating media, e.g. glycols [7].

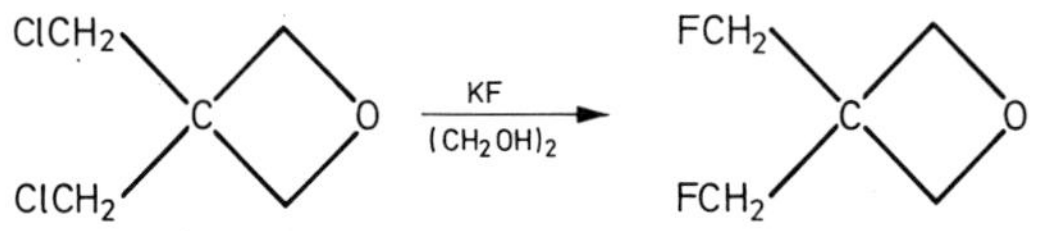

(5-2)

Examples of 3,3-disubstituted oxetanes obtained by these methods are listed in Table 5.1. More information on the synthesis of oxetanes, including those substituted in 2 and 4 positions, can be found in Ref. [8].

Table 5.1 Synthesis of oxetanes [7]

Substituents	Method of preparation, yield, %				b.p. °C
	a	b	c	d	
none	44	45			48
3,3-dimethyl-	57	54	66		80
3,3-diethyl-	90				129
3,3-bis(chloromethyl)-	90	90			80/10 mm Hg
3-chloromethyl- 3-hydroxymethyl-	86				64/0.1 mm Hg
3-chloromethyl- 3-ethyl-	82				75/34 mm Hg
3,3-bis(hydroxymethyl)-	70		70		128/0.04 mm Hg mp = 84°
3,3-bis(iodomethyl)-				89	mp = 50°
3,3-bis(cyanomethyl)-				89	138/0.04 mm Hg mp = 76.5°

a according to Scheme 5.1 where X = H, Y = Cl;
b according to Scheme 5.1 where X = $OCOCH_3$, Y = Cl;
c by elimination of CO_2 from cyclic carbonates;
d according to Scheme 5.2

5.2.1 Synthesis of 3,3-Bis(Chloromethyl)Oxetane(BCMO)

The classical method of BCMO synthesis is by Sørensen and Campbell [9]:

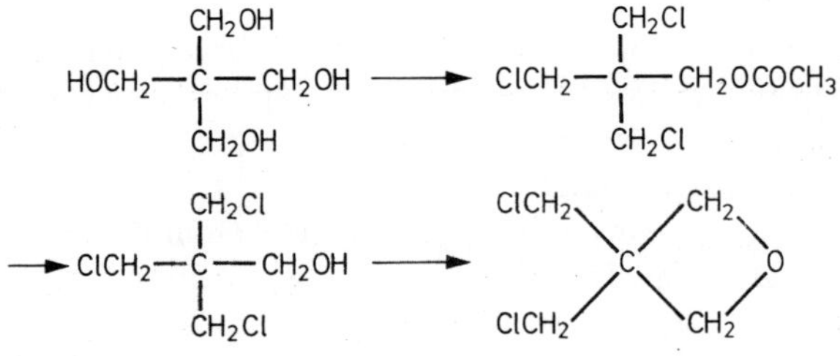

(5-3)

According to this procedure, pentaerithritol is treated with the mixture: $CH_3COOH/H_2O/HCl$ at 160 °C in an autoclave and the product is refluxed with CH_3OH. Crude trichloropentaerithritol is then refluxed with KOH in a CH_3OH/H_2O mixture.

BCMO is purified by distillation at reduced pressure, b.p. = 101 °C at 27 mm Hg or 62 °C at 4 mm Hg [9].

The purity of the monomer can be determined by GC or cryometry [10]. The refractive index can also be used to determine the purity. For polymerization grade monomer (purity over 99.8% according to GC) $n_D^{20} = 1.48583–1.48586$ [11].

5.3 Polymerization of Oxetanes

5.3.1 Initiation

Two initiating systems were most often used for the initiation of oxetane polymerization, namely BF_3 and its complexes with ethers, and the R_3Al/H_2O system.

It was established that the polymerization of oxetane (OX) cannot be initiated with BF_3 alone in the absence of water or other proton donating compounds [12,13]. The same phenomenon was observed later for the polymerization of BCMO where the addition of controlled amounts of water increased the overall polymerization rate [14,15]. Thus, the necessity of water cocatalysis is well documented for OX-BF_3 and BCMO-BF_3 systems, the rates and yields of initiation in these systems are however unknown.

Water plays also an important role in R_3Al initiated polymerizations, described in terms of a cationic process [16]. Kinetic studies on the BCMO-R_3Al-H_2O system revealed that the rate of polymerization depends on the concentration of added water, i.e., the rate increases until the ratio $[H_2O]/[(i\text{-}C_4H_9)_3Al]$ approaches unity and than decreases for $[H_2O] > [(i\text{-}C_4H_9)_3Al]$ [17,18].

When the concentration of $[(i\text{-}C_4H_9)_3Al]$ was changed at constant $[H_2O]$, similar effect was observed, namely the rate of polymerization increased until the 1:1 ratio was reached and then remained constant [18].

These results indicate that the true initiator is formed in this system by the reaction of alkylaluminum compounds with water. The initiator is most effective when the molar concentration ratio of both components is equal to 1. Considering that the molecular weights depend on water concentration, it was deduced that approximately ten molecules of water are required to start one chain. This means that an average initiator molecule results from the reaction of approx. ten molecules of H_2O with approx. the same number of AlR_3 molecules.

The initiation rate was low in this system, kinetic curves showed marked acceleration periods. Analysis of the initial periods of kinetic curves, considering the slow initiation-fast propagation process [19], yielded of an apparent rate constant of initiation $k_i = 1.6 \cdot 10^{-3}\ mol^{-1} \cdot l \cdot s^{-1}$ (C_6H_5Cl, 70 °C) [18] ($k_p = 8.5\ mol^{-1} \cdot l \cdot s^{-1}$ for the same conditions; cf. next paragraph).

Similar dependences were observed for the polymerization of 3-methyl-3-chloromethyloxetane (MCMO) initiated by the same catalytic system and studied by Kropachev and Aleksiuk [20–22]. The activity of the $(i\text{-}C_4H_9)_3Al—H_2O$ system was highest, however, for a 2:1 molar ratio of the components. Isobutane was formed in the reaction between R_3Al and H_2O in quantities corresponding to the following stoichiometry:

$$(i\text{-}C_4H_9)_3Al + H_2O \rightarrow (i\text{-}C_4H_9)_2Al{-}O{-}Al(i\text{-}C_4H_9)_2 + 2\,C_4H_{10} \quad (5\text{-}4)$$

Thus, it was assumed that species like $R_2AlOAlR_2$ are responsible for initiation in this system. The rate constant of initiation was calculated as $k_i = 1 \cdot 10^{-6}\ mol^{-1} \times l \cdot s^{-1}$ (C_6H_5Cl, 20 °C) [21]. It has to be stressed that molar ratios close to 2:1 can also be obtained for longer chain alkylalumoxanes.

Polymerization of BCMO in the solid state can also be induced by radiation [23–25]; this subject is discussed separately.

5.3.2 Propagation

Propagation in the polymerization of oxetanes can be represented as an S_N2-type reaction, the growing species being the tertiary oxonium ion:

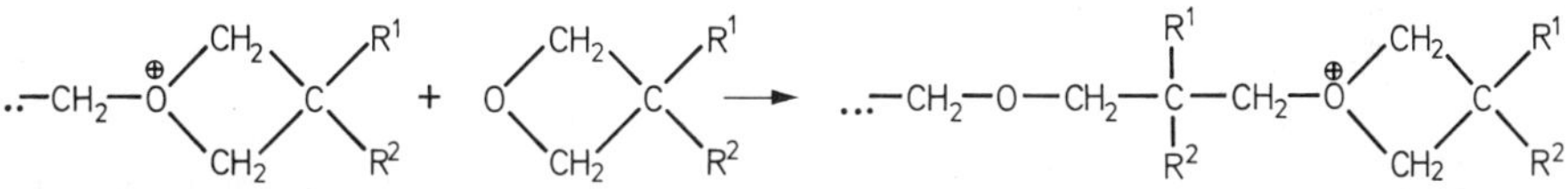

(5-5)

Saegusa and coworkers were first in attempting to determine the concentration of active species in these systems by using the "phenoxy end-capping" method [26]. Worsfold pointed out, however, that this method gives a sum of concentrations of both active (monomer holding) and inactive branched (polymer holding) oxonium ions formed by degradative chain transfer to polymer [27]:

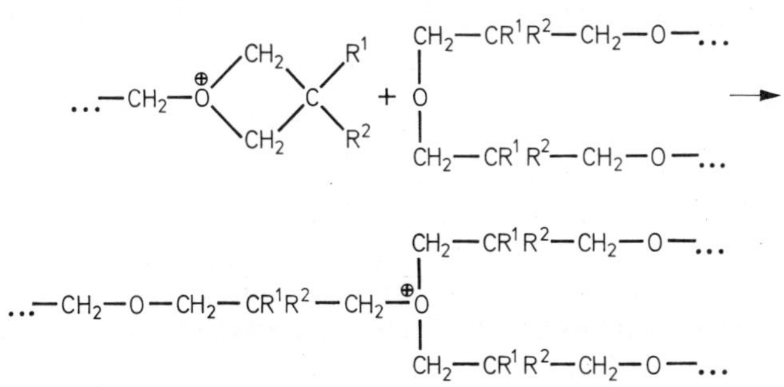

(5-6)

Thus, the values of the propagation rate constants reported by Saegusa [26]:

$$k_p(OX) = 5.7 \cdot 10^{-2}\ mol^{-1} \cdot l \cdot s^{-1}\ (CH_2Cl_2, -10°, \text{initiator: } BF_3)$$

$$k_p(DMO) = 6.8\ mol^{-1} \cdot l \cdot s^{-1}\ (CH_2Cl_2, -9°, \text{initiator: } BF_3)$$

may be the lowest estimates.

The kinetics of propagation in the polymerization of BCMO was studied by two of the present authors [18].

PolyBCMO is partly crystalline and its solubility in typical organic solvents is limited [28]. Thus, if polymerization is carried out at ambient temperatures, polymer precipitates out of solution. At higher temperatures (>50 °C) and low monomer concentrations (<0.3 mol · l^{-1}) polymerization in chlorobenzene solvent, however, is homogeneous over the entire conversion range.

Kinetic studies of propagation were carried out only for one system, namely for the polymerization of BCMO initiated with $(i\text{-}C_4H_9)_3Al/H_2O$ in C_6H_5Cl solvent over a temperature range of 55–90 °C. Analysis of the non-stationary kinetics led to a rate constant of propagation $k_p = 8.5 \text{ mol}^{-1} \cdot l \cdot s^{-1}$ (C_6H_5Cl, 70 °C) [18]. The apparent activation energy of propagation was $E_p = 25.2 \pm 8.4$ kJ · mol^{-1} (6.0 ± 2.0 kcal · mol^{-1}). In view of the relatively high temperature of the experiments and the relatively low polarity of chlorobenzene k_p value seems to represent the rate constant of propagation of ion-pairs.

5.3.3 Termination

Early kinetic studies indicated that polymerization of oxetanes proceeds essentially without transfer or termination, and that the only reaction leading to deactivation of growing species is degradative chain transfer to polymer [15].

Thus, cationic polymerization of BCMO is a typical and well documented example of cationic ring-opening polymerization, in which chain transfer to polymer leads to termination [15,18]. Chain transfer to polymer is a general phenomenon in cationic ring-opening polymerization of heterocycles. The extent this process affects the kinetics depends on the relative reactivity of strained (monomer holding) and unstrained (polymer holding) oxonium ion:

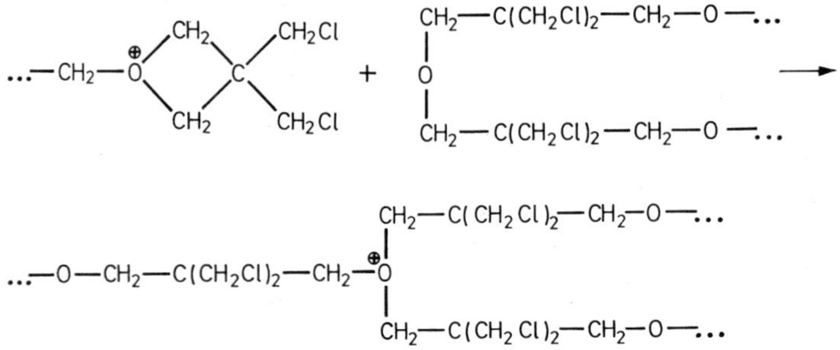

(5-7)

According to kinetic analysis the reactivity of the unstrained ion is so low that in polyBCMO it can be considered to be inactive.

This type of termination is kinetically distinguishable from other possible termination reactions such as spontaneous first-order decomposition of active centers or termination with monomer, because the rate of termination increases with conversion according to:

$$\frac{-d[P^*]}{dt} = k_t[P^*]([M]_0 - [M]_t) \tag{5-8}$$

where: [P*] — concentration of active species ($[M]_0 - [M]_t$) — concentration of polymer (in terms of monomer units).

A complete kinetic scheme, involving slow initiation, propagation, and termination by chain transfer to polymer, has been developed [29,30] and it was conclusively shown that it is applicable for BCMO polymerization. The pertinent kinetic equations were given in Vol. I (Adv. Polymer Sci. *37*). The rate constant of termination by chain transfer to polymer was $k_t = 1.2 \cdot 10^{-3}\ mol^{-1} \cdot l \cdot s^{-1}$ (C_6H_5Cl, 70 °C) [18].

5.3.4 Molecular Weights

Black and Worsfold measured the molecular weights of oxetane polymers prepared with $Ph_3C^+PF_6^-$ and $(C_2H_5)_3O^+PF_6^-$ as initiators [27]. Some representative results are shown in Table 5.2.

Table 5.2 Molecular weights in the polymerization of oxetane (CH_2Cl_2, −30 °C) [27]

[Oxetane] $mol \cdot l^{-1}$	$Ph_3C^{\oplus}PF_6^{\ominus}$ $\times 10^3\ mol \cdot l^{-1}$	$\overline{M}_n$ [a] calcd.	$\overline{M}_n$ (GPC)	$\overline{M}_w/\overline{M}_n$
0.513	0.544		26400	1.88
0.513	1.37		21400	1.83
0.513	1.87		22800	1.67
	$(C_2H_5)_3O^{\oplus}PF_6^{\ominus}$			
0.068	1.85	1800	3900	4.2
0.272	1.72	7270	12500	2.0
0.544	1.55	8800	19200	2.0
0.459	0.99	2700	34700	1.5

[a] Assuming living system

Molecular weights were higher than the calculated values assuming a living process; this may be explained by slow initiation, which was indeed proposed for $(C_2H_5)_3O^+PF_6^-$ initiated polymerization on the basis of kinetic measurements.

Table 5.3. Dependence of molecular weight of poly-3,3-bis(chloromethyl)oxetane) (BCMO) on the concentration of water, ($[BCMO]_0 = 0.31\ mol \cdot l^{-1}$, $[(i\text{-}C_4H_9)_3Al]_0 = 2 \cdot 10^{-2}\ mol \cdot l^{-1}$, solvent C_6H_5Cl, temp.: 70 °C) [18]

$[H_2O]_0$ $10^{-4}\ mol \cdot l^{-1}$	Conversion %	Molecular weight [a] 10^3
12.5	80	430
7.5	80	1500
3.5	80	2200

[a] Viscosity, from the dependence $\overline{DP}_n = 1230\ [\eta]^{1.95}$ determined for column-fractionated samples, $\overline{M}_n$ measured by osmometry in cyclohexanone solvent [31]

Cationic polymerization of BCMO proceeds without appreciable transfer except for degradative chain transfer to polymer, which, in fact, is a termination rather than transfer. Thus, molecular weights of polymers prepared in the absence of the purposely added chain transfer agents are high, some having a number average molecular weight up to $2 \cdot 10^6$. It was demonstrated that molecular weights were inversely proportional to the concentration of active initiator, which for the R_3Al/H_2O system was governed by the concentration of water (at low $[H_2O]/[R_3Al]$ratios). This is illustrated by the data of Table 5.3.

The same phenomenon was also observed for BF_3-initiated polymerization, although because of higher concentration of free water in the system the molecular weights were usually lower [15].

On the contrary, molecular weights do not depend on the concentration of BF_3 or R_3Al, providing their concentration is higher than that of water. In both systems molecular weights increase with increasing initial monomer concentration and for the latter system there is an almost linear relationship between molecular weight and conversion. Thus, in spite of rather complex kinetics, molecular weights are close to theoretical values calculated on the basis of the simple equation applicable for living systems:

$$\overline{DP}_n = \frac{[M]_0 - [M]_t}{[I]_0} \tag{5-9}$$

where the concentration of growing species as indicated above, is governed mainly by the concentration of water present in the system. Some pertinent data are given in Table 5.4 [18].

Table 5.4 Dependence of molecular weight of poly-3,3-bis(chloromethyl)oxetane (BCMO) on conversion $[BCMO]_0 = 0.15$ mol · l^{-1}, $[(i\text{-}C_4H_9)_3Al] = 8 \cdot 10^{-3}$ mol · l^{-1}, $[H_2O]_{0\,(add)} = 1 \cdot 10^{-2}$ mol · l^{-1}, C_6H_5Cl, 70 °C [18]

Conversion %	$\overline{M}_n \cdot 10^3$ (viscosity) [31]
20	131
44	232
60	344
70	465
98	606

In the polymerization of MCMO with the R_3Al/H_2O system, molecular weights do not depend on conversion. This is illustrated by the data in Table 5.5 [22].

Table 5.5. Dependence of molecular weight and molecular weight distribution of poly-3-methyloxetane (MCMO) on conversion $[MCMO]_0 = 0.55\ mol \cdot l^{-1}$, $[(i\text{-}C_4H_9)_3Al]_0 = 10^{-2}\ mol \cdot l^{-1}$, $[H_2O]_{0\,(added)} = 3 \cdot 10^{-3}\ mol \cdot l^{-1}$, C_6H_5Cl, 20 °C [22]

Conversion	$\overline{M}_n \cdot 10^3$ osmometry	$\overline{M}_w \cdot 10^3$ ultracentrifuge	$\overline{M}_w/\overline{M}_n$
4.6	610	820	1.29
21.6	720	720	1.30
45.8	700	820	1.30
89.0	760	820	1.27

5.3.5 Formation of Cyclic Oligomers in the Polymerization of Oxetanes

The general discussion of cyclization phenomena in cationic ring-opening polymerization is given in Sect. 3.

Formation of cyclic oligomers in the polymerization of oxetane has already been noted by Rose, who found that in addition to polymer, cyclic tetramer was also formed in quantities depending on the temperature [13]. Polymerization of OX with BF_3 initiator at —80 °C gave 4% tetramer while at 100 °C as much as 50% cyclic tetramer was isolated. This work was later reinvestigated, and cyclic trimer in addition to cyclic tetramer was isolated from the polymerization products; the quantities of both oligomers were dependent on the temperature and on the initiator used [32]. The effect of the initiator structure is illustrated by the data in Table 5.6.

Table 5.6 Formation of cyclic oligomers in the polymerization of oxetane; 1,2-dichloroethane, 70 °C [32]

Initiator	Mol %	% Conversion to cyclic	
		trimer	tetramer
$Et_3O^{\oplus}BF_4^{\ominus}$	0.14	9.7	20
$Et_3O^{\oplus}SbCl_6^{\ominus}$	0.15		1.0
$Et_3O^{\oplus}PF_6^{\ominus}$	0.13	1.0	1.8
$Et_3O^{\oplus}SbF_6^{\ominus}$	0.03	1.0	1.95
$HOSO_2CF_3$	0.24	5.4	2.9

Table 5.7 illustrates the effect of the temperature. The values given in Tables 5.6 and 5.7 represent final yields measured at complete monomer conversion; cyclic oligomers were not formed after the monomer was consumed. Isolated cyclic oligomers could not be reconverted to linear polymer (or monomer) under the typical polymerization conditions.

In the polymerization of 3,3-dimethyloxetane (DMO) initiated with $(C_2H_5)_3O^+BF_4^-$ in CH_2Cl_2 at 20 °C, Goethals and Bucquoque isolated and analyzed cyclic oligomers

Table 5.7 Formation of cyclic oligomers in the polymerization of oxetane; initiator: $Et_3O^{\oplus}BF_4^{\ominus}$, 0.14 mol %, solvent: 1,2-dichloroethane [32)]

Temperature	% Conversion[a] to cyclic	
°C	trimer	tetramer
—25	0	1.7
0	0	3.5
25	0	
35	0.8	3.9
50	3.7	16
70	20	9.7
100	20	16

[a] Concentration of oligomers were determined by GC

up to nonamer by GC-MS spectrometry [33)]. The typical cyclic oligomer distribution was as follows:

tetramer	92%	heptamer	0.2%
pentamer	7.1%	octamer	0.1%
hexamer	0.7%	nonamer	0.07%

The overall yield of cyclic tetramer was close to 20%. Cyclic tetramer (up to 20%) was also formed in the polymerization of 3,3-diethyloxetane under the same conditions [33)].

Cyclic oligomers were isolated also from the products of bulk polymerization of BCMO at 180 °C initiated with $(C_2H_5)_3Al—H_2O$ initiator [34)]. Extraction of the polymer with acetone gave 2% solids with m.p. = 122 °C. On the basis of molecular weight measurements and IR and ^1H-NMR spectra the product was identified as cyclic trimer.

5.3.6 Polymerization of BCMO in the Solid-State

BCMO can be polymerized in the solid state under the influence of ionizing radiation [23–25)]. This is a handy method of preparing polymer directly from monomer, in the absence of solvent and initiators. Polymerizations proceed to limited conversions (maximum 15–20% was observed at temperatures just below the melting point T_m = 18.5 °C). The rate increases with increasing temperatures (up to the melting point), irradiation time, dose rate and the size of the monomer crystals. Molecular weights are relatively high, $[\eta]$ = 0.5 dl/g (in cyclohexanone at 40 °C).

It is generally accepted that the solid state polymerization of BCMO initiated with γ-radiation proceeds by a cationic mechanism. Typical radical inhibitors affect neither the rate nor the final yield of polyBCMO. On the basis of the dependence of the polymer yield on the size of monomer crystals and the observed high degree of orientation in the polymers, it was originally suggested that the structure of the monomer crystal lattice directly imposes order and orientation upon the polymer [35)].

Wegner, however, established that radiation-induced solid-state polymerization of BCMO leads to a polymer morphology, which is incompatible with the so-called topochemical polymerization, i.e., a process in which monomer molecules are transformed into polymer without destruction of the crystal lattice [36]. Electron microscopy, X-ray analysis and electron diffraction studies, have shown that polymerization starts at the edges and imperfections of the monomer crystals and that amorphous polymer is formed initially. Further transition from the amorphous state leads to the thermodynamically unstable monoclinic β-form. Density measurements indicate that the polymer is only 45–50% crystalline. The density of the amorphous polyBCMO is 1.368 g/cm^3; the density calculated for the crystalline polymer from crystallographic data of the β-form is 1.456 g/cm^3. The density of the product of the radiation-induced solid-state polymerization is 1.41 g/cm^3 [36].

5.3.7 BCMO Copolymers

Problems related to the reactivity of oxetanes in copolymerization are discussed in Sect. 15 (Copolymerization). This paragraph briefly describes the preparation of block copolymers of BCMO by sequential polymerization.

It has been claimed that sequential polymerization of the corresponding monomers leads to AB block copolymers containing polyBCMO and polyTHF blocks [37]. Reactions were typically carried out in the following way: 3 ml (37 mmol) THF was polymerized in 7 ml heptane for 18 hrs at 0 °C with 0.02 mmol BF_3 and 0.01 mmol epichlorohydrin (promoter). 5 mmol BCMO (0.6 ml) was then added, the mixture cooled to −50 °C and the excess of unreacted THF and solvent were removed by distillation. The second stage of the synthesis was carried out at −30 °C for 3 hrs. After this time 1.02 g of polymer were isolated. The product was characterized by solubility; the increase of molecular weights during the second stage was not proven nor were GPC measurements performed. The difference in solubilities of the corresponding pair of homopolymers, however, is indeed large enough to assume the formation of block copolymer.

Another approach for the preparation of BCMO graft copolymers was based on the use of macromolecular initiators [38]. Polystyrene or polyTHF containing dioxolenium groups in the chain were prepared and used as initiators of BCMO polymerization as shown for polystyrene:

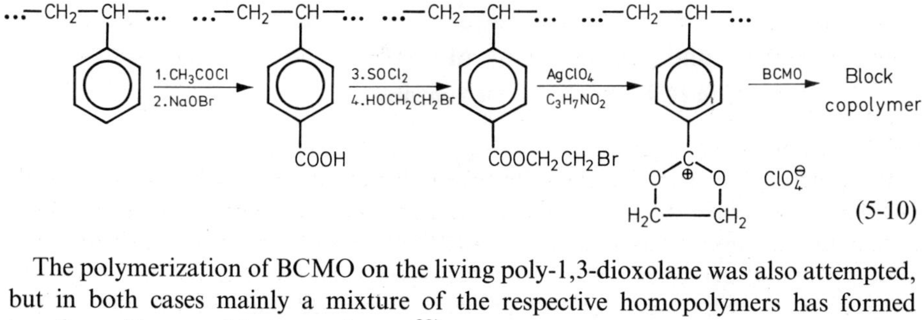

(5-10)

The polymerization of BCMO on the living poly-1,3-dioxolane was also attempted, but in both cases mainly a mixture of the respective homopolymers has formed together with some block copolymer [39].

5.4 Other Oxetanes

Goethals, in a review published in 1965, listed 33 cationically polymerized differently substituted oxetanes [8]. In most cases the corresponding monomer was synthesized, and its ability to polymerize was confirmed by routine experiments using BF_3 initiator. Visual observations were made (i.e. viscous oil, amorphous solids, crystalline) but no details as to yields and molecular weights of polymers prepared under truly reproducible conditions were given.

Recently Kops described in more detail the polymerization of a 2-substituted oxetane, 2-methyloxetane [40]. By the use of $(C_2H_5)_3O^{\oplus}PF_6^{\ominus}$ initiator and CH_2Cl_2 solvent at −80 °C, Kops obtained reasonable yields (30–60%) of polymers with $\overline{M}_n$ as high as $3 \cdot 10^5$.

According to ^{13}C-NMR analysis, the polymers were not uniform, containing both head-to-head and head-to-tail units, and the proportion of the latter increased with decreasing temperatures:

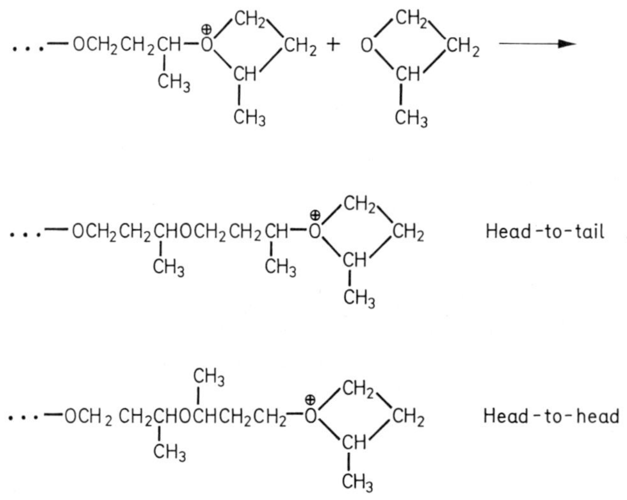

(5-11)

This result indicates that opening of the 4-membered ring is nonspecific and both $O{-}C^2$ and $O{-}C^4$ bonds can be broken.

5.5 Properties and Applications of Poly-3,3-Bis (Chloromethyl)Oxetane

PolyBCMO is one of the large-scale production polymers based on the cationic ring-opening polymerization. Production was started in the mid-fifties by Hercules Co. (Penton®) but it was discontinued in the late 70 s. There are also indications of its production in the USSR (Pentaplast®).

PolyBCMO exhibits an excellent combination of physical-chemical-processing characteristics rendering it a competitive thermoplastic material (engineering plastic).

In particular, it has a combination of relatively high melting point, outstanding thermal stability and unusually good chemical resistance (e.g. stable to conc. H_2SO_4 up to 120 °C) [28]. Thus, Penton has found numerous applications for the construction of chemical equipment.

PolyBCMO is a light-coloured thermoplastic with a crystalline melting point of 181 °C. Its mechanical properties are close to those of Nylon 6 (impact strength is poorer) as well as to those of low density polyethylene.

The crystalline polymer may exist in two forms α and β. The higher melting α-form, having an orthorhombic unit cell, is formed when the polymer crystallizes out of the melt or solution. The thermodynamically unstable β-form occurs when quenched amorphous polymer is annealed above the glass-transition temperature; annealing at 115 °C transforms the β-form into the α-form. The average size of lamellae formed in isothermal crystallization is 1000–2000 Å in lateral dimensions, with a thickness of 100 Å.

The principal thermal, electrical and mechanical properties of Penton are given in Ref. [28].

5.6 References

1. Rose, J. B.: in "The Chemistry of Cationic Polymerization" Plesch, P. H., Ed., Macmillan, New York, 1963
2. Dreyfuss, P., Dreyfuss, M. P.: in "Ring Opening Polymerization" Fritsch, K. C., Reegan, S. L., Eds. Marcel Dekker, New York, 1969
3. Dreyfuss, P., Dreyfuss, M. P.: in "Chemical Kinetics Vol. 15. Non-radical Polymerization", Bamford, C. H., Tipper, C. F. H. Eds, Elsevier, Amsterdam–Oxford–New York, 1976
4. Hirano, T., Nakayama, S., Tsuruta, T.: Makromol. Chem. *176*, 1897 (1975)
5. Koinuma, H., Hirai, T.: Makromol. Chem. *178*, 241 (1977)
6. Sawada, H.: Thermodynamics of Polymerization, Marcel Dekker Inc. New York, Basel, 1976, Chapter 7
7. Cyclic Monomers, Frisch, K. C. Ed., Wiley-Interscience, New York, 1972, p. 54
8. Goethals, E. J.: Industrie Chemique Belge, *30*, 559 (1965)
9. Sørensen, W. R., Campbell, T. W.: Preparative Methods of Polymer Chemistry, 2-nd, Ed. Wiley Interscience, New York, 1968, pp. 367–378
10. Hoare, F. E., Melia, T. P.: Polymer *3*, 277 (1962)
11. Penczek, I., Penczek, S.: J. Polymer Sci. *8*, 2465 (1970)
12. Farthing, A. C., Reynolds, R. J. W.: J. Polymer Sci. *12*, 503 (1954)
13. Rose, J. B.: J. Chem. Soc. *1956*, 542, 546
14. Farthing, A. C.: J. Chem. Soc. *1955*, 3648
15. Penczek, I., Penczek S.: Makromol. Chem. *67*, 203 (1963)
16. Saegusa, T., et al., Makromol. Chem. *53*, 203 (1962)
17. Kubisa, P., Brzeziński, J., Penczek, S.: Makromol. Chem. *100*, 286 (1967)
18. Kubisa, P., Penczek, S.: Makromol. Chem. *130*, 186 (1969)
19. Beste, L. F., Hall, H. K. Jr.: J. Phys. Chem. *68*, 269 (1967)
20. Aleksiuk, G. A. et al.: Vysokomol. Soed. *21*, 2142 (1979)
21. Aleksiuk, G. A. et al.: Polymer J. *13*, 23 (1981)
22. Aleksiuk, G. A. et al.: Polymer J. *13*, 33 (1981)
23. Okamura, S., Hayashi, K., Kitanishi, Y.: J. Polymer Sci. *58*, 925 (1962)
24. Chapiro, A., Penczek, S.: J. Chim. Phys. *59*, 696 (1962)
25. Hayashi, K., Nishii, M., Okamura, S.: J. Polymer Sci. *C4*, 839 (1963)
26. Saegusa, T.: J. Macromol. Sci., Chem. *A6*, 997 (1972)
27. Black, P. E., Worsfold, D. J.: Can. J. Chem. *54*, 3325 (1976)

28. Boardman, H.: in "Enclopedia of Polymer Science and Technology" Mark, H. F., Gaylord, N. G. Eds. Interscience, Vol. 9, p. 668
29. Penczek, S.: Bull. Acad. Pol. Sci. Ser. Sci. Chem. *18*, 47, 53 (1970)
30. Penczek, S.: Makromol. Chem. *134*, 299 (1970)
31. Brzeziński, J.: Polimery (Poland) *9*, 367 (1964)
32. Dreyfuss, P., Dreyfuss, M. P.: Polymer J. *8*, 81 (1975)
33. Bucquoye, M. R., Goethals, E. J.: Makromol. Chem. *179*, 1681 (1978)
34. Arimatsu, Y.: J. Polymer Sci. *4*, 728 (1966)
35. Bamford, C. H., Eastmond, E. C. H.: Quart. Rev. *23*, 271 (1969)
36. Keggenhoff, B., Wegner, G.: J. Polymer Sci., Pol. Phys. Ed. *11*, 759, 769 (1973)
37. Saegusa, T., Matsumoto, S., Hashimoto, Y.: Macromolecules *3*, 381 (1970)
38. Yamashita, Y., Chiba, K.: Polymer J. *4*, 200 (1973)
39. Yamashita, Y., Okada, M., Hirota, M.: Angew. Makromol. Chem. *9*, 136 (1969)
40. Kops, J., Spaangard, H.: Pol. Bull. *4*, 505 (1981)

6 Oxolanes (Tetrahydrofurans)

6.1 Introduction

Synthetic applications of the cationic polymerization of tetrahydrofuran (THF) are related to two groups of products, namely high-molecular-weight polymer and lower-molecular-weight products with controlled $\overline{DP}_n$ and well defined end-groups which are being used as telechelics.

Substituted derivatives of THF, except 2-methylTHF, do not polymerize (cf. Chapt. 2).

6.2 High-Molecular-Weight Polytetrahydrofuran

Synthesis of high polymers by cationic ring-opening polymerization of heterocycles is hampered by chain transfer which may occur by:

— back-biting or end-to-end closure (intramolecular chain transfer);
— scrambling by intermolecular chain transfer,
— proton transfer (formation of double bonds at the chain end).

All these reactions proceed to a larger or smaller extent in the cationic polymerization of *any* heterocycle. In the polymerization of THF the rate of these reactions is reduced and the polymerization becomes a living one.

The small contribution of macrocyclization in the polymerization of THF is due to two effects: The formation of smaller cyclic oligomers like dimer and trimer is thermodynamically unfavorable (in contrast to the formation of e.g. 1,4-dioxane in the polymerization of ethylene oxide) because 10- and 15-membered cyclic ethers are strained. It is also kinetically hampered since the rate of chain transfer to polymer which leads to ring formation is low due to the lower basicity of the polymer units than that of THF.

Proton transfer in THF polymerization at moderate temperatures is negligible and other sources of termination such as anion instability can be avoided by a proper choice of polymerization conditions.

The synthesis of high-molecular-weight polyTHF became possible after the discovery of the living nature of THF polymerization. This discovery started with the work of Meerwein [1] and the observations were then formulated in terms of a living process simultaneously and independently by Rosenberg et al. [2], Dreyfuss and Dreyfuss [3] and Vofsi and Tobolsky [4] (cf. Adv. Pol. Sci. *37*). Further development at DuPont, BASF and other companies led eventually to commercialization of polyTHF diols. Later Smith et al. [5–7] used superacid derivatives for THF polymerization. These pioneering works showed the way toward the preparation of high-molecular-weight polyTHF and for the synthesis of telechelics.

High purity THF can be obtained by distilling from the blue "solution" of potassium-sodium alloy (at least at the laboratory scale), the CH_2Cl_2 solvent preferably purified over a sodium mirror and the initiator (e.g. $(CF_3SO_2)_2O$) purified by distillation. The reacting solution is best prepared under vacuum. The expected $\overline{DP}_n$ is given by

$$\overline{DP}_n = \frac{[THF]_0 - [THF]_e}{[I]_0 - [I]_e}, \tag{6-1}$$

where I is initiator.
The desired end-groups can be introduced after the living polymer-monomer equilibrium is reached. The preparation of polyTHFs, terminated at both ends with primary amino, bromine, isocyanate, and thiol end-groups, has been described by Smith and Hubin [5].

Properties of the high-molecular-weight polyTHF, which to our knowledge has not been commercialized, are given in the reviews on polyTHF by Dreyfuss and Dreyfuss [3,8].

PolyTHF is a crystalline (up to 80%) material [9], melting at 42 °C, although higher values (up to 60 °C) have also been reported and it behaves as an elastomer above this temperature.

Telechelic polyTHF oligodiols (DuPont, BASF, Quacker Chem.) have found important applications, mostly as soft segments in multiblock elastoplastic materials such as polyurethanes (e.g. Spandex®) and polyesters (Hytrel®).

6.3 Telechelic PolyTHF: α,ω-Diols

PolyTHF that found practical applications have $\overline{M}_n$ ranging from 1000 to 2000 and carry two hydroxyl end-groups per macromolecule:

$$HO\text{+}CH_2CH_2CH_2CH_2O\text{+}_nH \tag{6-2}$$

These oligomers are prepared by polymerization of THF with protonic acid initiators, and according to the patent literature and other available information, fluosulfonic acid has been used on a technical scale. A recent paper describes the successful application of H_2SO_4/SO_3 mixtures [11]. The diol of $\overline{M}_n = 1000$ is a white opalescent solid at 24 °C; it melts at 32 °C giving a syrupy liquid with a hydroxyl number of 112–116, an acid number less than 0.05, a carbonyl number less than 4.0 and a Ca content less than 5 ppm; Na and K are less than 10 ppm [8].

6.3.1 Special Features of Acid-Initiated Oligomerization of THF

Although the cationic polymerization of THF was studied extensively, especially during the last decade, and it is believed to be the best understood system among cationic ring-opening polymerizations at the present time, (cf. Adv. Polymer Sci. *37*, (1980) Sect. 4.2.) surprisingly little attention has been paid to the mechanism of polymerization initiated by protonic acids. This is probably due to the fact that these systems show some peculiarities which are purposely avoided in studies aimed at determining the general features of THF polymerization.

Initiating systems required for the preparation of THF oligomers carrying two OH end-groups must have the following characteristics:

— $\overline{M}_n$ in the range 1000–2000 should be accesible and easily controllable;
— the practical rate of polymerization should be sufficiently high;
— the polymer should carry strictly two OH end-groups per macromolecule;

These combined requirements are not easily attained in one initiating system. Below we will discuss advantages and deficiences of selected protonic acids.

6.3.1.1 The THF-CF_3SO_3H System

In the THF-CF_3SO_3H system the initiating acid is not completely consumed at the monomer-polymer equilibrium. As illustrated by the data shown in Table 6.1 the molecular weights of the products formed exceed many times the values calculated by:

$$\overline{DP}_n = \frac{[THF]_0 - [THF]_e}{[CF_3SO_3H]} \qquad (6\text{-}3)$$

Table 6.1. Molecular weights of polyTHF obtained with $HOSO_2CF_3$ and $C_2H_5OSO_2CF_3$ as initiators [12]

Reaction system (molar ratio)	Time, h	Conversion, %	$\overline{M}_n$ calcd.	$\overline{M}_n$ measured
$THF/CH_3NO_2/C_2H_5OSO_2CF_3$ 23.1/9.1/1.0	24	47	780	740 (HO- by ^{1}H-NMR)
$THF/CH_3NO_2/HOSO_2CF_3$ 23.1/9.1/1.0	24	40	680	10000 (HO- by IR)

This may be attributed either to kinetic (slow initiation) or thermodynamic (chain-coupling) reasons, as indicated in Scheme (6-4).

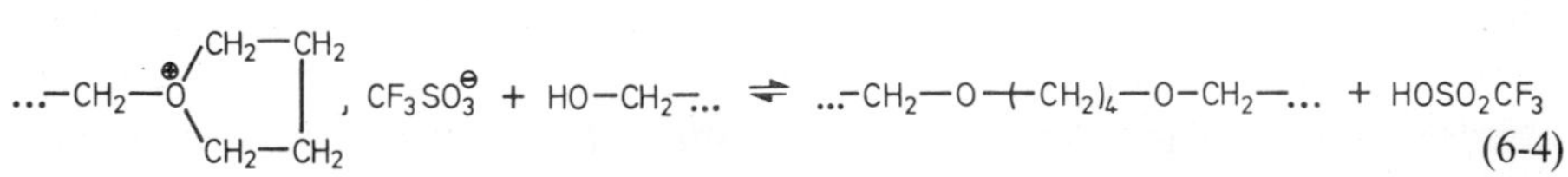

(6-4)

To describe such a system the equilibrium acid concentration has to be included in Eq. (6.3); thus:

$$\overline{DP}_n = \frac{[THF]_0 - [THF]_e}{[HOSO_2CF_3]_0 - [HOSO_2CF_3]_e} \qquad (6\text{-}5)$$

Pruckmayr explains his results [12] by assuming that the formation of high-molecular-weight polymer is due to chain-coupling (Scheme 6-4). In this case the acid concentration in the equilibrium governing the $\overline{DP}_n$ would depend on the equilibrium constant of Reaction (6-4).

However it was not proven, that the observed results are not of kinetic origin, i.e. due to slow propagation. In such a scheme, growth would proceed on active species formed only by a small fraction of acid, and at the moment monomer-polymer equilibrium was reached some amount of unconsumed acid would still remain.

Insufficient data exist to conclude which of the two possible schemes is valid; in fact the molecular weights are not simply given by the ratio of monomer to initiator and thus cannot easily be regulated. The preparation of oligomeric diols having $\overline{M}_n$ = 1000–2000 would require unreasonably high starting concentrations of CF_3SO_3H.

6.3.1.2 The THF-FSO_3H System

When fluosulfonic acid (FSO_3H) was used as the initiator of THF polymerization, the higher molecular weight polymer with $\overline{DP}_n \gg ([THF]_0 - [THF]_e/[FSO_3H]_0$ was formed initially. In the further course of the reaction, the molecular weight decreased and approached calculated values. HF is formed during the polymerization and as determined by ^{19}F-NMR its ultimate concentration equals that of FSO_3H [13]. The results of a typical polymerization are summarized in Table 6.2.

Table 6.2 Polymerization of THF with HSO_3F as initiator in CD_3NO_2 solvent at 25 °C. Molar ratio $THF/CD_3NO_2/HSO_3F$ = 11.6/4.35/1.0 [13]

Polymerization time h	Conversion %	$\overline{M}_n$
0.7	48.0	2100
1.9	52.0	1440
3.8	52.8	1250
4.8	53.4	950
24.0	55.5	910

The observed dependence of molecular weights on time was explained by assuming that in the presence of FSO_3H, polymerization proceeds initially by chain coupling [12] i.e., by the mechanism similar to that described for CF_3SO_3H:

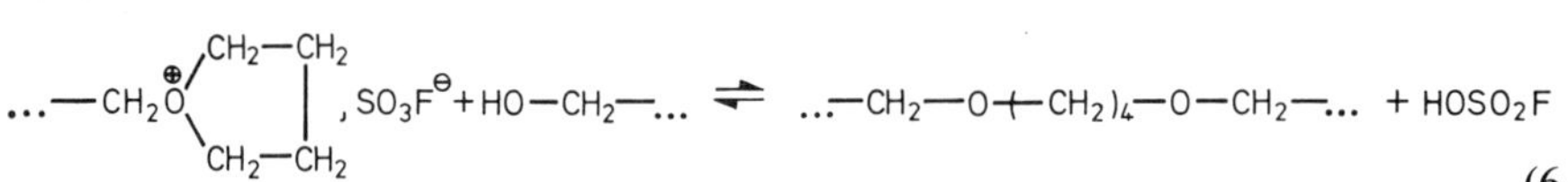

(6-6)

Since fluosulfonic acid is bifunctional, the acid or its esters can further react with hydroxyl end-groups:

$$\ldots\!-CH_2OH + FSO_3H \rightleftharpoons \ldots\!-CH_2OSO_3H + HF$$

$$\ldots\!-CH_2OSO_2F + HOCH_2-\ldots \rightleftharpoons \ldots\!-CH_2OSO_2OCH_2-\ldots + HF$$

(6-7)

As a result of these reactions, fluosulfonic acid and its esters are converted into the sulfonic acid derivatives. The equilibrium constant of the coupling reaction is different for fluosulfonic and sulfuric acid and redistribution of molecular weights occurs in the course of polymerization [13].

In these studies the molecular weights have been obtained by determining the hydroxyl end-group content in the polymers after hydrolysis.

Thus, in regard to final molecular weights one should distinguish between two types of coupling reactions:

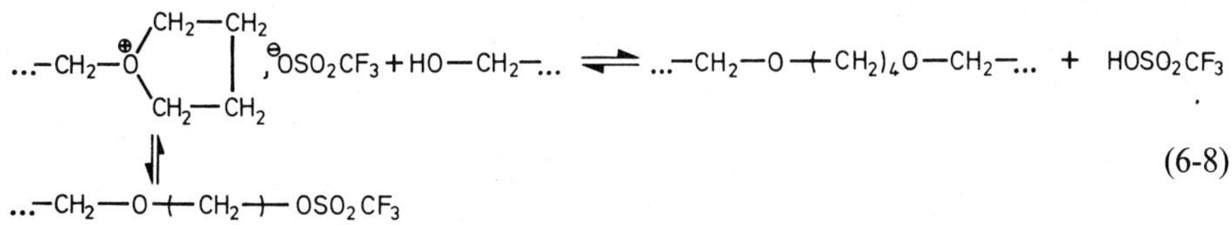

(6-8)

and

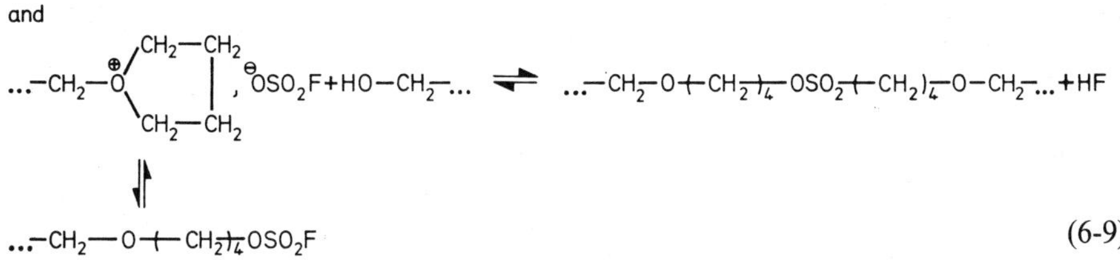

(6-9)

Reaction (6-8) dominates in CF_3SO_3H-initiated polymerization and leads to ether bonds which are unaffected by hydrolysis. Thus, this type of coupling leads to an irreversible increase of molecular weights. Reaction (6-9), yields ester bonds in the presence of HSO_3F and H_2SO_4 which undergo hydrolysis; thus, the coupling reaction in these systems does not affect the molecular weight after hydrolysis.

6.3.1.3 The THF-H_2SO_4 System

The molecular weights of polyTHF oligomers prepared in the presence of H_2SO_4 are given by the ratio of monomer to initiator concentrations (Eq. 6-3). Although the reaction analogous to Eq. (6-8) (expulsion of initiator, doubling the length of the macromolecule through ether bonds) is also possible with the sulfate ester, it has been shown by molecular weight measurements [11] that the contribution of this reaction is negligible in comparison with internal sulfate reactions. As already pointed out, macromolecules with the internal sulfate ester bonds are hydrolysed back during polymer work-up, leading to the degree of polymerization given by Eq. (6—3).

6.3.1.4 Rate of Polymerization in THF-Protonic Acid Systems

The rates of THF polymerization initiated by protonic acids having anions that are capable of forming covalent bonds, are affected by the existence of macroion-macroester equilibria (cf. Adv. Pol. Sci. *37* Sect. 5.1.2).

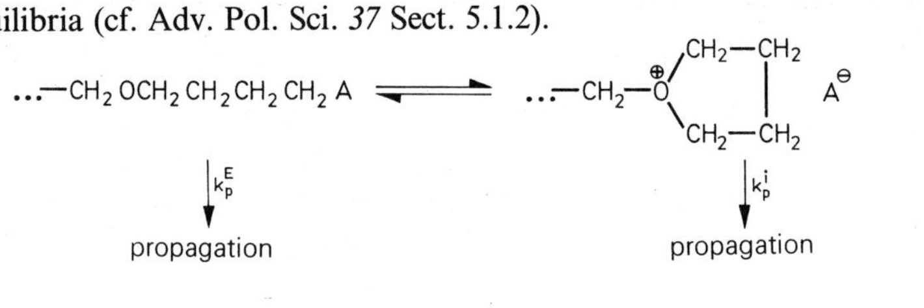

(6-10)

When $A^{\ominus} = CF_3SO_3^{\ominus}$, propagation by macroions is approximately 100 times faster than by macroesters [16]. Thus, only 2% macroions (of all of the growing species) counterbalance the contribution of growth by macroesters.

While k_p^E for the sulfate ester (e.g. ...—O—$(CH_2)_4OSO_3H$) is unknown, it is certainly much lower than that of the triflate ester (the higher the nucleophilicity of the parent anion the lower the reactivity of the ester).

Since the equilibrium constant of ion-ester interconversion is also unknown, the relative contribution of ionic and covalent growth cannot be predicted.

The rate of polymerization is very low even at high acid concentrations (up to 20 mol %) [11]. To increase the polymerization rate to acceptable values it is necessary to decrease the nucleophilicity of the sulfate anion. This can be done in various ways; one that is practical involves the addition of SO_3 [11]. Most likely the pyrosulfate anions thus produced increase the proportion of ions (as compared with sulfate anions) and/or yield more reactive macroesters. We are inclined to think that the first explanation is valid.

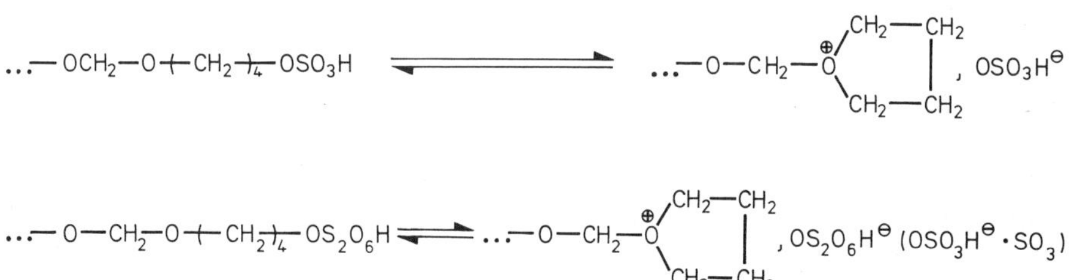

(6-11)

However, when pyrosulfate anions are used to boost the rates of polymerization, one has to use such high initial concentrations of the initiating system H_2SO_4/SO_3 that the highest $\overline{M}_n$ that can be reached is approx. 1000. Thus, a further increase of the rate is required. One practical solution is to add $HClO_4$ [11].

Under these conditions we have the following equilibria:

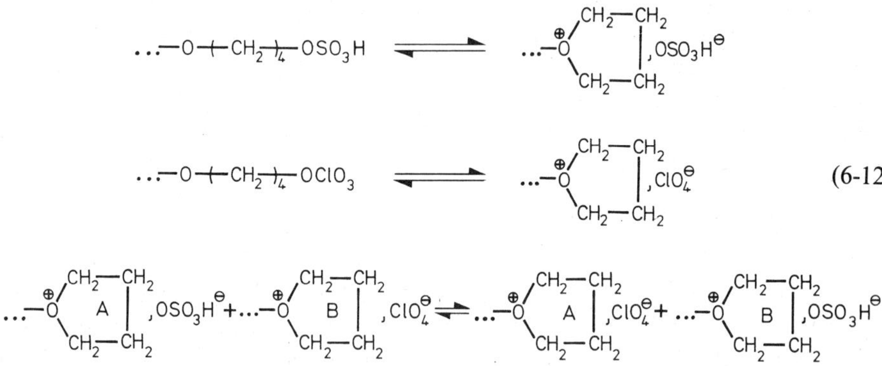

(6-12)

As indicated in Scheme (6-12), anion exchange gives every macromolecule a chance to exist as the perchlorate. Due to the low nucleophilicity of perchlorate anion, as compared to sulfate anion, the equilibrium between the perchlorate ester and the corresponding ion-pair is shifted more toward the ion-pair than the sulfate. This results in a higher overall rate of polymerization. A similar exchange has been observed in the THF-$CF_3SO_3^{\ominus}$-$SbF_6^{\ominus}$ system [18].

6.3.1.5 Conclusions

As indicated earlier, the initiating system should give products with desired $\overline{M}_n$s at a practically acceptable rate. The rate of THF polymerization is highest with CF_3SO_3H as initiator, however, the molecular weights are considerably higher than those given by the monomer to initiator ratio because some of the acid remains unconsumed.

Desired molecular weights may easily be attained in THF-H_2SO_4 systems; however, the polymerization rate is too low for practical purposes.

HSO_3F gives a higher rate of polymerization, but the serious disadvantage of this system is the liberation of highly corrosive HF.

To our knowledge, industrial processes are thus based on HSO_3F or H_2SO_4 initiators in combination with SO_3 or $HClO_4$ to increase the rate of polymerization.

6.4 THF Oligomers with Other Than —OH End-Groups

There are three ways of introducing the desired end-groups into polyTHF, namely by initiation, chain transfer or termination.

Initiation with acid chlorides complexed with metal halides (or silver salts) has been used [17], e.g.:

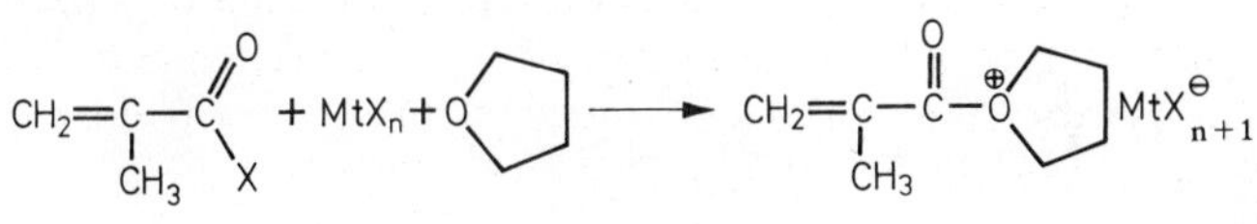

(6-13)

Other ionogenic groups can also be used, the major problem is quantitative initiation. Franta et al. studied the dependence of the extent of initiation vs. side reactions (e.g. $H^{\ominus}$ transfer leading to a double bond at the head-group) and found nearly quantitative initiation with the following compounds: $(C_6H_5)_2CH^{\oplus}SbF_6^{\ominus}$ [18], $C_6H_5CO^{\oplus}SbF_6^{\ominus}$ [19], $C_2H_5CO^{\oplus}SbF_6^{\ominus}$ [19], $SbF_6^{\ominus}O^{\oplus}C{-}C_6H_4{-}CO^{\oplus}SbF_6^{\ominus}$ [19], $SbF_6^{\ominus}O^{\oplus}C{-}(CH_2)_4{-}CO^{\oplus}SbF_6^{\ominus}$ [20], $(CH_3)_2CH^{\oplus}SbF_6^{\ominus}$ [19]. These salts were obtained in situ by reaction of the corresponding halide with silver salts:

$$RX + AgSbF_6 \rightarrow R^{\oplus}SbF^{\ominus} + AgX \qquad (6\text{-}14)$$

Some other salts, e.g. $(C_6H_5)_3C^{\oplus}SbF_6^{\ominus}$ (formed in situ), resulted predominantly in H^- transfer or proton elimination [19]. These facts have been rationalized (Adv. Polymer Sci. *37*, 1980 Sect. 3.2.1) when we discussed factors influencing the relative proportions of initiation by direct addition. These factors include first of all the ability of the second THF molecule to add to the oxonium ion formed by reaction between THF and the initiating carbenium ion (preformed or transient). Steric hindrance, like that observed in oxonium ions formed by triphenylcarbenium ions and THF, is a major factor preventing initiation by direct addition. It has been discussed in detail in the aforementioned series, the triphenylcarbenium ion gives exclusively $H^{\ominus}$ transfer and no initiation by direct addition.

Chain transfer as a method of introducing end-groups has mostly been studied by Rosenberg [21] and Heitz [22,23]. The latter author, in a series of papers based on the precise MWD analysis, studied the polymerization of THF in the presence of anhydrides (Ac_2O) as chain transfer agents. When Ac was $CH_2{=}CHCO$ or $CH_2{=}C(CH_3)CO$, polymerization carried out in the presence of catalytic amounts of protonic acids ($HSbF_6$ or CF_3SO_3H) gave polymers with

$$\overline{DP}_n = \frac{[THF]_0 - [THF]_e}{[Ac_2O] + [HA]} \tag{6-15}$$

The possibility of chain transfer by Ac_2O:

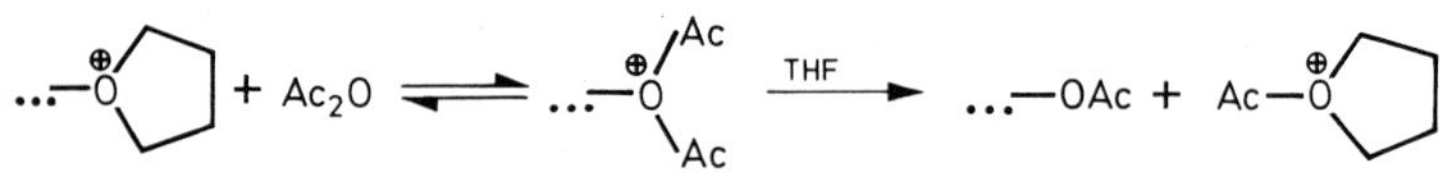

(6-16)

was rejected because similar processes in which $CH_3CO^{\oplus}SbCl_6^{\ominus}$ was used as initiator instead of protonic acid did not lead to low-molecular-weight polyTHF with AcO end-groups. Under these conditions higher molecular weight polyTHF is formed and Ac_2O remains unconsumed. Thus, protonic acid is required to catalyze the reaction of the anhydride with tetrahydrofuranium cations as well as the reaction with hydroxyl groups. We assume that the reactive intermediates are the $AcO^{\ominus}$ anion and acetylium cation present in minute but sufficient amounts, to accelerate telomerization:

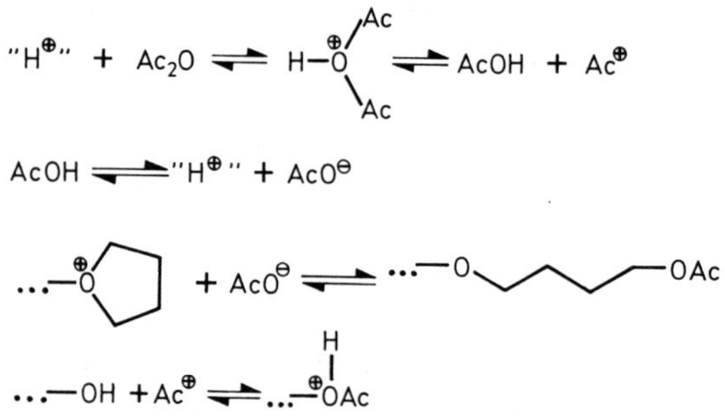

(6-17)

(where "$H^{\oplus}$" means solvated proton).

The macromolecules formed initially carry at one end an OH group and at the other end an OAc group. The OH end-groups are subsequently converted into OAc groups with $Ac^{\oplus}$ and protonic acid is regenerated.

Two additional observations support this mechanism. Polymerization initiated by protonic acid initially gives high-molecular-weight polymer (high viscosity of the mixture), equilibration occurs only after longer reaction times and the molecular weights decrease to the value given by Eq. (6-15). Analysis of the reaction mixture by GPC shows at the beginning of the reaction a series of double peaks, which indicates the presence of OH terminated oligomers in considerable amounts (cf. Fig. 2.1 in Sect. 2). After longer reaction times redistribution of molecular weights occurs; high polymer is converted into oligomers, but at the same time the proportion of the OH terminated oligomers decreases and becomes very low in the final product:

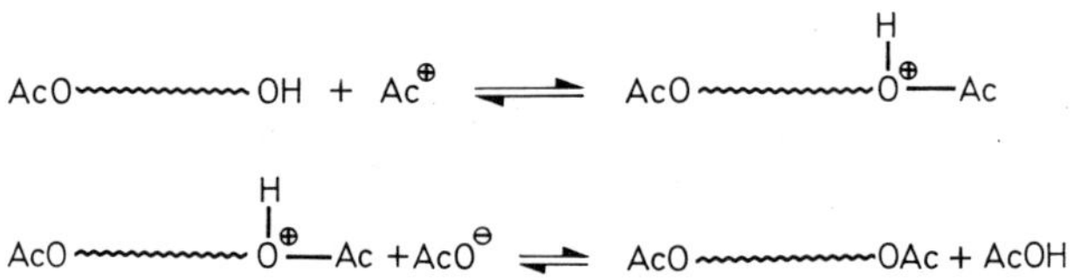

(6-18)

This observation indicates clearly that the formation of OAc-terminated oligomers proceeds not during the polymerization but after the polymerization, as shown in Scheme (6–17). Such a reaction mechanism was shown to operate for $Ac = CH_2 = CH—CO—$ (acrylic anhydride) and $Ac: CH_2 = C(CH_3)CO—$ (methacrylic anhydride) but was also suggested to operate for acetic anhydride [23]. Thus, earlier mechanistic interpretations of THF-$(CH_3CO)_2O$ systems based on the assumption that anhydrides are chain transfer agents should be abandoned.

Oligomers prepared by Heitz with acrylic anhydride (60% conversion, bulk process) had $\overline{M}_n$s ranging from 450 to 400 and functionalities from 1.8 to 2.

Chain termination can be used for preparing polyTHF with one or two active species per macromolecule. This procedure was used to prepare macromonomers with a polyTHF trunk, as well as block and graft copolymers (cf. Sect. 15).

Typical polyTHF macromonomers prepared by termination of mono- and dicationically living polyTHF are shown below:

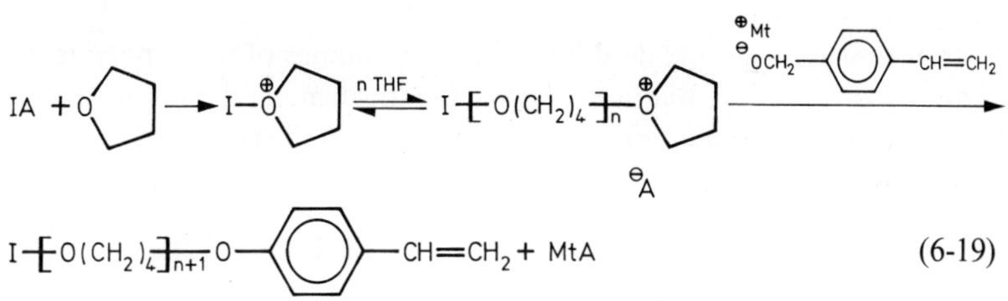

(6-19)

where I = (e.g.) an alkyl or acyl group, Mt = (e.g.) $Na^{\oplus}$, and $A^{\ominus}$ = (e.g.) $SbF_6^{\ominus}$.

Macromonomers of this or related structures have been prepared by Asami [24,25] and Franta [26].

Telechelics with a polyTHF trunk and two polymerizable end-groups have also been prepared [27,28] (cf. Chap. 15 on Copolymerization).

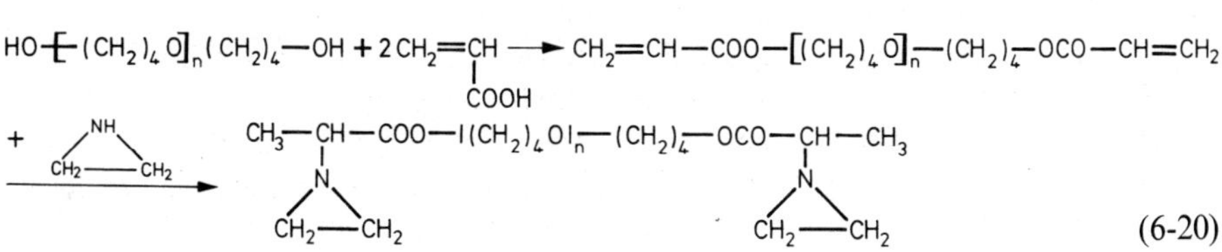

(6-20)

Star-shaped polyTHF can be fitted with desired head-groups by initiation with multioxocarbenium ions [29], and the desired end-groups by termination.

6.5 Copoly(THF-Oxirane) Diols

Linear strictly bifunctional oligomers of THF with molecular weights close to 1000 or 2000 and having two terminal OH groups are prepared in bulk with strong protonic acids (cf. Sect. 6.2). These oligomers are mostly used as soft segments in block copolymers made by polycondensation with diisocyanates or with 1,4-alkylene (ethylene, butylene) terephthalates.

A number of attempts has been made to improve some properties of THF oligomers. The major deficiency of these polymers is their ability to crystallize. The melting temperature of polyTHF is above room temperature (m.p. = 37 °C) which complicates handling and influences the properties of the final products. Recently a decrease of m.p. to 22 °C has been disclosed in a DuPont patent [30] that claimed lower melting ABA terpolymers of polyTHF (B) and poly(ethylene oxide). The synthesis starts on the terminal OH groups of polyTHF. Other attempts to lower the melting temperature have been based on the copolymerization of THF with various oxiranes. Since some of the applications include solid rocket propellant, unusual comonomers have also been tested, e.g. nitroglicydyl ether [31].

1,2-Propylene oxide PO was mostly used as comonomer in the preparation of bifunctional THF copolymers by direct copolymerization process. The required products must have a sufficiently low m.p., uniform structure and the functionality equal to two. It is also desirable to have the same OH end groups, either primary (preferentially) or secondary.

In this section we first describe the major findings of the copolymerization of THF with 1,2-propylene oxide and then a practical method published by Futamura [32] for the synthesis of the bifunctional copolymer of THF and PO.

6.5.1 Structure and some Properties of THF Copolymers

THF-PO copolymers obtained by various methods differ in several properties.

According to Futamura [32] copolymer prepared at the initial concentration of both monomers ($[M_1]_0 + (M_2]_0)$ close to 7 mol · l^{-1} at 0 °C in $C_2H_4Cl_2$ solvent (approx. 55 Vol. %) has an average functionality (F_n) of 2.0. Using $BF_3 \cdot O(C_2H_5)_2$ catalyst (0.08 mol %) and 4 mol % of 1,4-butandiol, copolymers containing 12–75% PO and $\overline{M}_n$ from 500–5000 were obtained. $\overline{F}_n$ was determined by comparing the $\overline{M}_n$ measured by GPC, and calculated from OH end-group content. The concentration of OH end-groups was determined by ^{13}CNMR using the α carbon resonances:

(6-21)

In off-resonance decoupled spectra these C atoms give a distinguishable doublet and a triplet, respectively. The ...—CH_2CH_2OH end-group (THF as a terminal unit) was absent. The ratio between secondary and primary end-groups was independent of $\overline{M}_n$ and was close to 2:1.

The copolymer prepared by Hammond et al. [33], under essentially the same conditions, had different characteristics. GPC analysis showed bimodal molecular weight distribution, besides a "high-molecular-weight" polymer an oligomer fraction was also formed. The latter consisted mainly of cyclic PO tetramer but mixed (PO and THF) tetramers and pentamers were also identified. The $\overline{M}_n$ determined by GPC was low (usually below $\overline{M}_n = 500$). The OH end-groups were determined by acetylation with acetic anhydride in pyridine with $HClO_4$ catalyst.

The $\overline{M}_n$ calculated from the OH end-groups content (assuming two OH groups per macromolecule) was about 50% higher than $\overline{M}_n$ by GPC, indicating that the end functionality of the copolymer was considerably less than 2.

Blanchard et al. have tested a variety of catalytic systems including $BF_3 \cdot OEt_2$ and $SnCl_4$ with diols and trityl salts [34,35]. Working under similar conditions as the previous two groups they prepared copolymers with $\overline{M}_n$ in the $\overline{M}_n = 500$–750 range (vpo). GPC analysis showed bimodal distribution with a low-molecular-weight polymer ($\overline{M}_n \sim 300$) also present.

Entelis et al. have studied the copolymerization of THF with three different oxiranes: PO, α-epichlorohydrin (ECH) and nitroglicydyl ether (NGE)[31]:

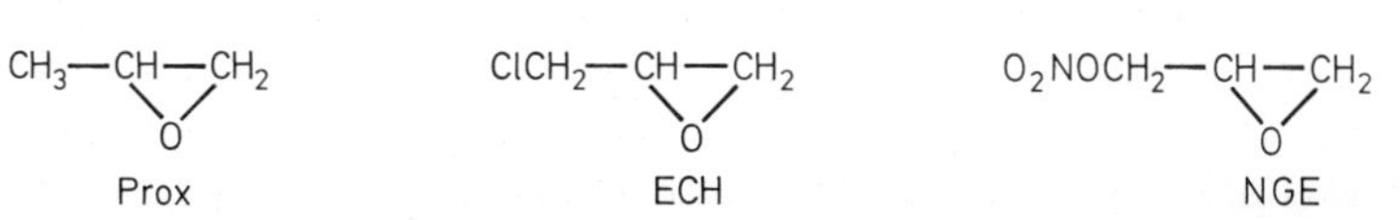

(6-22)

Using BF_3 catalyst, these authors have found a large proportion of cycles in all three systems (up to 40% with NGE), which decreased with an increase of $[THF]_0$. In the linear polymer a significant amount of double bonds was detected.

In bulk copolymerizations ($[PO] = 1.9\ mol \cdot l^{-1}$, $[THF]_0 = 10.7\ mol \cdot l^{-1}$), the concentration of cyclic PO tetramer did not exceed 4%. The proportion of cyclic oligomers may be further decreased i.e., to about 1%, by addition of propionic aldehyde in a concentration comparable to that of $[PO]_0$. Addition of propionic aldehyde also influences the molecular-weight distribution; in the presence and absence of aldehyde, $\overline{M}_w/\overline{M}_n \simeq 2$ and 4, respectively.

Besides hydroxyl end-groups fluoroalkyl and unsaturated end-groups were also detected in copolymer. These functions decrease the functionality of the desired diols. The introduction of low-molecular weight diols (propanediol, butanediol) as chain transfer agents decreases to negligible amounts the terminal unsaturation (up to 40% in the absence of diol; bromine number = 7, with $[BF_3]_0 = 2.5 \cdot 10^{-2}\ mol \cdot l^{-1}$) and fluoroalkyl end groups (up to 30% without diols). This can be ascribed to a change in the mechanism in the presence of relatively strongly basic hydroxyl compounds (cf. Sect. 3.2.7).

6.6 References

1. Meerwein, H., Delfs, D., Morschel, M.: Angew. Chem. *72*, 927 (1960)
2. Rosenberg, B. A., et al.: J. Polym. Sci. *C16*, 1917 (1964)
3. Dreyfuss, P., Dreyfuss, M. P.: Adv. Pol. Sci. *4*, 528 (1967)
4. Vofsi, D., Tobolsky, A. V.: J. Polym. Sci. *A3*, 3361 (1965)
5. Smith, S., Hubin, A. J.: J. Macromol. Sci. Chem. *7*, 1399 (1973)
6. Smith, S., Hubin, A. J.: US Pat. 3631199 (1973)
7. Smith, S., Hubin, A. J.: US Pat. 3436359 (1969)
8. Dreyfuss, P., "Polytetrahydrofuran", Gordon and Breach New York, 1982
9. Trick, G. S., Ryan, J. M.: Polym. Prepr. *7*, 92 (1966)
10. "Encyclopedia of Chemical Technology", Kirk, R., Othmer, B., Eds. New York, 1966, Vol. 10, p. 249
11. Matsuda, K., et al.: Polymer Prepr. *15*, 468 (1974)
12. Pruckmayr, G., Wu, T. K.: Macromolecules, *11*, 265 (1978)
13. Pruckmayr, G., Wu, T. K.: Macromolecules, *11*, 662 (1978)
14. Kubiczek, L., private communication
15. Baran, T., et al.: Makromol. Chem. in press
16. Matyjaszewski, K., Diem, T., Penczek, S.: Makromol. Chem. *180*, 1817 (1979)
17. Sierra-Vargas, J., Franta, E., Rempp, P.: Pol. Bull. *3*, 83 (1980)
18. Afshar-Taromi, F. et al.: Makromol. Chem. *179*, 849 (1978)
19. Zilliox, J. G., et al.: IUPAC International Symposium Mainz 1979, p. 56
20. Franta, E., et al.: J. Polym. Sci., Symp. *56*, 139 (1976)
21. Rosenberg, B. A., et al.: Dokl. Akad. Nauk SSSR, *175*, 365 (1967)
22. Kress, H. J., Heitz, W.: Makromol. Chem., Rapid Commun. *2*, 427 (1981)
23. Heitz, W., Stix, W., Kress, J.: IUPAC International Symposium Mainz 1979, p. 200
24. Asami, R., et al.: Polym. Bull. *2*, 713 (1980)
25. Asami, R., et al.: Polym. J. *13*, 141 (1981)
26. Sierra-Vargas, J., Franta, E., Rempp, P.: Makromol. Chem. *182*, 2603 (1981)
27. Brauer, G. M.: Polym.-Plast. Technol. Eng. *9*, 87 (1977)
28. Brit. Pat. 1044753 (1966)
29. Franta, E., et al.: J. Polym. Sci., Polymer Symposia *56*, 139 (1976)
30. US Pat. 4299993, 1981
31. Entelis, S. G., Korovina, G. U.: Makromol. Chem. *175*, 1253 (1974)
32. Mochel, V. D., Bethea, P. W., Futamura, S.: Polymer *20*, 65 (1979)
33. Hammond, J. M., Hopper, J. F., Robertson, W. G. P.: J. Polym. Sci. *9*, 265, 281, 295 (1971)
34. Blanchard, L. P., Singh, J., Baijal, M. D.: Can. J. Chem. *44*, 2679 (1966)
37. Blanchard, L. P., Asselin, J. C., Malhotra, S. L.: Polymer J. *7*, 326 (1975)

7 Cyclic Acetals

7.1 Introduction

Cyclic acetals are 5- and higher membered monomers with at least one unit in which two oxygen atoms flank a substituted or unsubstituted methylene group. If the methylene carbon is unsubstituted the monomers are cyclic formals.

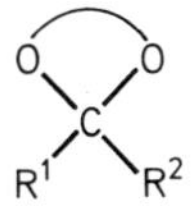

(7-1)

This group of monomers has a number of distinctive features, i.e.,

- high stability towards alkalies and low stability towards acids; thus, cyclic acetals polymerize only by a cationic mechanism;
- low basicity, usually lower than that of respective polymer units, a unique feature among the oxygen-containing heterocyclics;
- high reactivity of the $—O—CR_2—O—$ group both in strained (monomers) and unstrained (polymers) molecules; thus, under the conditions of polymerization the reaction rates with monomer participation (i.e. propagation) are comparable to those with polymer participation (e.g. transacetalization).

1,3,5-Trioxane (TXN), the cyclic trimer of formaldehyde, occupies a special position among cyclic acetals:

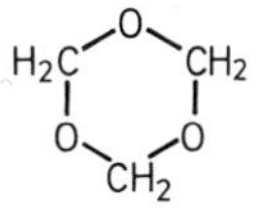

(7-2)

The cationic polymerization of TXN is used in large-scale industrial processes, therefore this system will be discussed separately in more detail.

7.2 Synthesis and Polymerizability of Cyclic Acetals

There are a few general methods of preparing cyclic acetals described in the literature [1, 2]. The most frequently used method is based on the reaction of the corresponding diols with aldehydes or ketones:

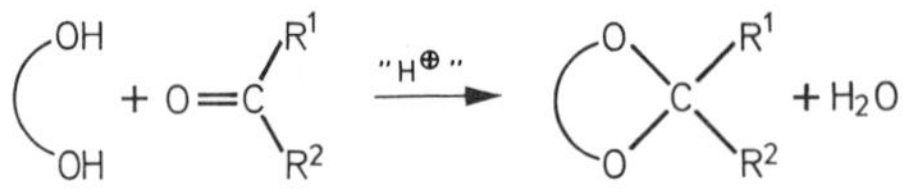 (7-3)

Some of these reactions are essentially quantitative and are even used in analytical chemistry (e.g. benzal of pentaerithritol)[3] or to introduce protective groups in the synthesis of nucleic acids or sugars (reactions of hydroxyl groups at adjacent carbon atoms in sugar with acetone or benzaldehyde)[4].

The conformations of cyclic acetals have been discussed by us in Adv. Polymer Sci. **37** (1980).

The scope of application of cyclic acetals for the synthesis of polymers is very much limited because substitution (particularly at the C-2) position, retards polymerization. Earlier observations have been recently complemented by the unsuccessful attempts directed toward the polymerization of a large variety of higher substituted cyclic acetals[5].

The general thermodynamics of polymerization of cyclic acetals and the influence of substitution are discussed in Chapt. 2 of this volume (Thermodynamics). It may suffice to state here, that the monomers used to date for polymer synthesis are mostly derivatives of 1,2-glycols or 1,4-glycols and formaldehyde (i.e. 5- and 7-membered formals). 6-membered formals (1,3-dioxane and its derivatives) are nonpolymerizable due to the thermodynamic restrictions.

Acetals are one of a few groups of monomers for which the best general polymerization conditions can be given. Thus, if the aim is the preparation of high-molecular-weight polyacetals we would advise to use <0.1 mol% trifluoromethanesulfonic anhydride well below 0 °C in bulk. If the polymerization proceeds violently under these conditions and the time for mixing is insufficient to obtain uniform solution, both temperature and initiator concentration should be decreased.

Some cyclic acetals may have a low ceiling temperature and polymerization may ensue at a lower temperature than applied. We recall here the polymerization of 4-phenyl-1,3-dioxolane (considered earlier nonpolymerizable) at —20 °C by Elias et al.[6]. The T_c of 4-phenyl-1,3-dioxolane is 20 °C[6].

In Table 7-1 we present major information on the polymerization of representative cyclic acetals.

In some works, aimed at the preparation of high molecular weight polymers, low molecular weight oligomers resulted. It is known that polyacetals are particularly prone to give cyclic oligomers by back-biting (cf. Sect. 3.2). Thus, particularly at lower initial monomer concentrations ($[M]_0$) a high proportion of macrocycles may be formed[24]; therefore it is advisable to use as high as possible $[M]_0$ to lower their proportion (not their concentration which does not depend on $[M]_0$).

It has been shown that although the concentration of cyclic oligomers is invariant at equilibrium (unaffected by $[M]_0$, $[I]_0$ or temperature, provided that high molecular weight polymer is also formed), their concentration may grow slower than that of linear polymer with monomer conversion[25]. This effect named "kinetic reduction" can be used to enhance the $\overline{M}_n$ of the linear polymer and its proportion.

Therefore, one should observe GPC traces (like Schulz et al.[9,21]) of the polymerizing mixture and stop the polymerization when the proportion of cyclic and linear polymers have reached the desired values (cf. Sect. 5).

Table 7.1. Polymerizability of cyclic acetals

Monomer	Properties	Polymerization	Ref.
1,3-dioxolane 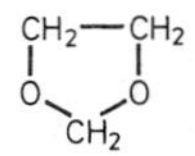	b.p. 74 °C	high-molecular-weight linear polymer accompanied by macrocyclics	this chapter
1,3-dioxane 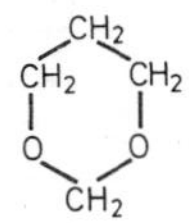		non-polymerizable	[7)]
1,3-dioxepane 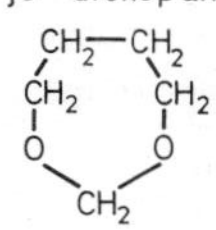	b.p. 119 °C	high-molecular-weight linear polymer	this chapter
1,3,5-trioxepane 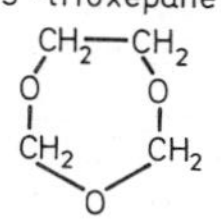	b.p. 50 °C/12 torr	waxlike polymers, $\overline{M}_n$ not given, for discussion of polymerization thermodynamics cf. Chap. 2	[9)]
1,3,6-trioxoxane 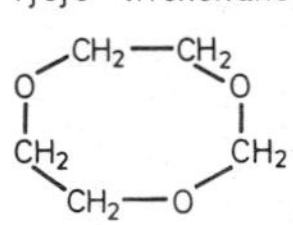	b.p. 60 °C/50 torr	$\overline{M}_n$ up to ~3000 obtained at 0 °C, soln. 2.4 mol · l^{-1} in CH_2Cl_2 with $2.6 \cdot 10^{-2}$ mol · l^{-1} of $(C_2H_5)_3O^+BF_4^-$	[10)]

Table 7.1. (continued)

Monomer	Properties	Polymerization	Ref.
1,3-dioxepin 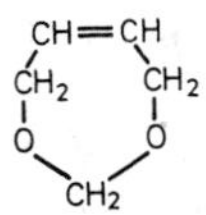	b.p. 127 °C	Polymers with [η] up to 0.3 dl · g^{-1} (measured in THF at 25 °C) bulk polymerization with 1 mol % of $BF_3 \cdot OEt_2$, 300–400 h	11)
1,3,5-trioxane 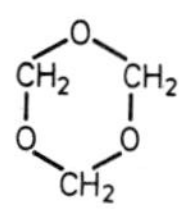	b.p. 114.5 °C, b.p. of azeotrope (70% TXN, 30% H_2O) 91.3 °C, m.p. 64 °C	high-molecular-weight linear polymers	this chapter
4-methyl-1,3-dioxolane	b.p. = 85–86 °C	$\overline{M}_n$ up to 2000 obtained at —20 °C, $[M]_e$ = 4.5 mol · l^{-1} (in bulk with 0.2 mol · l^{-1} $BF_3 \cdot OEt_2$) $\overline{M}_n$ up to 2500 at —75 °C (soln. 5 mol · l^{-1} in CH_2Cl_2 with $2 \cdot 10^{-2}$ mol · l^{-1} of $HClO_4$)	12) 13)
4-ethyl-1,3-dioxolane	b.p. = 110–111 °C	$\overline{M}_n$ up to 4000 obtained at —40 °C, $[M]_e$ = 4.75 mol · l^{-1} (in bulk with 0.2 mol · l^{-1} of $BF_3 \cdot OEt_2$)	14)
4-isopropyl-1,3-dioxolane	b.p. = 65–67 °C/100 torr = 129–130 °C/760 torr	traces of liquid oligomer obtained at —60 °C	14)
4-phenyl-1,3-dioxolane	b.p. = 120 °C/12 torr	$\overline{M}_w$ up to 50000 obtained at —350 °C *polymerization* (bulk with 0.02–2% mol of $HClO_4$) (Ceiling temperature in bulk >20 °C)	6)
2-methyl-1,3-dioxolane	b.p. = 82 °C	oligomers with $\overline{DP}_n$ up to 10, obtained at —75 °C (soln. 7.0 mol · l^{-1} in CH_2Cl_2 with $5.6 \cdot 10^{-2}$ mol · l^{-1} of $HClO_4$)	15)

2-vinyl-1,3-dioxolane	b.p. = 71 °C/64 torr	polymers with complicated structure resulting from parallel vinyl and ring-opening polymerization in the presence of cationic initiators	[16,17]
4,5-dimethyl-1,3-dioxolane	cis-4.5-dimethyl b.p. = 97.5 °C, trans-4,5-dimethyl b.p. = 102 °C	only cis-4,5-dimethyl gives at —40 °C low molecular weight oligomers with poor yield (bulk, $BF_3 \cdot OEt_2$) trans-4,5-dimethyl (and 4,4-dimethyl) do not polymerize at —60 °C (bulk, $BF_3 \cdot OEt_2$)	[18]
2-methyl-1,3-dioxepane	b.p. = 65 °C/90 torr	$\overline{M}_n$ up to 8000 (soln. 5.8 mol · l^{-1} in $C_2H_4Cl_2$ with $1 \cdot 10^{-2}$ mol · l^{-1} of $(C_2H_5)_3O^{\oplus}BF_4^{\ominus}$)	[19]
2,2-dimethyl-1,3-dioxepane	b.p. = 73–74 °C/104 torr	non-polymerizable at —78 °C (soln. 5 mol · l^{-1} in CH_2Cl_2 with $5 \cdot 10^{-2}$ mol · l^{-1} $BF_3 \cdot OEt_2$)	[19]
macrocyclic formals $\overline{\left(CH_2-CH_2-O\right)_n CH_2-O}$ (cyclic) n = 2–4	n = 3 b.p. = 56 °C/0.4 torr, m.p. = 27 °C, n = 4 b.p. = 51 °C/10^{-3} torr, m.p. = 23.5 °C	linear polymers $\overline{M}_n$ up to 40000 and cyclic oligomers (soln. in CH_2Cl_2 2.5 mol · l^{-1} with $2.5 \cdot 10^{-3}$ mol · l^{-1} of $HOSO_2CF_3$)	[9,22]

In some other investigations, whose aim was monomer synthesis, monomer could not be isolated and polymer was the only product [26].

$$HO(CH_2)_n{-}OH + C_4H_9{-}O{-}CH_2{-}O{-}C_4H_9 \rightarrow$$
$$\rightarrow \ldots [(CH_2)_n O{-}CH_2{-}O] \ldots + 2\, C_4H_9OH$$
$$\ldots [(CH_2)_n O{-}CH_2{-}O] \ldots \xrightarrow{\text{heating}} \overline{[(CH_2)_n O{-}CH_2{-}O]} \qquad (7\text{-}4)$$

Evidently synthesis conditions favoured polymer formation. Under these circumstances on can try to obtain a monomer by polymer degradation proceeding by depropagation. "Distillation" of the polymer in the presence of suitable non-volatile initiators may help. The low volatility of initiator is particularly important to prevent polymerization of the distilling monomer.

7.2.1 Preparation of High Molecular Weight Polyacetals

As already mentioned, the polymerization of 1,3,5-trioxane and bicyclic acetals (including anhydro sugars) is discussed separately. In this section we shall only give the conditions for preparation of high-molecular-weight polymers from the parent monomer of this class: 1,3-dioxolane (DXL). The majority of conclusions applies also to the 7-membered acetal 1,3-dioxepane (DXP).

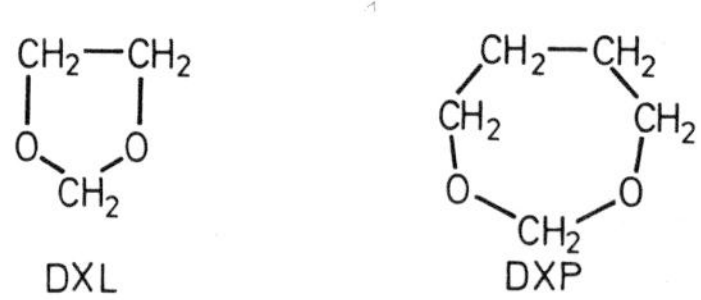

(7-5)

The preparation of high polymers under living polymerization conditions requires that the polymerization is carried out in the presence of counterions which do not lead to irreversible termination [27–29]. Noncomplex counterions, derivatives of strong protic acids (superacids), e.g. the $CF_3SO_3^{\ominus}$ anion, may react with cationic growing species. This reaction is reversible and affects only the polymerization kinetics but not the molecular weights [30]. These counterions are classified as stable.

Stable counterions thus include the most stable complex anions: $SbF_6^{\ominus}$ and $AsF_6^{\ominus}$ [27,31,32] as well as the noncomplex anion $CF_3SO_3^{\ominus}$. Counterions such as $SbCl_6^{\ominus}$, $BF_4^{\ominus}$ and even $PF_6^{\ominus}$ should be classified as unstable, because in their presence irreversible termination is frequently observed [27–29].

Carbenium salts [e.g. $(C_6H_5)_3C^{\oplus}A^{\ominus}$] initiate the polymerization of DXL by hydride-transfer which introduces an additional complication [28,29,33,34].

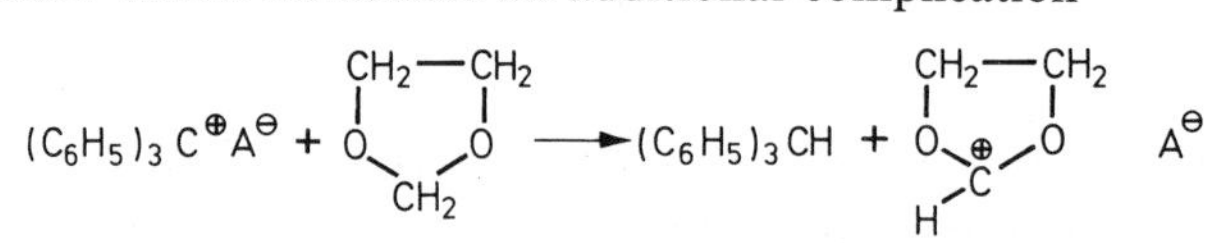

(7-6)

Taking the above discussed factors into consideration we may select the group of initiators which initiate quantitatively by addition, the terminationless polymerization of DXL and which are either commercially available or can easily synthesized:

- trifluoromethanesulfonic (triflic) acid $HOSO_2CF_3$, its anhydride $(CF_3SO_2)_2O$ and methyl or ethyl esters $ROSO_2CF_3$ [30] (all are commercially available).
- oxonium salts, e.g. $(C_2H_5)_3O^{\oplus}SbF_6^{\ominus}$ [27,32]
- oxycarbenium salts, e.g. $C_6H_5CO^{\oplus}SbF_6^{\ominus}$ or $CH_3CO^{\oplus}SbF_6^{\ominus}$ [31], a convenient synthesis method for these initiators is given in Ref. [35]
- 1,3-dioxolan-2-ylium(dioxolenium) salts:

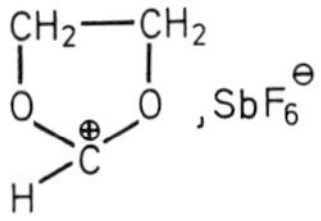

(7-7)

which may be prepared by reaction (7-6) and isolated as pure, crystalline products [36].

Polymerization of DXL initiated by these initiators and conducted at ambient or low temperatures (—78 °C to room temp.) in CH_2Cl_2 solvent (CH_3NO_2 may also be used) gives polymers with molecular weights close to the calculated values:

$$\overline{DP}_n = \frac{[DXL]_0 - [DXL]_e}{[initiator]_0} \tag{7-8}$$

These polymers carry well defined end-groups which may be incorporated by choosing the proper initiator and terminating agent [31,32]. Representative examples are collected in Table 7-2.

Table 7.2. Structure of poly-1,3-dioxolane[a] end-groups. Polymerization conditions $[DXL]_0 = 4.05\ mol \cdot l^{-1}$, CD_3NO_2 solvent, —15 °C, 12 h [31]

Initiator	Terminating agent	$\overline{DP}_n$ calcd. $\frac{[DXL]_0 - [DXL]_e}{[initiator]_0}$	$\overline{DP}_n$ (NMR)[b] found from ^1H analysis of the indicated end-group	
$C_6H_5CO^{\oplus}SbF_6^{\ominus}$ $4.7 \cdot 10^{-3}\ mol \cdot l^{-1}$	C_2H_5ONa	700	650 $C_6H_5COO—$	575 $—OC_2H_5$
$C_6H_5CO^{\oplus}SbF_6^{\ominus}$ $1.95 \cdot 10^{-3}\ mol \cdot l^{-1}$	$P(n\text{-}C_4H_9)_3$	1690	1380 $C_6H_5COO—$	
$(C_2H_5)_3O^{\oplus}SbF_6^{\ominus}$ $5.2 \cdot 10^{-3}\ mol \cdot l^{-1}$	$P(n\text{-}C_4H_9)_3$	750	1000 $C_2H_5O—$	920 $—\overset{\oplus}{P}(n\text{-}C_4H_9)_3$

[a] perdeuterated monomer was used for polymerization;
[b] end-groups identified by ^1H-NMR

It should be remembered, that cyclic acetal polymerization is always accompanied by cyclization. These aspects of polymerization are discussed in detail in Sect. 3 of this volume. Here, we wish only to stress that high-molecular-weight linear polymers are not the only products of polymerization carried out under conditions listed in Table 7-2, but some cyclic oligo- and polymers also form.

7.2.2 Preparation of Polyacetal-based α,ω-Diol Oligomers

It has recently been found that, if the polymerization of DXL or DXP with $HOSO_2CF_3$ initiator is carried out in the presence of 1,4-butanediol, low-molecular-weight linear oligomers terminated at both ends with OH groups are formed and the formation of the cyclic fraction is suppressed [37]. One molecule of macrodiol is formed per one molecule of used diol. Thus, molecular weights in the range of 1000–2000 can easily be obtained. Analysis based on the complexation of the HO groups with hexafluoroacetone and determination of the peak area and the chemical shift of the resulting complex by ^{19}F-NMR spectroscopy has shown that each macromolecule contains two hydroxyl end-groups and that only one type of end-groups ...—OCH_2CH_2OH exists. This means that the glycol used is not located at the chain end. The following scheme is proposed to account for the results [37]:

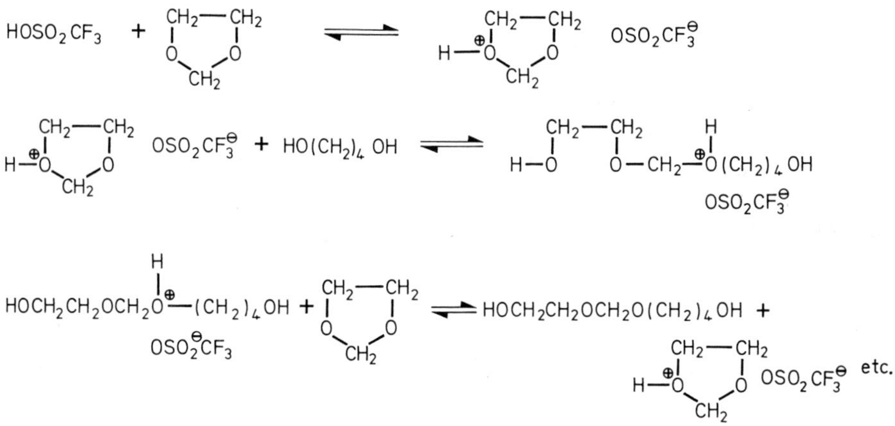

(7-9)

Table 7.3. Structure of products formed in the cationic polymerization of 1,3-dioxolane in the presence of 1,4-butanediol. Reaction conditions: 25 °C, 2–24 h [37]

Monomer mol · l⁻¹	$[HOSO_2CF_3]_0$ mol · l⁻¹	$[HO(CH_2)_4OH]_0$ mol · l⁻¹	$\overline{M}_n$ calcd[a]	$\overline{M}_n$ measured		
				^1H;NMR[b]	^{19}F-NMR[c]	vpo[d]
DXL 12.8 (bulk)	$4 \cdot 10^{-3}$	0.567	1300	1350	—	1600
DXL 12.8 (bulk)	$3 \cdot 10^{-2}$	0.586	1100	1400	900	1100
DXP 9.53	$8.9 \cdot 10^{-3}$	0.427	1800	1990	2080	1700

[a] $\overline{M}_n \text{ calcd} = \dfrac{[DXL]_0 - [DXL]_e}{[diol]_0} \times 74$;

[b] from the ratio of $-O\!\!-\!\!(CH_2)_2-O-$ units to $-O\!\!-\!\!(CH_2)_4-O-$ units;

[c] terminal OH group complexed with hexafluoroacetone [38]

[d] vapor pressure osmometry (tonometry)

The repetition of the reaction sequence shown in Scheme (7-9) leads to the following structure:

$$H\!-\!\!\left[OCH_2CH_2OCH_2\right]_n\!O(CH_2)_4O\!\left[CH_2OCH_2CH_2O\right]_n\!\!-\!H \qquad (7\text{-}10)$$

This mechanism is similar to that which seems to operate in the cationic oligomerization of oxiranes in the presence of compounds containing hydroxyl groups ("activated monomer polymerization"), discussed in detail in Chapter 4.

Typical reaction conditions and analysis of oligomer structures are listed in Table 7.3.

7.3 Polymerization and Copolymerization of 1,3,5-Trioxane

7.3.1 Introduction

Polymerization of 1,3,5-trioxane (TXN) gives linear polyoxymethylene (POM), a homopolymer of formaldehyde [39]. This is the only polyacetal made on the technical scale. Two methods are used for the industrial production of stable, high-molecular-weight POMs. This is either the anionic polymerization of formaldehyde or the cationic copolymerization of the cyclic trimer of formaldehyde: TXN with ethylene oxide or 1,3-dioxolane (DXL):

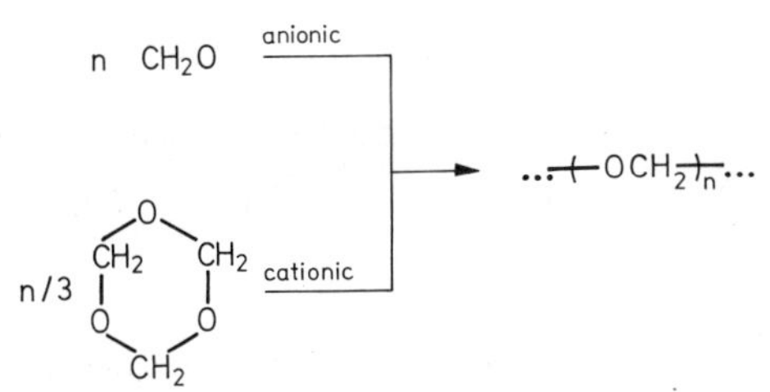

(7-11)

In this chapter only the cationic polymerization and copolymerization of TXN are discussed.

The cationic polymerization of TXN has a long history. Polyformaldehyde was first obtained in this way by Staudinger as early as in the 1920s (anionic polymerization was applied at the same time) [40]. However, both anionic polymerization of formaldehyde as well as the cationic polymerization of TXN lead, in the presence of water, to polymers containing hemiacetal end-groups. It became apparent much later that in order to get a polymer that could be kept in the molten state for a time long enough for extrusion or injection molding, it is necessary to block the hydroxyl end-groups to prevent hemiacetal degradation by unzipping. It took more than 30 years until this has been understood, stabilization methods were developed, and laboratory-curiosity polyacetals became large-scale industrial polymers. In the late 1950s DuPont commercialized the end-capped homopolymer of formaldehyde (Delrin®)[41] while Celanese and Hoechst as well as a few other companies developed the cationic bulk copolymerization of TXN with small amounts (2–5%) of comonomer (Celcon®,

Hostaform®)[42]. Copolymerization in solution of hydrocarbons with gradual polymer precipitation also has been developed on the technical scale [43].

7.3.2 Preparation, Purification and Some Properties of 1,3,5-Trioxane

1,3,5-Trioxane (TXN) is usually prepared by treating a 60–65% aqueous solution of formaldehyde with acid (e.g. 2% of sulfuric acid) and distilling off the azeotrope of TXN with water (70% of TXN, b.p. 91.3 °C). The trimer can be extracted from the aqueous distillate with e.g. methylene dichloride or trichlorobenzene.

A recommended method of TXN purification involves recrystallization from water, diethyl ether or methylene dichloride, vacuum sublimation or distillation [44]. TXN is inert towards alkalies; thus, basic drying agents may be used to remove water. On the large scale TXN is conveniently dried by distillation of the crude trimer with solvent which forms ternary azeotropes with high water content (e.g. n-heptane)[45].

The high-purification in laboratory requires refluxing over sodium-potassium alloy followed by distillation under nitrogen [46]. It is advisable to keep the distilled product in the molten liquid state, to prevent premature spontaneous polymerization due to crystallization-melting cycles.

Table 7.4 Physical properties of 1,3,5-Trioxane [47]

Property	Value
Melting point	62 °C
Vapor pressure at 25 °C	13 mm Hg
Boiling point	115 °C
Heat of vaporization	460 J/g (110 cal/g)
Heat of sublimation	630 J/g (150 cal/g)
Specific gravity at 65 °C (liquid TXN)	1.17 g/cm^3

Some properties of TXN are listed in Table 7.4. Pure TXN is colorless, crystallizes in needles or rhombohedral form. It is soluble in alcohols, ketones, organic acids, esters, phenols, aromatic hydrocarbons and chlorinated hydrocarbons. Solubility in aliphatic and cycloaliphatic hydrocarbons is limited at room temperature but rises markedly with temperature. At 20 °C water dissolves about 20% wt of TXN, and solubility in water increases with increasing temperature [47].

TXN is flammable. Toxicity has not been demonstrated [47].

7.3.3 General Features of the Polymerization

TXN polymerizes easily in the presence of cationic initiators, and polymerization does not occur in the presence of anionic or radical initiators [48]. Claims of radical-cationic polymerization have been made but the evidence is not convincing.

A typical initiator for TXN polymerization is boron trifluoride, usually in the form of its complexes with ethers, esters or amines. Such stable, easily available complexes are conveniently handled and can be purified by distillation [49].

The patent literature discloses a large variety of other initiators including different Lewis acids, stable carbenium, oxonium or diazonium salts, and derivatives (esters or anhydrides) of strong protonic acids. However, BF_3 seems to maintain its superior position in industrial applications.

Recently low concentrations of triflic acid and its derivatives (particularly the anhydride) were shown to give high molecular weight homo- and copolymers (with DXL) in high yield. It is remarkable that in order to obtain comparable yields it was sufficient to use almost 10^2 times less anhydride than $BF_3 \cdot O(C_2H_5)_2$ [50].

As it has been mentioned the polymerization of TXN may be carried out in bulk or in solution [47,48,51].

In the bulk process, molten TXN is polymerized in admixture with comonomer in the 60–120 °C range (above the melting and below the boiling point of TXN). Several problems may occur during this bulk process. The heat of polymerization of TXN is close to 17 kJ/mol (4.0 kcal/mol) [48] and this heat has to be removed. Thus, a special reactor design is required. Several examples are given in the patent literature and one of the possible solutions is screw extruders or related apparatures permitting adequate temperature control, efficient mixing and particle size control.

Solution polymerization can be carried out in an inert solvent, such as saturated hydrocarbons (e.g. cyclohexane), over a wide range of temperatures. To achieve reasonable rates of polymerization, temperatures close to 60 °C are recommended. The polymer, which is insoluble in the monomer or the solvent, precipitates as a fine powder during the early stages of polymerization. The precipitated particles may aggregate which affects the properties of the mixture (the system solidifies and cannot be efficiently mixed) and renders isolation of the polymer by filtration difficult.

The aggregation of the polymer particles may be prevented by addition of a small amount of polyether [e.g. poly(ethylene oxide)] to the system [52].

In this process, polymer is obtained as a fine, free flowing powder with uniform particle size which facilitates its separation from the reaction medium by filtration.

Thermal stability of polyacetal is achieved by incorporating into the backbone $-CH_2-CH_2-$ units, i.e., disrupting the sequence of $-CH_2-O-$ units susceptible to unzipping. This is done, by copolymerization of TXN with other cyclic ethers or acetals, preferably with 1,3-dioxolane or ethylene oxide ($<5\%$ mol) [42,53].

TXN also may be polymerized in the solid state, this process is initiated either chemically or by radiation.

7.3.4 Preparation of High Molecular Weight Polymers of 1,3,5-Trioxane

Several methods of TXN polymerization have been studied and some of them are described in the patent literature. In this section we will give a detailed description of solution polymerization in which the association of precipitating polymer particles (which is a main shortcoming of the solution process) is avoided [52]. A typical procedure is as follows [52]: A 4-neck flask, fitted with a nitrogen inlet, thermometer, stirrer and condenser is charged with 3.6 g polyEO ($\overline{M}_n \sim 15000$) (preventing agglomeration of particles), 120 ml cyclohexane ($[H_2O] < 20$ ppm), 144 ml (168 g = 1.87 mol) of molten, freshly distilled TXN ($[H_2O] < 30$ ppm, $[CH_2O] < 10$ ppm, [HCOOH] < 10 ppm), and 7 g (0.094 mol) 1,3-dioxolane (DXL). The flask is thermostated at 60°. To initiate the polymerization, 0.1 %mol (with respect to TXN, i.e. 0.14 g) $BF_3 \cdot O(n\text{-}Bu)_2$ is added under constant stirring. Polymerization ensues after a short induction period, the temperature rises to 70–75° and polymer precipitates. The reaction is terminated after 60 min by adding a solution of NH_3 in CH_3OH, the polymer is filtered off, washed with warm water and methanol, and then dried at 70 °C. A typical yield is 70–80%.

The polymerization parameters influence, sometimes dramatically, the properties of the final product. First of all let us compare the behaviour of the system in the presence and in the absence of polyEO. In the absence of this polyether the reaction mixture containing 45% wt polyacetal solidifies and cannot be mixed anymore (at least at laboratory scale). In the presence of polyEO even a mixture containing 60% wt solid polymer is fluid and can be efficiently mixed.

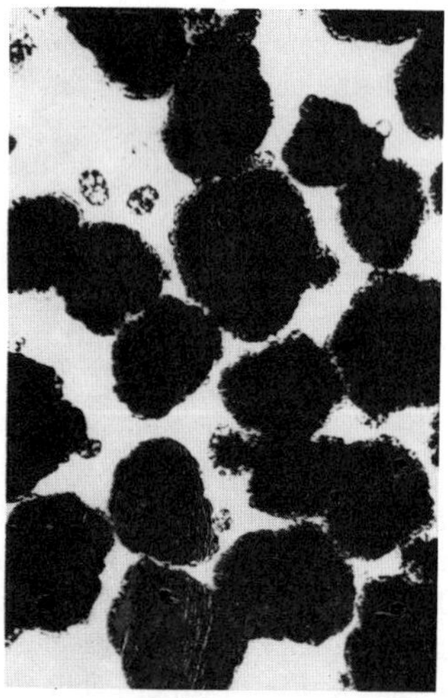

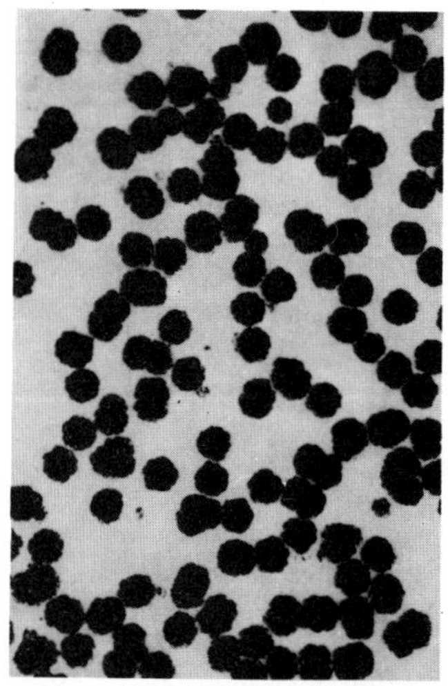

Fig. 7.1. Particles of copolymer obtained without poly(ethylene oxide); magnification 500x [1,3,5-trioxane] = 7.0 mol · kg^{-1}, [1,3-dioxolane] = 0.35 mol · kg^{-1}, $[BF_3 \cdot O(n\text{-}Bu)_2] = 2 \cdot 10^{-3}$ mol · kg^{-1}, conversion 71% [52)]

Fig. 7.2. Particles of copolymer obtained in the presence of 1.5% wt of poly(ethylene oxide) ($\overline{M}_n$ = 15,000); magnification 500x [1,3,5-trioxane] = 7.0 mol · kg^{-1}, [1,3-dioxolane] = 0.35 mol · kg^{-1}, $[BF_3 \cdot O(n\text{-}Bu)_2] = 4.0 \cdot 10^{-3}$ mol · kg^{-1}, conversion 86% [52)]

The reason for this behaviour is a difference in the shape and size of particles precipitating in the presence and in the absence of polyEO as shown in Figs. 7.1. and 7.2 respectively. Particles obtained in the presence of polyEO are of more regular shape and smaller in diameter. The amount of polyEO required for the modification of particles is shown in Fig. 7.3. The average particle diameter decreases with increasing polyEO concentration up to 1–1.5% and then remains practically constant. The changes of the particle size in the course of polymerization in the presence and absence of polyEO are shown in Fig. 7.4.

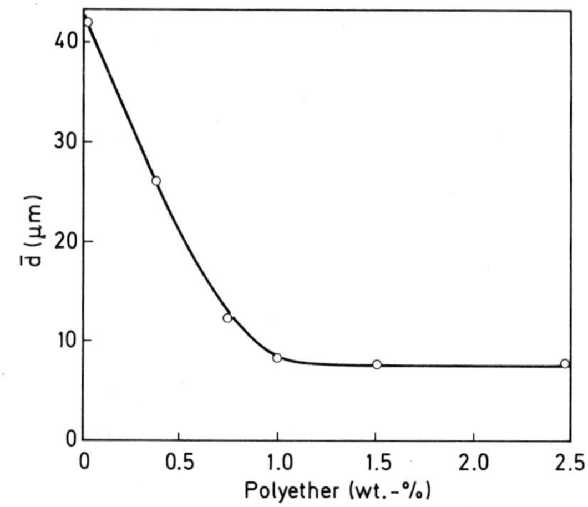

Fig. 7.3. Dependence of the average diameter of particles on the concentration of poly(ethylene oxide); Conditions as in Fig. 7.2 [52]

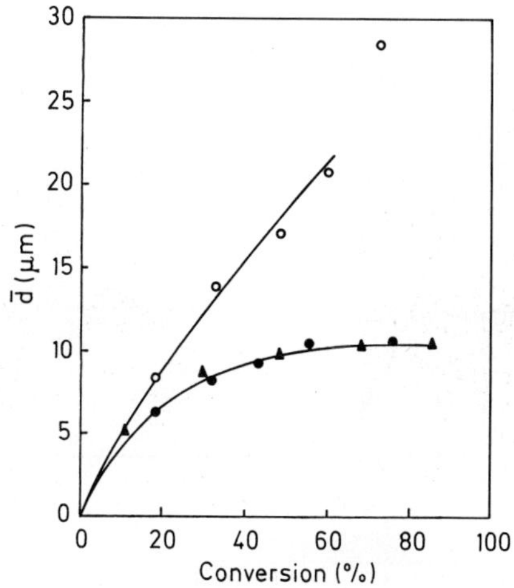

Fig. 7.4. Influence of poly(ethylene oxide) addition upon the dependence of the average particle size on conversion [52]. ○ — no poly(ethylene oxide) added; ▲ — 1% wt. of poly(ethylene oxide) added; ● — 1.5% wt. of poly(ethylene oxide) added. Conditions as in Fig. 1 and 2, respectively [52]

These results indicate that polyEO forms a protective layer on the surface of the growing polyacetal particles and at a certain particle size this layer effectively prohibits further growth of the individual particles. Consequently, new particles have to be formed which again may grow only to a certain size. From the amount of poly-EO and the total surface of particles, it was calculated that the particles are covered with a 4 Å-thick, i.e. monomolecular, layer of polyEO. The particles cannot aggregate because of repulsion of the external polyEO layers. The polyEO can not be separated from the polyacetal by extraction, thus it is probably, chemically bound. It was assumed that the chemical bond is formed by slow segmental exchange, a process similar to transacetalization. Instead of randomization, long polyEO blocks will be incorporated into polyacetal chains:

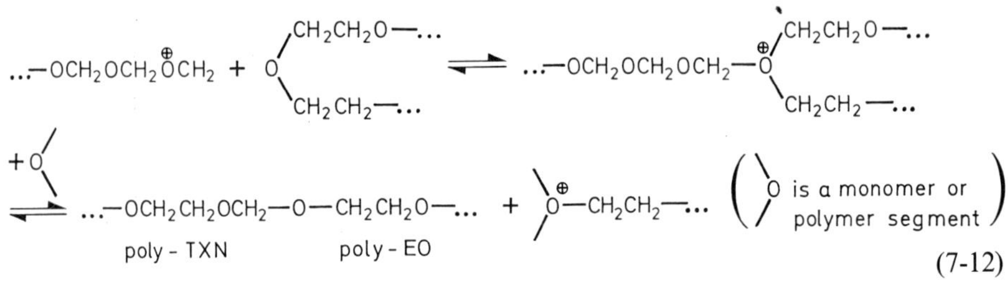

(7-12)

Since polyEO may interact with the growing species, it may affect not only the shape and distribution of particles but also the rate of polymerization and the molecular weight. The rate of polymerization decreases in the presence of polyEO.

From the practical point of view a disadvantage of this process is longer reaction times and the advantage is easier temperature control. The presence of polyEO leads also to a decrease of the molecular weight of the product and decreases the content of the thermally unstable fraction, as shown in Fig. 7.5.

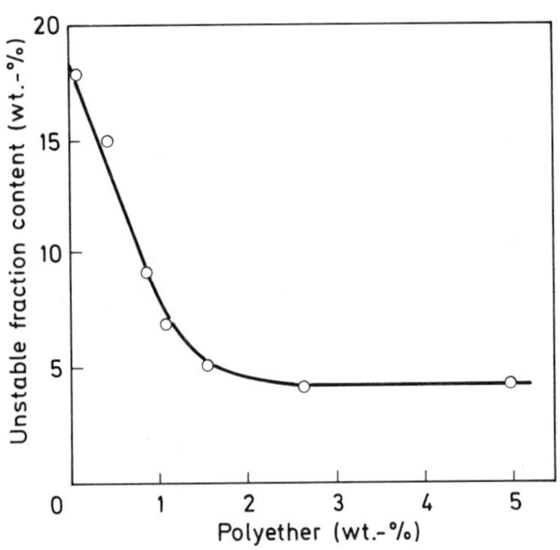

Fig. 7.5. Influence of poly(ethylene oxide) ($\overline{M}_n = 15000$) addition on the content of the unstable fraction in copolymer, [1,3,5-trioxane] = 7.0 mol · kg^{-1}, [1,3-dioxolane] = 0.375 mol · kg^{-1} [52)]

7.3.5 Mechanistic Aspects of 1,3,5-Trioxane Polymerization Pertinent to Polymer Synthesis

In spite of hundreds of papers describing the phenomenology of TXN homopolymerization and TXN-DXL copolymerization (rates, $\overline{M}_n$) on polymerization variables (cf. for example Refs. [24,48,58)]), little is known as to the elementary reactions involved. Progress was hampered by the insolubility of the polymer and crystallization that occurs during polymerization. Another difficulty is the complexity of thermodynamics of TXN polymerization.

This section is not an exhaustive review of all of the papers published; instead we will try to present a coherent view of cationic TXN polymerization needed to comprehend phenomena taking place during polymer synthesis.

7.3.5.1 Initiation

Boron trifluoride in form of its complexes is commonly used as the initiator of TXN polymerization. The existing hypotheses concerning the initiation mechanism may be divided into two categories:

a) direct initiation:

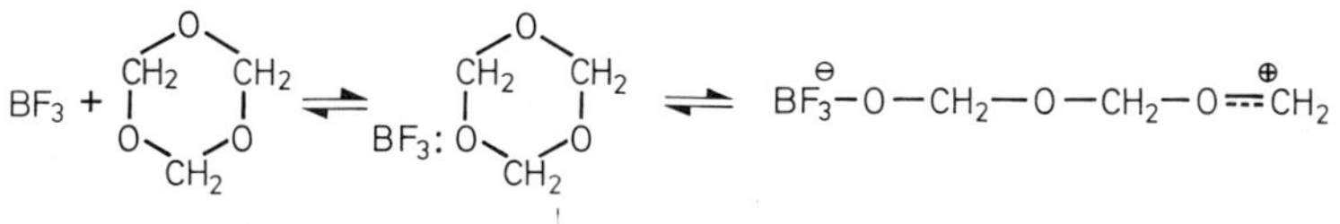

(7-13a)

b) initiation by coinitiator

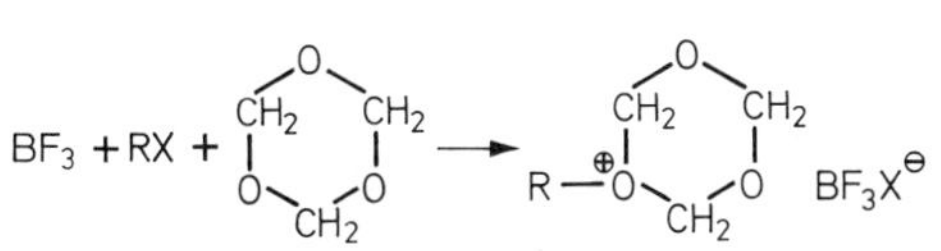

(7-13b)

where RX = H_2O, ROH or other suitable compounds, as discussed below.

Kern and Jaacks, who first studied this problem thoroughly, expressed the opinion that the polymerization of TXN can be initiated by BF_3 even in the absence of protogenic compounds [54]. Later, on the basis of new observations, they weakened their position and concluded that it was impossible to decide whether initiation can occur in the absence of coinitiator [55].

Since then, two reports have appeared claiming that polymerization of carefully dried TXN cannot be initiated by BF_3 complexes alone [56,57]. According to Collins, TXN dried over a sodium-potassium alloy to such an extent that the concentration of water is below the detection level of the Karl-Fischer method [57], does not undergo polymerization for 48 hrs in bulk at 70 °C in the presence of $1.25 \cdot 10^{-3}$ mol $\cdot$ l^{-1} $BF_3 \cdot O(n\text{-}C_4H_9)_2$. This may be due to the low ring strain and low basicity of TXN; complexes of BF_3 with more basic and more strained oxiranes may open giving zwitter-ions (cf. Sect. 4.2.). However, upon addition of water in concentration comparable to that of the BF_3 complex the polymerization of TXN starts. The same result was reported earlier for the solution polymerization in cyclohexane [56].

According to Higashimura et al. the mechanism of initiation may be different in higher polarity media [58]. While it was not clearly shown that polymerization of TXN was not initiated by $BF_3 \cdot OEt_2$ alone in nonpolar media (n-hexane or benzene), strong accelerating effect of water was indeed demonstrated. This effect was much less pronounced in the more polar ethylene dichloride and disappeared completely in the highly polar nitrobenzene.

An additional evidence in favour of different contributions of various initiation mechanisms in nonpolar and polar solvents was derived from studies on the thermal stability of polymers carrying hemiacetal (hydroxyl) end-groups. Under comparable conditions, polymers prepared in n-hexane decomposed 5–10 times faster than those obtained in nitrobenzene [58]. This indicates that the hydroxyl end-group content is higher in the former polymer which is consistent with the mechanism of Eq. (7–13b).

In summary, there are strong evidences in favour of coinitiation at least in nonpolar media. Of course, coinitiator like water (if present) will also be active in polar media (even if weak coinitiation by a polar solvent would also proceed).

Attempts to analyze initiation in more quantitative terms, are difficult. In systems in which water acts as the coinitiator, the concentration of the protonic acid, being the true initiator, is usually unknown.

$$\begin{aligned} &BF_3 \cdot OR_2 + H_2O \rightleftharpoons (H^{\oplus}BF_3OH^{\ominus} + R_2O) \\ &\rightleftharpoons R_2O^{\oplus}\text{-}H \quad BF_3^{\ominus}OH \end{aligned} \tag{7.14}$$

The concentration of the true initiator in BF_3-initiated polymerization may be approximated by comparing the rate of polymerization in the presence of BF_3 complex and that initiated by perchloric acid. In the latter system the concentration of the potentially active initiator is equal to the concentration of acid. Some pertinent data are given in Table 7.5 [59].

The results indicate that although the concentration of $HClO_4$ is 300 times lower than that of the BF_3 complex, the rate of polymerization initiated with $HClO_4$ is still higher. The same was observed for triflic acid and its derivatives [50] (cf. Table 7.6).

Table 7.5 Copolymerization of 1,3,5-trioxane (50% wt) with 1,3-dioxolane (50% wt) (total 100 g) in bulk, 60 min [59]

Initiator, mol · 10^5	Temperature, °C	Yield, %
$HClO_4$	40	34
0.07	50	86
	60	94
	70	93
$BF_3 \cdot OBu_2$	40	15
21.3	50	43
	60	76
	70	83

Table 7.6 Copolymerization of 1,3,5-trioxane (7 mol · l^{-1}) with 1,3-dioxolane (0.35 mol · l^{-1}) in cyclohexane at 60 °C [50]

Initiator	Initiator concentration mol · l^{-1}	Polymerization time, min	[η] $g^{-1} \cdot cm^3$	% Yield	
				Stable fraction	Unstable fraction
$(CF_3SO_2)_2O$	$1.2 \cdot 10^{-3}$	30	261	82.0	7.2
$CF_3SO_2OC_2H_5$	$5.6 \cdot 10^{-3}$	66	135	79.8	8.0
$(C_6H_5)_3C^{\oplus}SbF_6^{\ominus}$	$3 \cdot 10^{-3}$	40	203	76.5	7.9
$BF_3 \cdot OBu_2$	$4 \cdot 10^{-2}$	30	160	78.0	6.0

Although the overall rate may be affected by other factors, it seems that the large difference in rates reflects the differences in the true concentration of growing chains (i.e. efficiency of initiation).

The rate of polymerization depends also on the nature of the complexing agent. The rate of polymerization decreases in the following order with different complexing agents:

$$BF_3 \cdot (EtOH)_2 > BF_3 \cdot (AcOH)_2 > BF_3 \cdot OEt_2 \tag{7-15}$$

Even if we accept Eqs. (7-13) a and b as the main processes which lead to the initiating species, other possibilities may still exist. For instance, the following reaction was proposed to explain that BF_3 complexes alone may initiate polymerization in polar

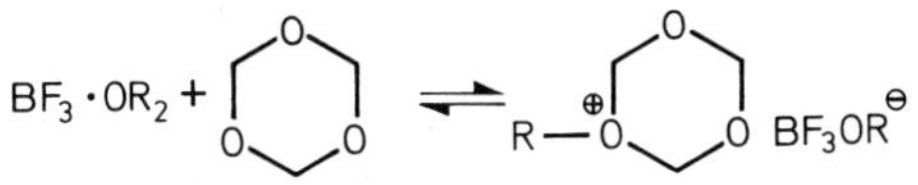

(7-16)

Similar reactions are known to proceed with reasonable rates, and although this ionization cannot be important in nonpolar media (polymerization does not occur in the absence of water), it may become important in the polar media [60].

Moreover, as pointed out by Enikolopyan et al., as soon as polymer is formed, it may react with BF_3 [similar to Reaction (7-16)] and it may become an important source of active species [61,62]:

$$\ldots\text{—O—CH}_2\text{—O—}\ldots + BF_3 \rightleftharpoons \ldots\text{—}\overset{BF_3}{\ddot{\text{O}}}\text{—CH}_2\text{—O—}\ldots \rightleftharpoons \ldots\text{—O—}BF_3^{\ominus} + \overset{\oplus}{\text{CH}_2\text{=}}\text{O—}\ldots \tag{7-17}$$

Such an accelerating effect of linear acetals (methylal or ethylal) has indeed been demonstrated in the homogenous polymerization of 1,3-dioxolane initiated by BF_3 complexes [63].

Evidently no single mechanism may represent the initiation of TXN polymerization with the most commonly used initiator: $BF_3 \cdot OEt_2$.

Depending on conditions, one or a few mechanisms may operate simultaneously, and initiation may proceed as a set of parallel reactions.

No attempt has been made to measure the concentration of growing species during polymerization, thus the rate and efficiency of initiation is unknown in these systems.

7.3.5.2 Propagation

It has been realized from the very beginning of systematic studies on polyoxymethylenes that the polymerization of TXN cannot be treated separately from crystallization [64–67], that occurs during polymerization. Three aspects of propagation will be discussed:

a) kinetics and chemistry of propagation
b) thermodynamics of propagation
c) phenomena related to the heterogenity of the system, particularly crystallization.

Although these phenomena are closely related and affect each other, for the sake of clarity they will be discussed separately as far as possible.

7.3.5.2.1 The Chemistry of Propagation

Understanding of the mechanism of propagation requires knowledge of the structure of the growing species, the mechanism of monomer addition and the rate constant of this process.

The structure of active species in the homogenous polymerization of cyclic acetals (e.g., DXL) was discussed in detail in Adv. Pol. Sci. **37**, Sect. 4.1.4. In short, the problem may be stated as follows. The polymerization of cyclic acetals involves an equilibrium between oxonium and carbenium (carboxonium) ions:

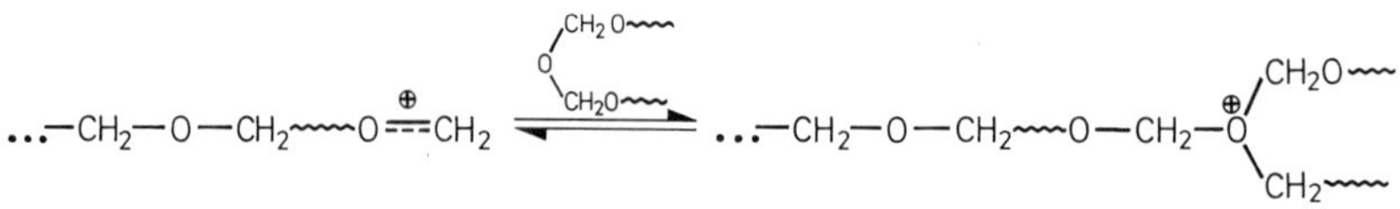
(7-18)

The ligands surrounding the oxonium center can be a part of a cyclic monomer, can belong to the own macromolecule, or can be part of some foreign cyclic or linear macromolecule. Thus, the oxocarbenium ion may be transformed into an oxonium ion by intramolecular reaction involving oxygen atoms of its own chain, or by intermolecular reaction involving oxygen atoms of a foreign polymer or a monomer molecule. The relative concentrations of these ions are governed by the corresponding equilibrium constants. Their role in propagation is affected by their concentrations and reactivities. Thus, at least one equilibrium constant and two rate constants should be known to decide to what extent are carbenium or oxonium ions responsible for growth. Such data are unavailable for TXN. Quantitative information exists only for the polymerization of DXL, and the dominant role of oxonium ions was demonstrated [68]. TXN is, however, considerably less basic than DXL [69], thus the equilibrium of Eq. (7-18) is shifted further to the left and correspondingly, the concentration of carboxonium ions should be higher.

Experimental results [54,70] indicate that in the polymerization of TXN the concentration of carboxonium ions in equilibrium with oxonium ions is not negligible.

According to work in this laboratory, oxonium ions do not abstract hydride anions. The presence of hydride abstraction is strong evidence for carbenium (alkoxycarbenium) ions in the system [33]. Since hydride transfer occurs during TXN polymerization [70,54], carbenium ions must be present. Unless the relative rates of hydride transfer and propagation are exactly known, however, it is impossible to estimate the proportion of carbenium ions in the system by using this approach.

7.3.5.2.2 Crystallization of the Growing Polyoxymethylenes

Polyoxymethylene is insoluble below 100 °C in common solvents and molten TXN. Thus, both in bulk or in "solution" polymerization, polymer precipitates from the reaction medium. Polyoxymethylene solutions can be obtained only well above 100 °C in strongly hydrogen bonding solvents [71,72]. TXN oligomers, even with low polymerization degrees, are insoluble in the reaction media; according to Staudinger formaldehyde oligomers become insoluble when their $\overline{DP}_n$ reach 40 [65]. The $\overline{DP}_n$ at which precipitation occurs depends on the temperature and solvent. As soon as the solubility limit is reached, the oligomer precipitates out of solution in the form of chain-folded crystals. Aspects of TXN polymerization related to crystallization, were extensively studied by Wegner. The present review follows the reasoning presented in Wegner's papers [73-75].

Almost immediately after addition of initiator to solutions of TXN, small hexagonal crystals are formed with a diameter from 1000 to 2000 Å (0.1–0.2 μm) and a height of 50–100 Å. Such typical chain-folded crystals are normally formed when oxymethylene oligomers or polymers are crystallized from solution. At first, these crystals grow only in diameter, but with increasing conversion the originally smooth surfaces develop growth spirals, and the crystals start to grow also in thickness. The mechanism of crystal growth is illustrated in Figs. 7.6 and 7.7.

Lateral growth involves conventional propagation; the growing species located at the surface of a side-wall react with monomer which, is incorporated into the

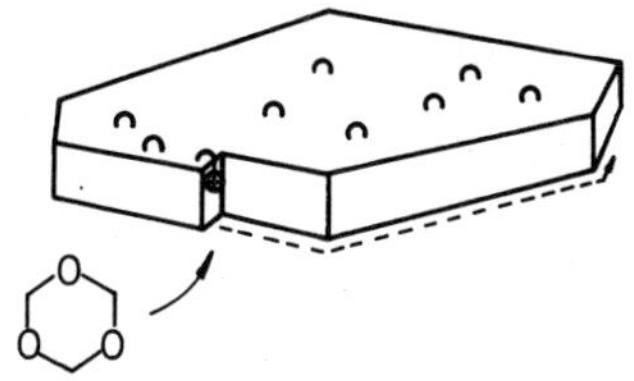

Fig. 7.6. Model for lateral crystal growth of polyoxymethylene crystals[74]

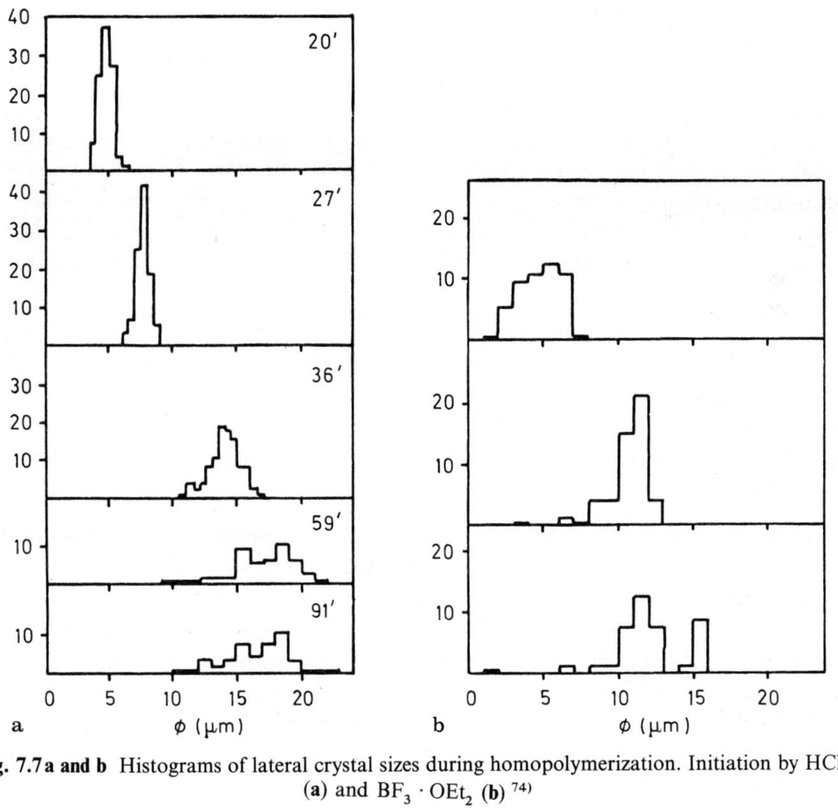

Fig. 7.7 a and b Histograms of lateral crystal sizes during homopolymerization. Initiation by $HClO_4$ (**a**) and $BF_3 \cdot OEt_2$ (**b**) [74]

crystal lattice. Less obvious is the mechanism of growth of crystal thickness. Wegner assumes monomer insertion into the chain folds activated by strain induced by these folds:

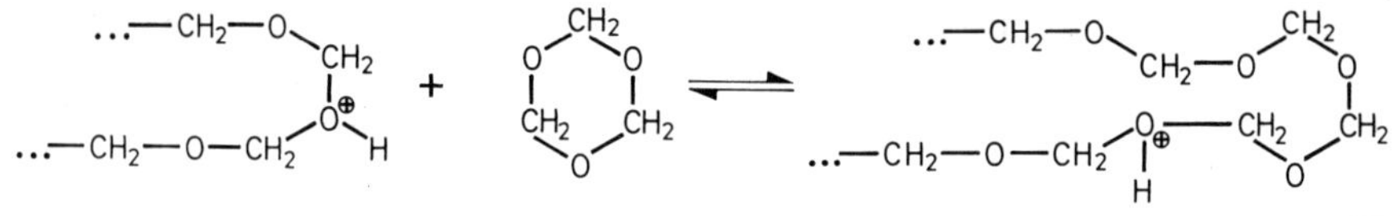

(7-19)

The term "insertion" describes, in our understanding, the end result but not the actual process; insertion does not have to proceed in one chemical step involving a four-center mechanism, but may occur stepwise by chain breaking, monomer addition, and recombination of the cation with a terminal hydroxyl group:

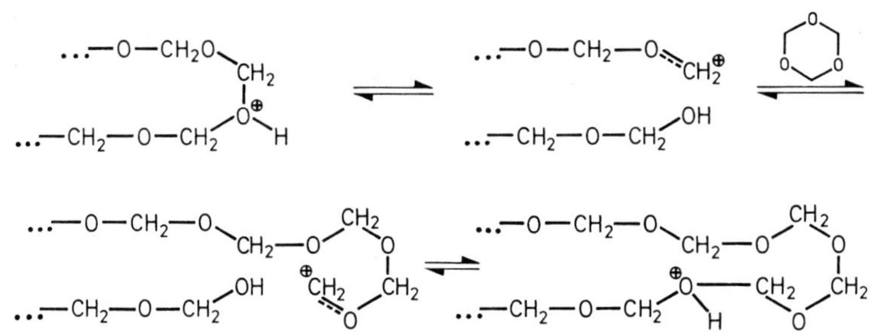

(7-20)

This process continues until the crystal reaches a thickness of several thousand Å. Crystal growth does not stop even after equilibrium conversion (polymer-monomer equilibrium) is reached. This is explained in terms of the Ostwald ripening. Solubility of particles depends on their size; thus smaller particles tend to redissolve (by depropagation) in favour of further growth of already larger ones. This is illustrated by the data shown in Table 7.7.

As shown in the last column of Table 7.7, the degree of crystallinity increases from 79% to almost 89% with increasing crystal size. This is already close to 95%,

Table 7.7 Growth of polyoxymethylene crystals during cationic polymerization of 1,3,5-trioxane in nitrobenzene at 35 °C [122)]

Time h	Conversion %	$\overline{M}_n \cdot 10^3$ [a]	Crystal[b] diameter μm	Crystal[c] thickness μm	Crystallinity[d] %
0.5	20	17.5	14	0.01	79.1
1.0	55	40	27	0.03	84.3
3.0	75	50	37	—	88.2
6.0	75	53	57	0.18	88.2
12.0	75	55	70	0.20	88.8

[a] viscosimetry; [b] light microscopy; [c] electron micrography; [d] density

the value given by Iguchi for extended chain polyoxymethylene crystals, obtained in the form of 1–3 μm thick and 30–80 μm long needles. These were prepared in the polymerization of TXN (0.32 mol · l^{-1}) with BF_3 (1.6 · 10^{-3} mol · l^{-1}) initiator and H_2O (4 · 10^{-5} mol · l^{-1}) cocatalyst in cyclohexane at 50 °C at 40% yield after 20 hrs [76].

Thus, by Wegner's scheme, typical chain-folded crystals are formed first and grow in both directions. Due to an increase in the crystal thickness, the distance between the folds becomes so large with conversion that the system assumes a mixed extended-chain-folded crystal morphology. This is evidenced by the high crystallinity and a high melting point (186.5 °C) of polyTXN prepared in this way. Characteristic properties of chain-folded and chain-extended polyTXN crystals (prepared in various ways) are compared in Table 7.8.

Table 7.8 Polyoxymethylene crystals

Form	Method of preparation	Mp °C	Density g/cm³	Crystallinity %	Type	Ref.
Needles 1–3 mm thick 30–80 mm long	Soln polymerization cyclohexane, 50 °C		1.493–1505	95	Extended chain	[76]
	After reprecipitation from HFA		1.42	67.5	Folded	
Hexagonal 0.2 mm thick	Soln polymerization nitrobenzene, 35 °C	186.5		89	Folded with large distances between folds	[75]
	Delrin, as supplied		1.42	62.3	Folded	[76]
	After recrystallization from benzene	175		83.3	Folded with increasing distances between folds	[76]

A characteristic feature emerges: Crystals formed in solution or by melt polymerization are more perfect than those formed upon crystallization of the preformed polyoxymethylene from solution or melt. If extended chain crystals with >95% crystallinity are reprecipitated from hexafluoroacetone sesquihydrate, the degree of crystallinity decreases to 67.5% [76].

Moreover, if polyoxymethylene is recrystallized from nitrobenzene and the usual chain-folded lamellae of ca. 100 Å thickness is treated with TXN in the presence of BF_3, more perfect extended-chain-type crystals are formed. These facts confirm the conclusion formulated by Wunderlich, that post-crystallization leads to chain-folded crystals because of kinetic restrictions, while crystallization in the polymerizing system gives thermodynamically more stable extended-chain crystals [77]. The formation of the "thermodynamic product" in polymerization is due to the growth of crystals at equilibrium polymerization conditions. Thermodynamically less stable crystals may "redissolve" as a result of depropagation, and crystals may thus grow further under equilibrium conditions. This (at least partly) eliminates kinetic restrictions and leads to the formation of thermodynamically more stable extended-chain crystals.

The situation is still more complex, when like in commercial polymers of TXN, the solid phase contains two types of units (most frequently ethylene oxide units from ethylene oxide or 1,3-dioxolane comonomers). Comparison of comonomer contents in the crystalline region of "as-polymerized" and recrystallized samples shows that segregation occurs in recrystallized samples, leading to a higher fraction of comonomer in the amorphous than in the crystalline phase [78]. The density and melting points decrease with increasing content of the comonomer units in both "as-polymerized" and recrystallized samples as shown in Fig. 7.8.

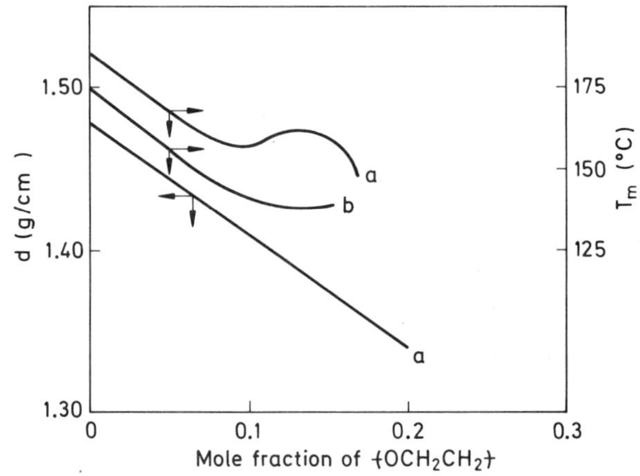

Fig. 7.8. Dependence of density and melting point of 1,3,5-trioxane-1,3-dioxolane copolymers on the mole fraction of 1,3-dioxolane units. a — as polymerized, b — recrystallized [78]

7.3.5.2.3 Thermodynamics of Propagation

Propagation in TXN polymerization is reversible but cannot be described by the equation:

$$n\ \boxed{(CH_2O)_3} \rightleftharpoons -\!\!(CH_2O)\!\!-_{3n} \qquad (7\text{-}21)$$

The growing polyoxymethylene chain may yield upon depropagation not only TXN but also formaldehyde and higher cyclic oligomers: 1,3,5,7-tetraoxocane (TTXN) and 1,3,5,7,9-pentaoxadecane; these species were identified in the polymerization mixture [79]. Formaldehyde and cyclic oligomers may reversibly add to the growing chain. Thus the complete thermodynamic description of this system requires a knowledge of the thermodynamic parameters of all individual steps. Major contributions in this area are due to Al. Al. Berlin and Enikolopyan et al. [80–86], and Romanov et al. [87–89].

The equilibrium concentration of pentamers and higher oligomers is very low. Consequently, these species will not be discussed further.

Thus, we are left with the following reversible reactions:

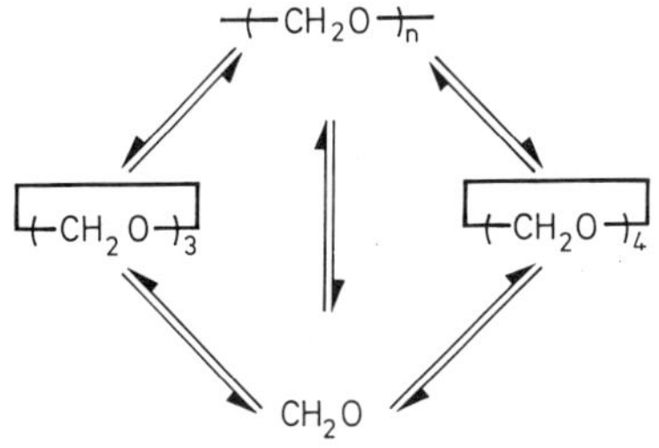

(7-22)

Moreover, in real systems (bulk or solution polymerization), polymerization proceeds in the crystalline state. Thus, thermodynamic parameters of phase transition are superimposed on thermodynamic parameters of chemical reactions (cf. Chap. 2 of this volume).

According to Eq. (7-22) the equilibrium concentration of each species is established during polymerization, and for each species a ceiling temperature exists above which the polymer is thermodynamically unstable with respect to this particular species. In practice this means that polymer decomposes thermally directly from the solid state and yields that entity with the lowest ceiling temperature. Thus, polyoxymethylene, although obtained from 1,3,5-trioxane, depropagates to formaldehyde. We will first discuss the equilibria involved in Scheme (7-22) as if these were for homogenous systems to get an insight into the thermodynamics of the chemical process alone, and then we will consider the thermodynamics of the phase transition to describe the thermodynamics of heterogenous polymerization of TXN.

The thermodynamic parameters of the reactions involved in Scheme (7-22) may be either calculated from thermochemical data or measured for a suitable model system. The first approach was used extensively by Enikolopyan's

Table 7.9 Thermodynamic parameters of polyoxymethylene formation

Monomer	Standard State	ΔH		ΔS⁰		Method	Ref.
		kJ/mol	(kcal/mol)	J/mol · K	(e.u.)		
Formaldehyde	gg	−55.2	(−13.2)	−149	(−35.6)	c	85)
	ss (NB)[a]	−33.4	(−8.0)	−76.5	(−18.3)	c	85)
	cc	−38.9	(−9.3)	−38.0	(−9.1)	c	85)
1,3,5-Trioxane	gg	−18.4	(−4.4)	−67.3	(−16.1)	c	85)
	ll	−9.2	(−2.2)	−9.2	(−2.2)	c	85)
	ss (NB)[a]	−8.4	(−2.0)	−5.8	(−1.4)	c	85)
	ss (B)[b]	−7.9	(−1.9)	−27.6	(−6.6)	e	87)
	cc	−6.3	(−1.5)	−2.9	(−0.7)	c	85)
1,3,5,7-Tetraoxocane	gg	−25.1	(−6.0)	−48.5	(−11.6)	c	85)
	ll	−10.9	(−2.6)	−7.1	(−1.7)	c	85)
	ss (NB)[a]	−9.2	(−2.2)	−12.1	(−2.9)	c	85)
	cc	−3.8	(−0.9)	+12.5	(+3.0)	c	85)

[a] nitrobenzene; [b] benzene; [c] calculated; [e] equilibrium studies

group [80–86], whereas Romanov's [87–89] group studied the distribution of cyclic and linear oligomers in the homogenous reaction of polyoxymethylene oligomers $(CH_3(O{-}CH_2)_nO{-}CH_3$ $(n < 9)$ with cationic initiators. Both approaches gave essentially similar results which are summarized in Table 7.9.

Inspection of the data given in Table 7-9 reveals that even in homogenous systems, polymerization of TXN (and TTXN) would still be feasible. Although the enthalpy of polymerization for ll or ss transformation (the meaning of the abbreviations are given in Chap. 2) has a relatively small negative value (which reflects the low ring strain of the 6- or 8-membered ring), this is partly counterbalanced by the low entropy of polymerization.

Calculated from the results given above, the hypothetical equilibrium monomer concentrations in a homogeneous polymerization is [90]:

$$
\begin{aligned}
[\mathrm{TXN}]_e &= \sim 2.5\ \mathrm{mol \cdot l^{-1}}\ (\mathrm{CH_2Cl_2},\ 20^\circ) \\
&= \sim 1.3\ \mathrm{mol \cdot l^{-1}}\ (\text{benzene},\ 20^\circ) \\
&= \sim 2.5\ \mathrm{mol \cdot l^{-1}}\ (\text{benzene},\ 55^\circ)
\end{aligned}
$$

In the real system, in which the dissolved or liquid (molten) TXN is directly converted into crystalline polymer, the over-all thermodynamic parameters are equal to the sum of ΔH and ΔS of the chemical reaction and the phase transition (cf. Chap. 2). It should be remembered, however, that polymerization and crystallization can be separated only formally.

For the transition:

$$
(\mathrm{CH_2O})_n\ (\text{dissolved in benzene}) \rightarrow (\mathrm{CH_2O})_n\ (\text{crystalline}) \tag{7-23}
$$

$\Delta H_{sc'} = -20$ kJ/mol (-4.8 kcal/mol) and $\Delta S_{sc'} = -41$ J/mol °K (-10 e.u.) [89]. These values were calculated by assuming that the polymer is 100% crystalline. Although this assumption is justified (the crystallinity of polyoxymethylene usually exceeds 50%, for example it is 55–60% for the commercial Hostaform copolymer) [91], the exact values of $\Delta H_{sc'}$ and $\Delta S_{sc'}$ for polymerization to the crystalline polymer are a sum of sc and cc′ transitions, and thus depend on the crystallinity of the polymer obtained under a set of given conditions.

Comparison of thermodynamic parameters of the chemical reaction and phase transition shows that the contribution of phase transition to the over-all values of thermodynamic parameters is larger than that of chemical reaction (ring-opening of TXN). Thus, TXN polymerization may be described as a process for which crystallization provides the driving force. ΔH and ΔS values for the polymerization of TXN, formaldehyde and TTXN involving phase transitions are listed in Table 7.10.

Let us now briefly discuss the practical consequences of these results. As shown in Chap. 2:

$$
\ln\,[\mathrm{M}]_e = \frac{\Delta H_0}{RT} - \frac{\Delta S_0}{R} \tag{7-24}
$$

It is instructive to calculate by Eq. (7-24) (using the values in Tables 7.9 and 7.10) the equilibrium concentrations of formaldehyde, TXN, and TTXN for the hypothetical

Table 7.10 Thermodynamic parameters of polyoxymethylene formation [85)]

Monomer	Standard State	ΔH		ΔS°	
		kJ/mol	(kcal/mol)	J/mol °K	(e.u.)
Formaldehyde	gl	−62.3	(−14.9)	−162	(−38.7)
	gc	−69.8	(−16.7)	−178	(−42.6)
	lc	−46.4	(−11.1)	−85.5	(−20.7)
	sc (NB)[1]	−78.2	(−18.7)	−100	(−24.3)
1,3,5-Trioxane	gl	−50.1	(−12.0)	−115	(−27.5)
	gc	−62.7	(−15.0)	−155	(−37.0)
	lc	−21.7	(−5.2)	−48.9	(−11.7)
	sc (NB)[1]	−21.3	(−5.1)	−53.1	(−12.7)
	sc (C)[2]	−19.6	(−4.7)	−50.1	(−12.0)
1,3,5,7-Tetraoxocane	gl	−67.7	(−16.2)	−112	(−26.8)
	gc	−83.2	(−19.9)	−165	(−39.5)
	lc	−26.3	(−6.3)	−46.0	(−11.0)
	sc (NB)[1]	−25.9	(−6.2)	−41.8	(−10.0)
	sc (C)[2]	−30.1	(−7.2)	−56.4	(−13.5)

1 — Nitrobenzene, 2 — CH_2Cl_2

polymerization to dissolved polymer, and for the real polymerization to crystalline polymer. This comparison illustrates the effect of polymerization in the crystalline state. The corresponding values are given in Table 7.11. Thus, the contribution of the thermodynamic parameters of phase transition leads to a considerable decrease in $[M]_e$ values; this is particularly so with respect to TXN. It follows from the data in Table 7.11 that, if one starts, for instance, from a 30% solution of TXN (~3.3 mol $\times l^{-1}$) then in homogenous solution only ~55% conversion would be possible whereas polymerization in the crystalline state would yield ~96% conversion. As it was stressed in Chapter 2, this is due to the polymerization taking place in the crystalline state. Otherwise, if polymer merely precipitated and growth occurred as under homogenous conditions, $[M]_e$ would be the same for both homogeneous and heterogeneous systems.

Table 7.11. Calculated concentration of monomers in equilibrium with polyoxymethylene; solvent: nitrobenzene, temp. = 25 °C

Standard state	Monomer	Equilibrium concentration mol/l
ss (i.e. from dissolved monomer to dissolved polymer)	Formaldehyde	0.013
	1,3,5-Trioxane	1.41
	1,3,5,7-Tetraoxane	0.10
sc′ (i.e. from dissolved monomer to crystalline polymer)	Formaldehyde	0.003
	1,3,5-Trioxane	0.11
	1,3,5,7-Tetraoxane	0.004

Table 7.12. Measured concentrations of monomers in equilibrium with polyoxymethylene

Standard state	Monomer	Solvent	Temp. °C	Equilibrium concentrations, mol/l	Ref.
sc'	Formaldehyde	CH_2Cl_2	30	0.06	92)
sc'	1,3,5-trioxane	Benzene	30	0.05	93)
		$C_2H_4Cl_2$	30	0.13	93)
		Nitrobenzene	30	0.19	93)
		Nitrobenzene	25	0.2	71)
		$C_2H_4Cl_2$	60	0.4	104)
		CH_2Cl_2	25	0.14	85)
		Nitrobenzene	25	0.10	85)
		n-Decane	25	0.003	85)
		Nitrobenzene	74	0.35	83)
		n-Decane	74	0.005–0.01	83)
sc'	1,3,5,7-tetraoxane	CH_2Cl_2	25	0.0075	83)
		Nitrobenzene	25	0.0065	85)
		n-Decane	25	$5 \cdot 10^{-4}$	85)
		Nitrobenzene	74	0.02	83)
		n-Decane	74	$<10^{-3}$	83)

The values of equilibrium monomer concentrations measured by different authors under different conditions are somewhat scattered. This is mainly due to experimental problems. Nevertheless, they are quite close to values obtained by calculations based on thermochemical data.

As follows from the above data, the equilibrium monomer concentration depends rather strongly on the nature of the solvent used. Possible reasons are discussed in detail in Chap. 2 (Thermodynamics).

The thermodynamics of TXN polymerization also influences the polymerization kinetics. Kern [92] has proposed that induction periods, frequently observed in the cationic polymerization of TXN, are due to the build up of the equilibrium formaldehyde concentration, which has to occur before polymer can be formed. According to other authors, it is not formaldehyde but TTXN that must be formed prior to polymerization [93]. We conclude that formation of polymer may require that the equilibrium concentration of any "monomeric" species which is in equilibrium with polymer is reached first.

Thus, one should know not only the equilibrium constants, but also the rate constants in the scheme:

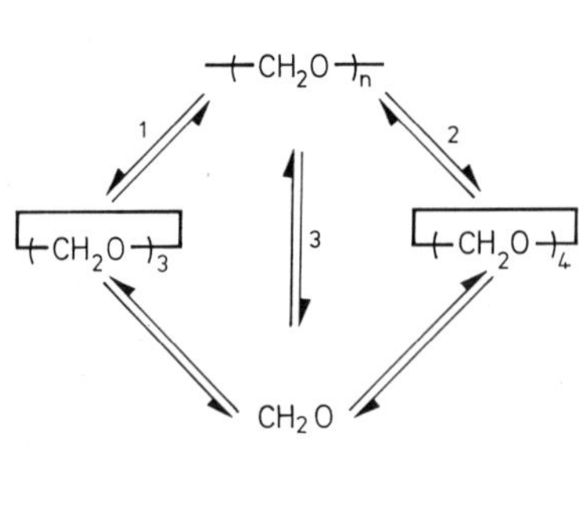

(7-25)

Polymer formed initially from TXN, is thermodynamically unstable with respect to formaldehyde and TTXN because their concentrations (zero at the beginning of the reaction) are below $[M]_e$. If the rates of reactions (2) and (3) are comparable with that of reaction (1), then, polymer formed from TXN will depropagate to formaldehyde and TTXN and this process will continue until the equilibrium concentrations of formaldehyde and TTXN are reached. Consequently, at this stage TXN will be consumed, however, polymer will not be formed.

Thus, in principle, we should distinguish between the induction period for the consumption of monomer (which may be of kinetic origin) and the induction period for the formation of polymer (which may be of thermodynamic origin). In practice, the latter is usually observed, as the beginning of the polymerization is generally regarded as the moment at which polymer starts to precipitate.

Kern and Jaacks found that addition of formaldehyde at the beginning of the reaction leads to a considerable shortening of the induction period, and used this observation to substantiate their hypothesis [92]. Studies of others, however, did not confirm this phenomenon [93].

Moreover, induction periods do not need to be only of thermodynamic origin. For example, the cationic polymerization of formaldehyde with $BF_3 \cdot OEt_2$ initiator proceeds without induction periods and initiation is "instantaneous" [94], thus the shortening of the induction period by formaldehyde may be to some extent due to the higher initiation rate. Recently, Wegner pointed out that nucleation of crystal growth may also lead to an induction period in TXN polymerization [75]. Analysis of the kinetics of nucleation, based on the theory of nucleation, led to the conclusion that the length of the induction period should be inversely proportional to the initiator concentration. Such a linear dependence was indeed observed for $HClO_4$-initiated polymerization.

More recently, Dimonie and Negulescu reported that the use of the BF_3-acrylonitrile complex instead of $BF_3 \cdot OEt_2$ as initiator but otherwise under identical conditions eliminates the induction period in the polymerization of TXN in benzene solvent at 35 °C [95]. The proposed interpretation may require confirmation.

An interesting analysis of the influence of thermodynamics on the kinetics of TXN polymerization is given in a series of papers by Enikolopyan et al. [96–98]. These authors consider polymerization to be homogeneous (in the kinetic sense) in contrast to the above-discussed polymerization in the crystalline state. The equilibrium between crystalline and dissolved polymer is considered (only the fragment of the chain that carries the active end needs to be dissolved) and on this basis some kinetic peculiarities of TXN polymerization are explained.

7.3.5.3 Chain Transfer Reactions

The importance of chain transfer in the polymerization of TXN is illustrated by the fact that the molecular weights of polymers cannot be increased much above 10^5 although kinetic chain lengths frequently reach 10^6.

Chain transfer may occur with any compound possessing reactive hydrogen atoms be these impurities or purposely added to the system. The most common impurity is water, a very active chain transfer agent [99,100]. Alcohols, carboxylic acids, acetals, some aromatic compounds (m-xylene, mesitylene, anthracene), and halogenated hydrocarbons may also function as effective chain transfer agents in cationic TXN polymerization.

Chain transfer with water involves formation of the corresponding oxonium ion with the subsequent release of a proton which initiates a new chain:

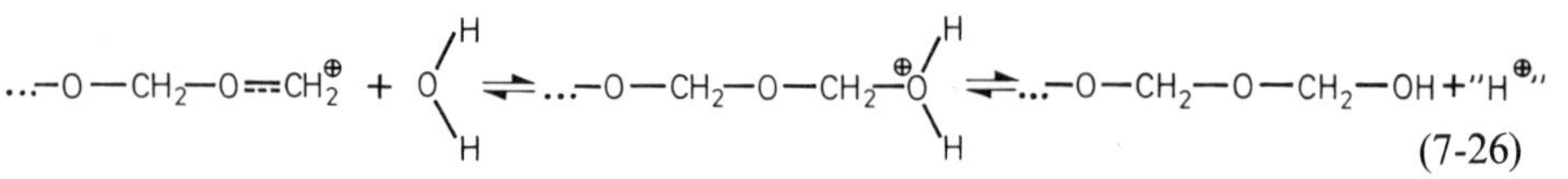

(7-26)

Even in most rigorously purified systems, molecular weights corresponding to the length of the kinetic chains cannot be attained. Search for the cause of chain shortening led to the discovery of hydride transfer in TXN polymerization [70].

As the result of hydride transfer from the monomer to the macrocation, the polymer would acquire a $-OCH_3$ end group and the 1,3,5-trioxane-2-ylium cation formed would initiate a new chain:

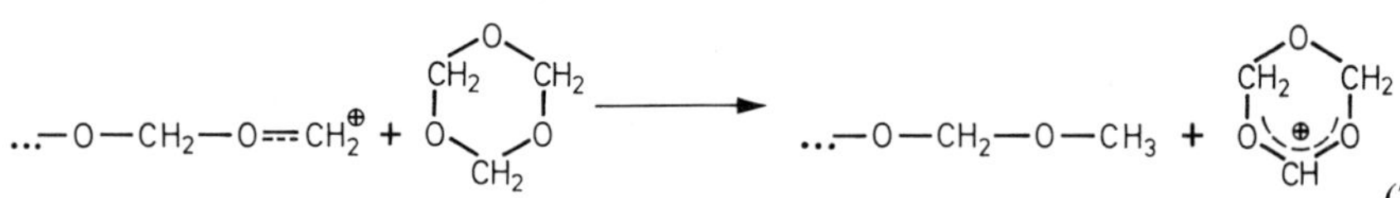

(7-27)

A similar reaction may also occur with participation of a polymer segment:

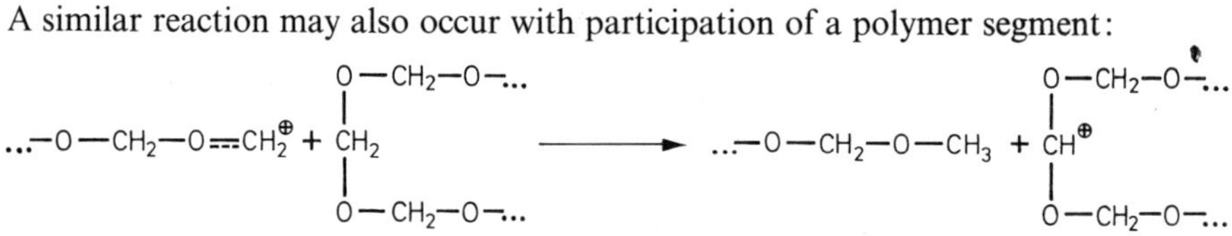

(7-28)

The stable OCH_3 end-groups, that are formed in this process, may be responsible for the observation that the polyoxymethylene contains up to 25% of a thermally stable fraction [48]. One has to remember, however, that thermal stability may at least partially be due to the inevitable presence of macrocyclics [101].

Model studies with $Ph_3C^{\oplus}$ as the hydride-ion acceptor, have demonstrated that hydride transfer proceeds much faster from the linear acetals ($CH_3OCH_2OCH_3$ as a model for the polyoxymethylene chain unit) than from TXN. The rate constant for the former reaction was $k = 6.4 \cdot 10^{-4}\ mol^{-1} \cdot l \cdot s^{-1}$ (CH_2Cl_2, 25 °C), whereas that for the reaction with TXN was below $10^{-5}\ mol^{-1} \cdot l \cdot s^{-1}$ under the same conditions [102].

These data would indicate that, as suggested by Weissermel [48], hydride transfer from polymer followed by cleavage of the newly formed cation predominates:

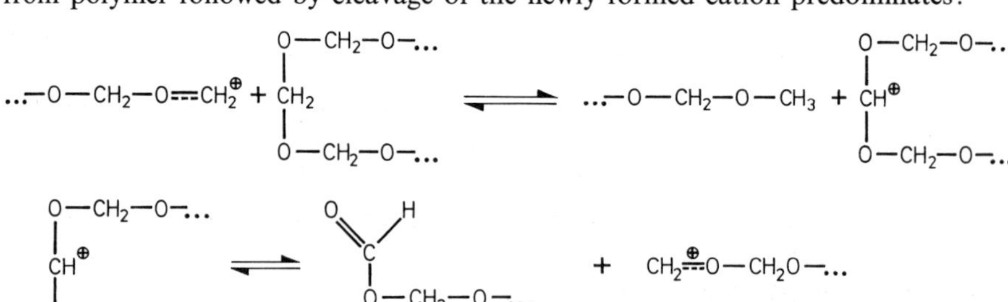

(7-29)

Hydride transfer from polymer may lead to branching:

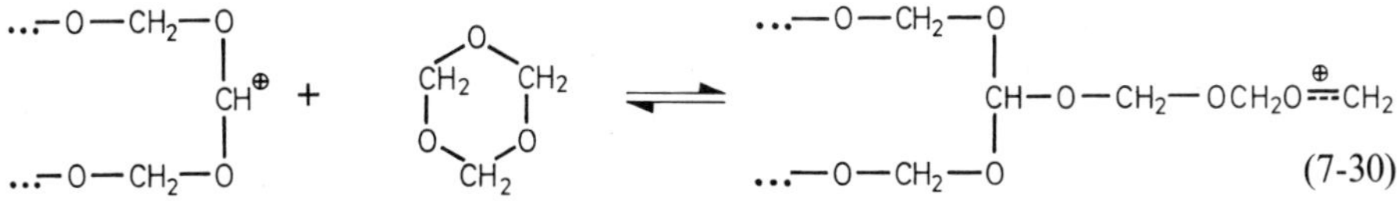

(7-30)

Indeed, branches formation was postulated [103], although on the basis of indirect evidence, for the polymerization at 80 °C. At higher temperatures, the reaction shown in Scheme (7-29) (involving cleavage) predominates.

Hydride transfer from monomer (assumed but not proven), would lead to the same end-groups [104]:

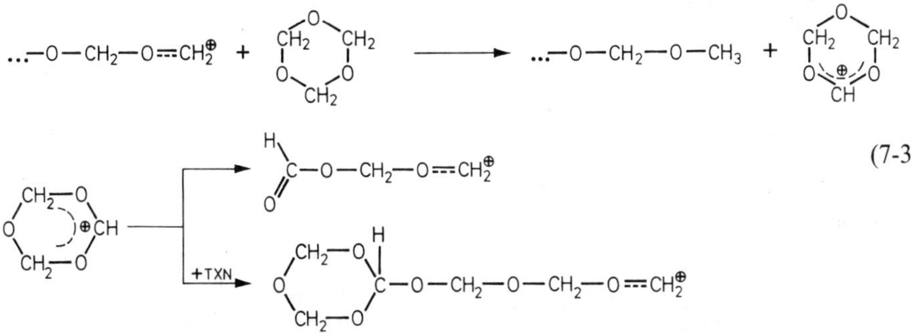

(7-31)

In this case, however, some cyclic orthoester end-groups should also be present, although they may be cleaved to formate end-groups during the polymerization [105]. The formation of these end-groups was postulated only on the basis of indirect evidence [104].

7.3.5.3.1 Chain Ttransfer to Polymer-Transacetalization

Chain transfer to polymer is a general phenomenon in ring-opening cationic polymerization of heterocyclic monomers and its basic mechanistic and kinetic features have been discussed in Vol. I Adv. Polymer Sci. 37 (1980), Sect. 5.3.

In some systems, mostly in the polymerization of cyclic sulfides and amines, the formation of "polymeric" onium ions is irreversible, i.e. the reactivity of these structures is so low that further propagation on these species does not proceed:

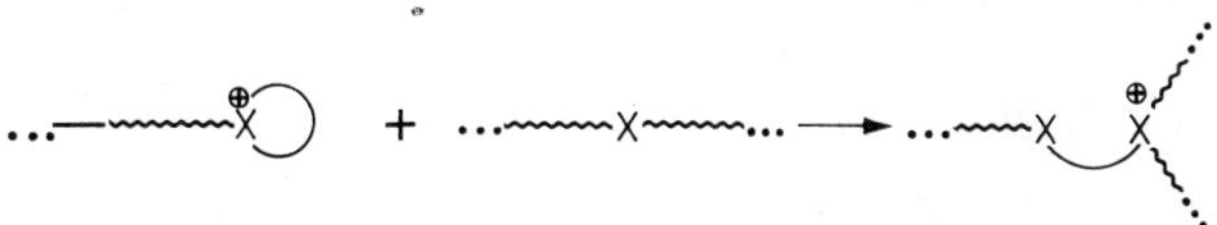

(7-32)

Such a process is thus equivalent to kinetic termination. In the homogenous cationic polymerization of cyclic acetals (e.g. DXL), chain transfer to polymer proceeds efficiently, because polymer is more basic than monomer. In this case, however, the polymeric oxonium ions remain active and may further propagate:

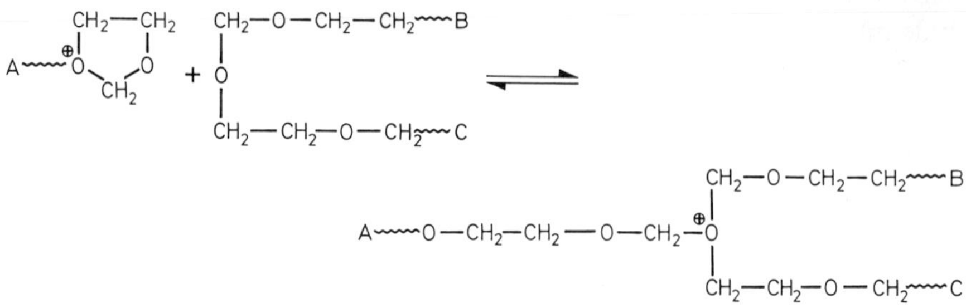

(7-33)

In Reaction (7-33) the two ligands, denoted by letters A and B, are identical, thus attack of an incoming monomer (or polymer) may occur with equal probability, on either carbon atom of the oxonium center. Attack on the methylene group in fragment A leads to the re-formation of the original polymer segment B–C, however, attack on B, leads to the new polymer segment A–C. If the polymer chain is composed of only one type of units this process has no practical consequences, except for affecting the molecular weight distribution.

If, however, B— is structurally different from A∿∿∿C, for example B∿∿ and A∿∿∿C contain a different number of comonomer units, then such a process, repeated many times, would lead to randomization.

Chain transfer to polymer, leading to transacetalisation is believed to be responsible for the formation of thermally stable TXN-DXL copolymers. The concept of producing stable formaldehyde polymers by introducing a small number of $—CH_2—CH_2—$ units, i.e., interrupting the $—CH_2—O—$ repeat units demands that the units which stop depropagation should be uniformly and randomly distributed along the polymer chain. Then, if the chain breaks for any reason at any unit, unzipping giving CH_2O would stop at the first $—CH_2—CH_2—$ unit.

$$\begin{aligned}&\ldots —CH_2—O—CH_2—O—CH_2—CH_2—O\!\left(CH_2—O\right)_n CH_2—OH \rightarrow\\&\ldots —CH_2—O—CH_2—O—CH_2—CH_2—OH + n + 1\ CH_2O \qquad (7\text{-}34)\end{aligned}$$

The low content of the unstable fraction (defined as the fraction of polymer, which is thermally degraded under the standard conditions of measurement) in copolymers of TXN with small amounts (2–5%) of DXL is an evidence that this condition is fulfilled, i.e. that the ethylene oxide units are uniformly distributed in the chain.

This cannot be achieved, however, by copolymerization alone, because the rate of consumption of both comonomers is not the same. [53,59]. As shown in Fig. 7.9 DXL is consumed much faster and the copolymer formed initially is considerably enriched by DXL units [53]. For example, bulk copolymerization of TXN with DXL (1:1 by weight) initiated with $BF_3 \cdot Et_2O$ yields at 15% conversion a product that contains 92% DXL units [53].

According to Jaacks this DXL-rich copolymer, formed during the first stage, is soluble in the reaction medium. Even when the DXL to TXN ratio was much smaller ($4.5 \cdot 10^{-2}$ mol · l^{-1} DXL, 2.5 mol · l^{-1} TXN in CH_2Cl_2), soluble copolymer containing more than 50% of DXL units formed initially and over 50% of DXL has already been consumed when the conversion of TXN was still negligible. Only then

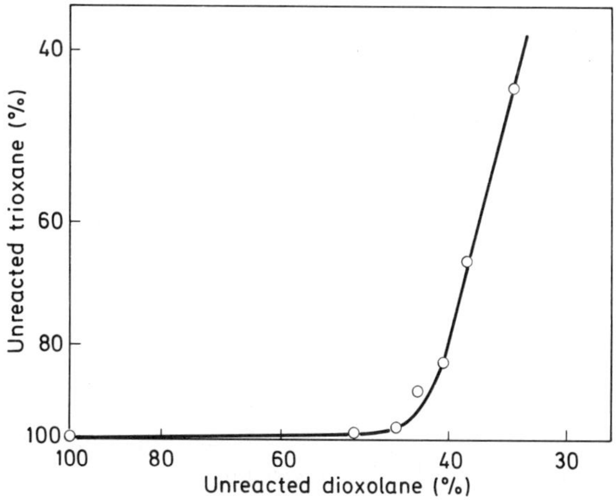

Fig. 7.9. Rate of monomer consumption in the trioxane-dioxolane copolymerization [1,3,5-trioxane] = 2.5 mol · l^{-1}, [1,3-dioxolane] = 4.5 · 10^{-2} mol · l^{-1}, [$SnCl_4$] = 1.7 · 10^{-2} mol · l^{-1}, CH_2Cl_2, 30 °C [53)]

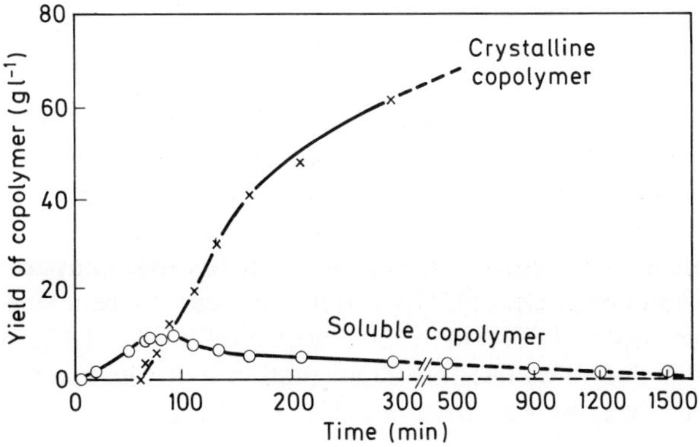

Fig. 7.10. Formation rates of soluble and insoluble trioxane-dioxolane copolymers, [1,3,5-trioxane] = 2.8 mol · l^{-1}, [1,3-dioxolane] = 0.28 mol · l^{-1}, [$SnCl_4$] = 2.5 · 10^{-2} mol · l^{-1}, CH_2Cl_2, 30 °C [53)]

did the insoluble copolymer composed almost entirely of TXN units start to precipitate. In the course of the further reaction, the yield of insoluble polymer continuously increased, while the concentration of the soluble copolymer decreased and finally became negligible. Typical conversion vs. time curves are shown in Fig. 7.10.

Evidently a mechanism must be available by which DXL, which initially yields soluble copolymer, is not only transferred into the insoluble TXN-rich copolymer, but is randomly distributed within this copolymer.

The most plausible explanation is that randomization occurs via transacetalization. Jaacks, however, questioned this explanation arguing that by a single transfer step a long block of soluble copolymer may be incorporated into the chain growing in the crystalline phase [53)]. This block must further be cleaved many times before complete randomization occurs and such a process would have to proceed in the crystalline phase. An alternative mechanism was proposed assuming depolymerization of the soluble copolymer due to different equilibrium concentrations of DXL and formaldehyde (formed from TXN) with respect to the dissolved and crystalline polymer (cf. Sect. 7.3.5.2.3). DXL formed by depropagation from the DXL rich soluble copolymer is repolymerized and introduced into the TXN rich insoluble polymer.

Although this is a logical mechanism and some arguments against a transacetalization mechanism have been substantiated, a number of experimental evidences favour the transacetalization mechanism put forward by Weissermel et al. [106].

7.3.5.4 Termination

Weissermel et al. view the bulk polymerization of TXN in the absence of impurities as a terminationless process, indicating that hydride transfer to the apparently still living active centers may be observed even several hours after the completion of reaction [48]. There are no published experimental results contradicting this point of view. It was argued, however, that although the growing species may be living, some of them may become unavailable for further propagation (buried) as a result of occlusion within the polymer particles. [107, 108]. This would lead to a more pronounced retardation than would be expected from monomer depletion. Such a termination mechanism is in a good agreement with the crystallographic description of the polymerization process. Aggregation of the growing crystals may prevent monomer to reach the growing sites.

The kinetics of polymerization, has been treated in terms of deactivation of growing species by embedding in the crystalline polymer, by Enikolopyan [110]. The equations derived fit well with the experimental findings, i.e., monomer conversion as a function of time and starting initiator concentration.

Active species may be deactivated by reaction with certain impurities, solvents or counterions [109]. For example, Enikolopyan et al. presented kinetic evidence indicating that the polymerization of TXN in the presence of $SbCl_6^-$ in CH_2Cl_2 solvent proceeds with termination due to counterion fragmentation and subsequent "deactivation" of the liberated $SbCl_5$ by side reactions [110]:

$$\ldots -OCH_2^{\oplus}SbCl_6^{\ominus} \rightleftharpoons \ldots -OCH_2Cl + SbCl_5 \qquad (7\text{-}36)$$

These and similar processes were first discussed by us for the polymerization of DXL with $SbCl_6^{\ominus}$ [29].

7.3.6 Solid-State Polymerization of 1,3,5-Trioxane

TXN undergoes solid-state polymerization by chemical-induced processes as well as by radiation. That TXN may undergo polymerization in the solid state was observed as early as in 1930 [111], but for a long time this observation was not used. In 1954 Mesrobian et al. published the first paper on the radiation-induced polymerization in the solid state (acrylamide polymerization) [112] and the solid-state polymerization of TXN was "rediscovered" a few years later [113]. Radiation-induced solid-state polymerization of TXN gained importance and it was believed that it could lead to the large-scale production of polyTXN. This process was studied at the pilot plant scale in France and in Japan.

The understanding of the mechanism of this reaction is important because there is no definite borderline between the solid-state and solution or bulk polymerization of TXN. In all these processes the polymer forms a solid phase from the beginning of polymerization and growth proceeds inside and/or on the surface of the solid particles. Thus, the difference between "solid-state" and "solution" processes stems from the different mobility and availability of monomer molecules being incorporated into the crystal lattice in the former and being in a liquid phase in the latter system.

7.3.6.1 Pecularities of Solid-state Polymerization

Chapiro, in his review article on radiation-induced polymerization, described solid-state TXN polymerization as the process in which polymerization may proceed without significant molecular diffusion [114)]. This conclusion was based on the reasoning that in the crystal lattice TXN molecules are piled one on top of the other with a distance between two molecules of 4.3 Å. If the rings open, a slight twisting of the molecule brings the active ends into a close proximity with its two neighbours. A stack of monomer rings is thus converted gradually into a polymer helix. Since the dimensions of crystalline polyoxymethylene are close to those of the monomer, large single crystals of polymers will result [114)].

This picture is slightly oversimplified, however, because it has been found that polymer formed during solid-state polymerization of TXN crystallizes as a helix with 1.81 $-CH_2O-$ units per turn, while the TXN molecule has 3 such units per ring [115)]. In the TXN crystals the distance between adjacent molecules is equal to 4.18 Å while three CH_2O units in a polymer helix occupy 5.77 Å [116)].

Also the cross section of a TXN molecule is much larger than the cross section of polyoxymethylene helix which is only 17.3 Å [116,117)].

Thus, polymer formation requires rearrangements in all a-, b-, and c-axis directions which in turn requires a relatively high mobility under the polymerization conditions [115)].

Indeed, a rather straightforward correlation was found between the rate of polymerization and mobility of the monomer within a crystal lattice by broad-line NMR [118)]. TXN polymerizes rapidly only above 40 °C and at this temperature a very narrow line is present in the centre of the ^{1}H-NMR spectrum, indicating considerable molecular motion. It was found also, that the maximum polymerization rate is reached at a temperature just below the melting temperature of TXN, implying again that extensive molecular mobility is required [119,120)].

These results indicate that RISS[1] polymerization of TXN is not a simple topochemical process, in which monomer molecules are converted into polymer units without leaving their lattice positions. It was suggested that polymerization may proceed according to a solid→gas→solid scheme, i.e. TXN sublimes from the monomer crystal lattice into cavities formed as a result of crystal lattice rearrangement and then adds to the growing end from the gas phase [120)].

---

[1] RISS: Radiation-Induced Solid-State

A more detailed discussion of the influence of the crystal lattice, the importance of lattice imperfections, phase transitions and molecular mobility on the solid-state TXN polymerization is given in review papers [121,122].

7.3.6.2 Radiation-induced Solid-state (RISS) Polymerization of 1,3,5-Trioxane

The development of solid-state polymerization was largely due to the use of high-energy radiation such as γ-rays [115,123], x-rays [124], electron beams [125,126] or α-particles [116]. γ-Rays are used most frequently. It is generally accepted that γ-ray-induced polymerization of TXN proceeds by a cationic mechanism. Ions or radical-ions are formed by electron transfer from TXN, the loss of hydrogen atoms or the heterolytic cleavage of the ring [127].

The early data on the rates, conversions and molecular weights are doubtful because RISS TXN polymerization is extremely sensitive to impurities (especially water). For carefully purified monomer, 80% conversions were reported for the temperature range 45–55 °C [121]. However, data published on RISS TXN polymerization contain many conflicting observations and as it is stated by Kovacs et al. in a recently (1981) published paper: "Our results also show once again the unexpectedly important role played by experimental conditions and that unidentified and therefore uncontrollable factors are involved" [129].

7.3.6.3 Chemically-initiated Solid-state Polymerization of TXN

These processes are carried out with crystals of TXN. An alternative is polymerization of crystals suspended in nonsolvent. Typical cationic initiators like BF_3 [130,131] or $SnCl_4$ [132,133] are most frequently used.

In suspension polymerization, initiator (e.g. $BF_3 \cdot OEt_2$) is dissolved in the suspending medium. In the polymerization of crystals alone, gaseous BF_3 (usually diluted with N_2) is used.

The characteristic feature of chemically initiated solid-state polymerization of TXN is the decrease of molecular weight with conversion. This is illustrated by the data in Table 7.13. It was concluded that high-molecular-weight polymer may be produced only on the surface of a crystal. At a later stage, when polymer is formed in the inner part of the crystal, lower-molecular-weight polymer results. The molecular weight increases with increasing temperature (cf. Table 7.13), and depends also on

Table 7.13 Solid-state polymerization of 1,3,5-trioxane (TXN). $[TXN]_0 = 2.0$ mol $\cdot l^{-1}$ (suspension in n-hexane) $[BF_3 \cdot OEt_2]_0 = 1 \cdot 10^{-2}$ mol $\cdot l^{-1}$ $[H_2O]_0 = 1.2 \cdot 10^{-3}$ mol $\cdot l^{-1}$ [131]

Sample	Temp. °C	Time, min	Conversion, %	$[\eta]^a$ dl $\cdot g^{-1}$
1	40	0.5	4.0	3.06
2	40	2.0	8.0	2.18
3	50	0.5	8.0	3.77
4	50	2.0	16.0	2.53
5	50	20	42.1	1.75

^{a} $[\eta]$ measured at 60 °C in solution of p-chlorophenol containing 2% α-pinene

solvation. This was again explained in terms of increased mobility of monomer molecules within the crystal lattice.

Essentially the same results were obtained for $SnCl_4$-initiated suspension polymerization [132]. The molecular weight decreased with increasing conversion and as in the BF_3 initiated polymerization increased after addition of a small amount of a good solvent [131].

7.3.7 Copolymerization of 1,3,5-Trioxane

Thermally stable polyacetals are prepared by cationic TXN copolymerization with a small amount of comonomer (DXL, EO). Thus, the mechanism of TXN copolymerization was more frequently studied than its homopolymerization.

In regard to the copolymerization process, we have to describe the rates of incorporation of comonomers and distribution of different units in the resulting copolymer (microstructure). In this section we briefly discuss characteristic features of copolymerization of TXN with cyclic formals and then with other oxacyclic and some vinyl monomers.

7.3.7.1 Copolymerization of 1,3,5-Trioxane with 1,3-Dioxolane

Some peculiarities of this system have already been described in connection with transacetalization reaction. As indicated by Jaacks [53], TXN-DXL copolymerization cannot be analysed by using the conventional Mayo plot because all propagation reactions are reversible and propagation is accompanied by chain transfer to polymer (transacetalization). Jaacks developed a method which, although based on some simplifying assumptions, gives the reactivity ratio r_1. The final equation

$$\log \frac{[M_1]_t}{[M_1]_0} = r_1 \log \frac{[M_2]_t}{[M_2]_0} \qquad (7\text{-}37)$$

is valid for $[M_1] \gg [M_2]$, with $M_1 = TXN$, $M_2 = DXL$

Further analysis of the copolymerization process showed that the fast initial conversion of DXL yielded copolymers soluble in the reaction medium (CH_2Cl_2). The DXL content in the product was above 50 % mol. At the point where a break occurred on the conversion curve a second type of copolymer began to form; it contained 98 mol % of oxymethylene units and crystallized out of solution. These observations were rationalized by the assumption that the relative reactivities of the comonomers (reactivity ratios) are different for the reaction occurring in solution and in the crystalline phase (cf. Sect. 7.3.5.2.2). The initial composition of copolymer and the distribution of units are continuously changing due to transacetalization. Thus, the alleged "reactivity ratios" determined for this copolymerization have no physical meaning.

7.3.7.1.1 Analysis of the Sequence Distribution in 1,3,5-Trioxane-1,3-Dioxolane Copolymers

Thermal stability of the TXN-DXL copolymers depends critically on the distribution of the DXL units (cf. Sect. 7.3.5.3.1). Thus, methods for the determination of copolymer microstructure based on NMR were developed, mostly by Schulz [134], and

Table 7.14. ^{1}H-NMR chemical shifts of oxyethylene and oxymethylene units in polyacetals

Monomer	Repeating unit of polymer	Chemical shifts ppm δ	Conditions	Ref.
1,3-dioxolane	$-O\underset{E}{\underline{CH_2CH_2}}O\underset{M}{\underline{CH_2}}-$	3.65 — E 4.65 — M	$CHCl_3$, 25 °C 60 MHz	10)
1,3,5-trioxepane	$-O-\underset{E}{\underline{CH_2-CH_2}}-O-\underset{M}{\underline{CH_2}}-O-\underset{M}{\underline{CH_2}}-$	4.89 — $M\underline{M}M$ 4.83 — $M\underline{M}E=E\underline{M}M$ 4.77 — $E\underline{M}E$ 3.73 — $M\underline{E}M$ 3.68 — $M\underline{E}E=E\underline{E}M$	$CDCl_3$, 25 °C 60 MHz	134)
1,3-dioxolane-trioxane copolym.	$(O-\underset{E}{\underline{CH_2-CH_2}}-O-\underset{M}{\underline{CH_2}})_n(O\underset{M}{\underline{CH_2}})_3$	4.73 — $M\underline{M}M$ 4.66 — $E\underline{M}M=M\underline{M}E$ 4.59 — $E\underline{M}E$ 3.59 — $M\underline{E}M$	CCl_4, 25 °C 100 MHz	135)
1,3-dioxepane	$-O-\underset{B}{\underline{CH_2CH_2CH_2CH_2}}-O-\underset{M}{\underline{CH_2}}-$	4.56 — $B\underline{M}B$ 3.49 — $M\underline{B}M$ 1.63 — $M\underline{B}M$	CCl_4	7)
1,3-dioxolane	$-O-\underset{E}{\underline{CH_2-CH_2}}-O-\underset{M}{\underline{CH_2}}-$	4.78 — $E\underline{M}E$ 3.73 — $M\underline{E}M$	CCl_4	7)

Opitz [135]. Studying the NMR spectra of polymers and copolymers of TXN, DXL and 1,3,5-trioxocane, modeling the various distributions of oxymethylene (M) and oxyethylene (E) units in the resulting polymers, these authors were able to determine the concentrations of MMM, MME and EME triads. The respective chemical shifts are given in Table 7.14. Thus, the chemical shifts of M in triads is not constant, and depends on the neighbouring units. The difference, however, between MMM and EMM triads is only ~0.05 ppm δ for the same TXN-DXL copolymer, which is insufficient to distinguish between these triads with a lower resolution apparatus.

^{13}C-NMR spectroscopy is much more suitable for structural studies. Extensive studies of the structures of homo- and copolymers of cyclic acetals were published by Schulz [134]. By using a shift reagent, 1,1,1,2,2,3,3-heptafluor-7,7-dimethyl-4,6-octadionatoeuropium ($Eu(fod)_3$), the pentad sequences, and with higher $Eu(fod)_3$ concentration the heptad sequences, were observed for both M and E units. Signals observed in the spectrum of TXN-DXL copolymer, are given in Table 7.15.

Table 7.15 ^{13}C-NMR spectra of 1,3,5-trioxane-1,3-dioxolane copolymer in CD_3Cl, shift reagent: $Eu(fod)_3$, $[Eu(fod)_3]/[-CH_2O-] = 0.00536$ [134]

Chemical shift ppm δ	Oxymethylene pentades	Oxymethylene heptades
96.57	MEMEM	EMEMEMM
96.05		MMEMEMM
92.53		EMMMEME
	MMMEM	MMMMEMM
93.30		MMMMEME
		EMMMEMM
92.01	MEMME	MMEMMEM
		MMMMMMM
88.86	MMMMM	MMMMMME
		EMMMMME
88.55	EMMMM	MEMMMMM
		MEMMMME
88.15	EMMME	MEMMMEM
		Oxyethylene heptades
69.13		MEMEMEM
		MEMEMME
68.29		MEMEMME
and		EMMEMME
67.60		EMMEMMM
		MMMEMMM

7.3.7.2 Influence of Monomer Content and Distribution on TXN Copolymer Properties

The thermal stability of TXN copolymers depends not only on the overall content of the foreign units but also on their distribution. Kučera attempted to define in quantitative terms, the relationship between the content of oxyethylene units (from DXL) in the copolymer and the amount of the unstable fraction [136]. The following equation was proposed, assuming that the stabilizing units are distributed statistically:

$$S = (1 - 2(1 - y_{DXL})/(1 + y_{DXL})\,\overline{DP}_n) \cdot 100\% \qquad (7\text{-}38)$$

where S denotes % of stable fraction and y_{DXL} the fraction of DXL units in the copolymer.

The agreement between the values calculated according to Eq. (7-38) and found experimentally was quite satisfactory, as shown in Table 7.16.

Table 7.16. Dependence of the content of the stable fraction[a] in the 1,3,5-Trioxane–1,3-Dioxolane copolymers[b] on the proportion of 1,3-Dioxolane units and the degree of polymerization [137]

$\overline{DP}_n$	1,3-dioxolane % wt	Stable fraction, % wt	
		measured	calculated
360	1.28	71.8	64.8
997	1.18	89.5	84.5
963	1.22	89.5	84.5
1267	1.22	92.0	88.0
775	3.06	94.0	92.1
1550	2.50	96.0	95.1
1100	3.88	97.0	95.6
1225	4.12	98.0	96.3

[a] on the basis of the weight loss at 222 °C;
[b] TXN and DXL purified over sodium and copolymerized with $\sim Si^+HSO_4^-$ initiator [137]

Table 7.17. The content of the unstable fraction in the 1,3,5-Trioxane-1,3-Dioxolane and 1,3,5-Trioxane–1,3-Dioxepane copolymers [138]

Comonomer in the feed % mol on TXN	Initiator $BF_3 \cdot (n\text{-}C_4H_9)_2O$ ppm on TXN	Yield %wt	$[\eta]$[a] 100 dl · g^{-1}	Unstable[b] fraction %wt
DXL				
2.4	800	65–67	1.72–1.81	6.6–8.3
5.4	940	72–78	1.55–1.70	6.1–6.8
7.5	1000	72–77	1.50–1.60	5.4–6.6
DXP				
3	740	72–78	1.70–1.87	6.8–7.6
5	870	83–85	1.93–2.08	3.3–4.9
7	1000	85	2.17	2.9

[a] determined at 90 °C for 0.2% solution of copolymer in the tetrachloroethane-phenol (3:1) mixture containing 0.2% α-pinene;
[b] on the basis of the weight loss at 150 °C

Table 7.18. Melting temperatures of 1,3,5-Trioxane copolymers[a] [139)]

Comonomer	% mol on TXN	Melting temperature	
		DTA	dilatometry
—		179	180
DXL	2	176	176
	4	170	172
	6	168	169
	7	164	165
1,3-dioxane	1	176	175
	2	174	173
	4	169	170
	6	166	165
	8	163	163
DXP	1	178	177
	3	173	173
	6	165	166
	7	163	162
	9	160	160
ECH	2	175	176
	4	172	172
	5	168	169
	7	165	166
	9	163	165

[a] Copolymerization in cyclohexane (50% soln.) with $BF_3 \cdot OEt_2$ (0.05 to 0.2 mol%) at 55 °C for 3–20 h

In contrast to Kučera's data Fejgin et al. [138)] found that the content of the unstable fraction does not change significantly when the proportion of DXL units increases above 2.5 mol %. Typical results are shown in Table 7.17. Similar results were obtained for the radiation-induced polymerization; copolymer prepared from a 95:5 mixture of TXN and DXL still contained 12% of the unstable fraction.

These discrepancies may be due, at least in part, to the different methods of copolymerization and different chances of copolymers to reach the statistical distribution, and/or to the different determination methods of the unstable fraction content.

Incorporation of comonomer units into the polyoxymethylene chain leads to a considerable decrease of the melting temperature [139)]. The effect of comonomer structure and concentration on the melting temperature is shown in Table 7.18.

With further increasing the DXL content, the melting temperature decreases to 32 °C for a DXL content of 72% and then rises again, as shown in Table 7.19 [91)].

7.3.7.3 Copolymers of TXN with Other Heterocyclic Monomers

TXN is copolymerized with DXL or EO to prepare thermally stable polyoxymethylene. Although there have been several attempts to replace these with other comonomers (some of them, e.g. DXP, indeed offer some advantages) [140)] it seems that DXL and EO retain their superior position among cyclic comonomers acting as "zipper jammers".

Table 7.19. Melting temperatures of 1,3,5-trioxane-1,3-dioxolane copolymers [91]

DXL %	[η] dl/g	Spec. vol. cm^3/g	M_p °C
12.7	1.48	0.7195	154
21.4	1.38	0.7332	143
30	0.97	0.7458	135
43.8	0.56	0.7600	100
46.0	0.77	0.7697	
52.4	0.70	0.7815	54
63.0	0.56	0.7869	48
64.9	0.70	0.7890	34
72	0.47	0.7865	32
79	0.61	0.7777	37
85	0.63	0.7773	40
90	0.62	0.7814	43
95	0.47	0.7826	44
100	1.11	0.7812	62

There has been a considerable amount of work directed towards the preparation of new copolymers of TXN. TXN easily copolymerizes with heterocyclic and some vinyl monomers. The patent literature discloses hundreds of attempted copolymerizations. We will not cover this literature extensively because, as we have stressed above, the best solution for practical purposes is copolymerization with DXL or EO, or terpolymerization with diepoxides when a higher modulus is desired [141].

One may prepare new products differing in properties from the basic polyacetal by functionalization of polyTXN. For example, the following approach may be used:

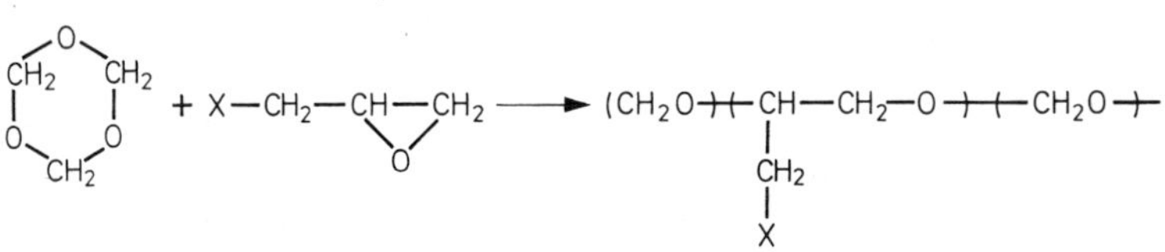

(7-39)

where X is a reactive group which may further be transformed. Using this method, polyTXN ionomers were prepared by reaction of thioglycolate anions with TXN-ECH copolymer [142].

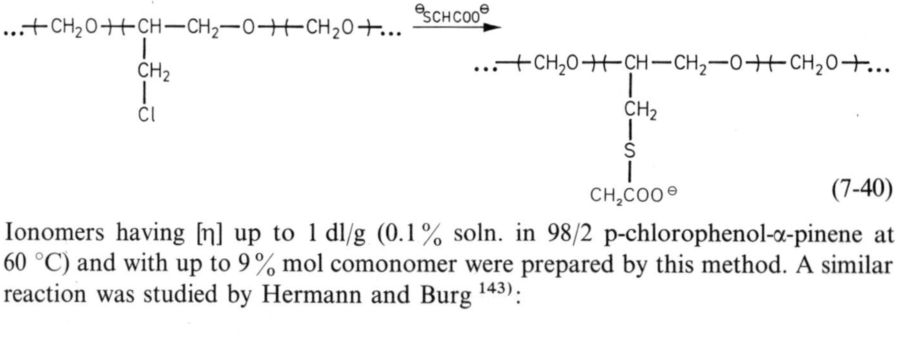

(7-40)

Ionomers having [η] up to 1 dl/g (0.1% soln. in 98/2 p-chlorophenol-α-pinene at 60 °C) and with up to 9% mol comonomer were prepared by this method. A similar reaction was studied by Hermann and Burg [143]:

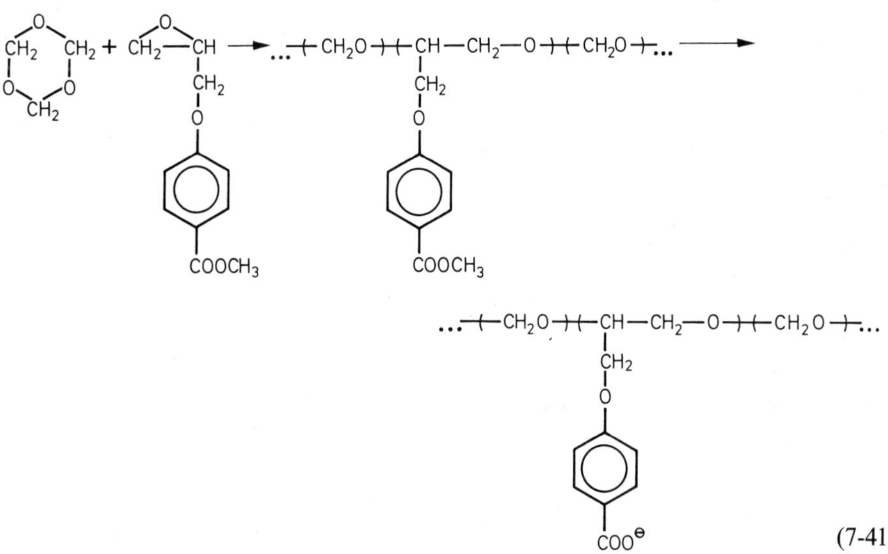

(7-41)

With the same aim of preparing polyTXN ionomers Vogl and MacKnight studied the copolymerization of TXN with ethyl glycidate[144,145]. The copolymers were subsequently converted to the corresponding polyacids:

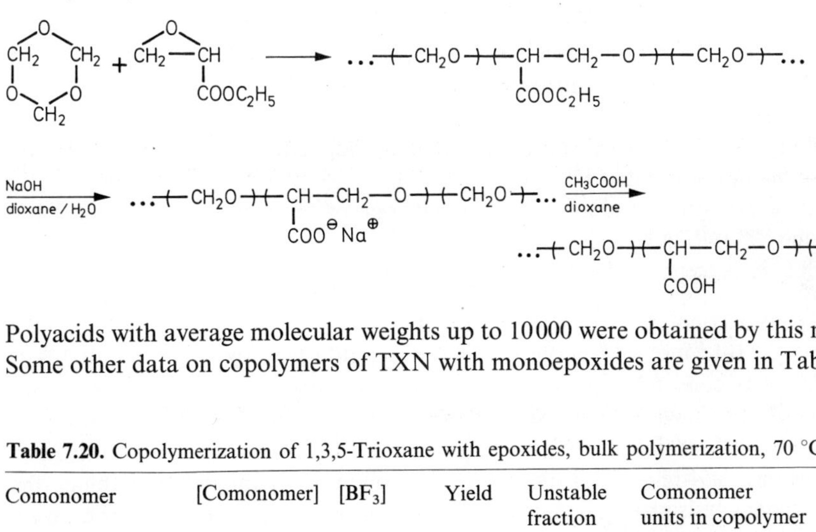

(7-42)

Polyacids with average molecular weights up to 10000 were obtained by this method. Some other data on copolymers of TXN with monoepoxides are given in Table 7.20.

Table 7.20. Copolymerization of 1,3,5-Trioxane with epoxides, bulk polymerization, 70 °C[106)]

Comonomer	[Comonomer] %mol	$[BF_3]$ %mol	Yield %wt	Unstable fraction %wt	Comonomer units in copolymer %wt	$[\eta]^a$ $dl \cdot g^{-1}$
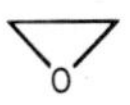	10	$6 \cdot 10^{-2}$	95	5	5	0.9
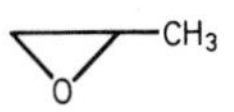	10	1	75	10	3	0.3
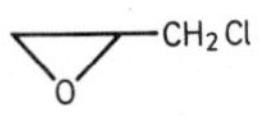	10	$6 \cdot 10^{-2}$	95	5	5	0.7

[a] 0.5 g of polymer in 88:2 butyrolactone-diphenylamine at 140 °C

Copolymerization of TXN with epoxides, particularly with diepoxides, was also used for the preparation of polyTXN with increased stiffness. Thus, terpolymerization of TXN, DXL and diepoxides (0.1–0.25% mol) (commercial epoxy resins) gives polymers with considerably higher moduli than conventional TXN-DXL copolymers [146]. For example, the addition of 0.25% mol Epidian 5® resin (a Polish product) leads to an increase in the modulus from 2300 MN/m^2 to 2880 MN/m^2. These products have been proposed as components in mixtures with conventional TXN-DXL copolymers.

Although TXN has most often been copolymerized with oxiranes, high-molecular-weight copolymers were also obtained in copolymerization with oxetanes [147] and lactones [148].

7.3.7.4 Copolymerization of 1,3,5-Trioxane and Other Cyclic Acetals with Vinyl Monomers

Kern et al. found that, contrary to most heterocyclic monomers, TXN can be copolymerized with styrene (St) to random copolymers [149]. Since that time the cationic copolymerization of TXN (and also DXL) with styrene has been studied by several groups but unfortunately GPC analysis was not commonly used and the only proof that copolymers were formed are based on solubility studies and on the presence of 4-phenyl-1,3-dioxane among the hydrolysis products of the products. Let us summarize the facts concerning TXN-styrene copolymerizations:

- in the presence of cationic initiator (e.g. BF_3) both TXN and St polymerize from mixed charges, i.e. concentration of both monomers in the charge decreases simultaneously [150].
- solubility studies exclude the presence of homopolymers; turbidity curves do not show discontinuities [151]
- when the resulting product is hydrolyzed, one of the products is 4-phenyl-1,3-dioxane [149] (4PDXL)
- 4PDXL is present in the copolymerizing mixture [150].

Higashimura copolymerized TXN with St: ([TXN] = [St] = 1.5 mol · l^{-1}, [BF_3 × OEt_2] = 2 · 10^{-3} mol · l^{-1} at 30 °C in CH_2Cl_2) and found by GLC that after 1 hr 25% TXN was consumed but only ~3% methanol-insoluble polymer was formed [150]. In addition to the unreacted TXN, the solution contained ~7% 4-phenyl-1,3-dioxane and ~5% tetraoxocane; the remaining ~10% could not be accounted for.

The presence of 4-phenyl-1,3-dioxane in the polymerizing mixture and among the copolymer hydrolysis products indicates that styrene is incorporated into the growing polyTXN chain and that attack occurs on the β-carbon atom of the double bond:

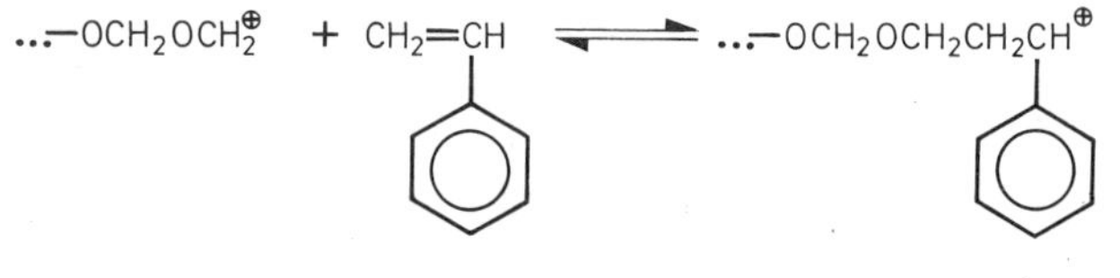

(7-43)

4-Phenyl-1,3-dioxane may be formed by back-biting:

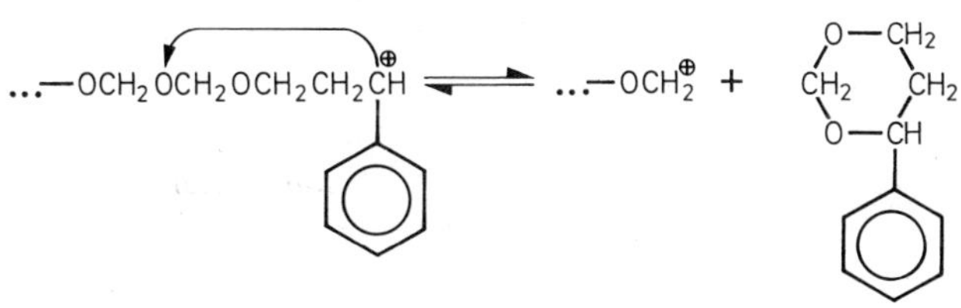

(7-44)

In essentially the same way 4-phenyl-1,3-dioxane may be formed by hydrolysis provided that St units are connected at both sides to oxymethylene units:

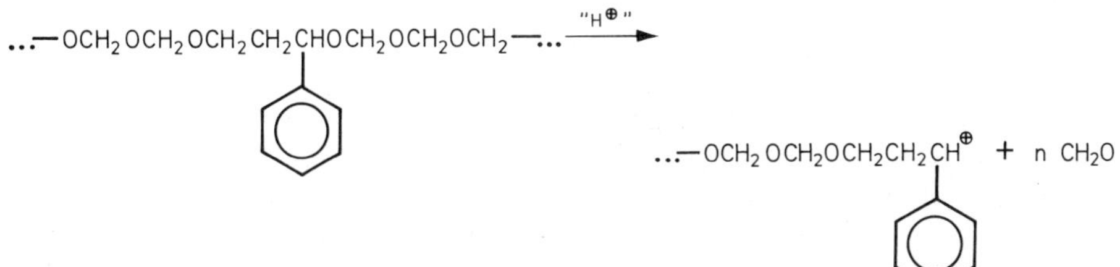

(7-45)

4-Phenyl-1,3-dioxane could not be formed if attack on the growing polyoxymethylene chain occurred at the α-position:

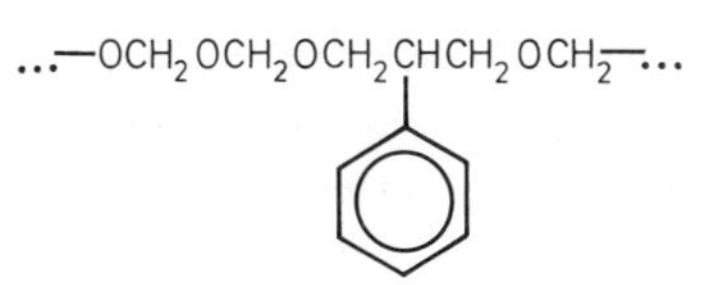

(7-46)

The copolymerization of TXN and St was analysed in a number of papers [150,151] in terms of conventional reactivity ratios without paying attention to the proper characterization of copolymers and other factors discussed in this volume (cf. Chap. 15). Some additional information comes from the studies of similar systems, i.e. DXL-St copolymerization [152]. Also in this case product characterization mainly involved solubility studies, although Yamashita et al. claimed that ^{1}H-NMR spectra confirmed that the product is indeed a true copolymer. This claim was based on a rather limited analysis of ^{1}H-NMR spectra, however, and was not confirmed by analysis of spectra of related models. Copolymerization conditions were as follows: $[DXL]_0 = 0.7 - 3.7 \text{ mol} \cdot l^{-1}$, $[St]_0 = 4.5 - 2.7 \text{ mol} \cdot l^{-1}$, $[BF_3 \cdot OEt_2] = 2{,}5 \cdot 10^{-2} \text{ mol} \cdot l^{-1}$, 25 °C, in toluene. After 2–8hrs, from 2% to 9% product with $[\eta] = 0.12 - 0.32 \text{ dl g}^{-1}$ (viscosity determination conditions not specified) was obtained.

The polymers contained, depending on the monomer ratio in the feed, from 26 to to 85 mol% DXL units and from 15 to 76 mol% St units. On the basis of overall product composition it was concluded that sequences containing up to 4 St units are present. Yamashita et al. reported the formation of copolymers with $\overline{M}_n$ as high as 332000 [153]. These products were obtained at 72% yield using $[DXL]_0/[St]_0 = 10$, however, copolymer formation was not proven. At lower $[DXL]_0/[St]_0$ ratios both yields and molecular weights were drastically reduced; thus using $[DXL]_0/[St]_0$ 1:1 a copolymer with $\overline{M}_n = 3520$ was obtained at 43% yield. Additional and conclusive proof that the products obtained were true copolymers was obtained from analysis of the hydrolysis products. In each DXL unit followed by a St unit, the acetal bond (OCH_2O) is replaced with an ether bond (OCH_2CH_2):

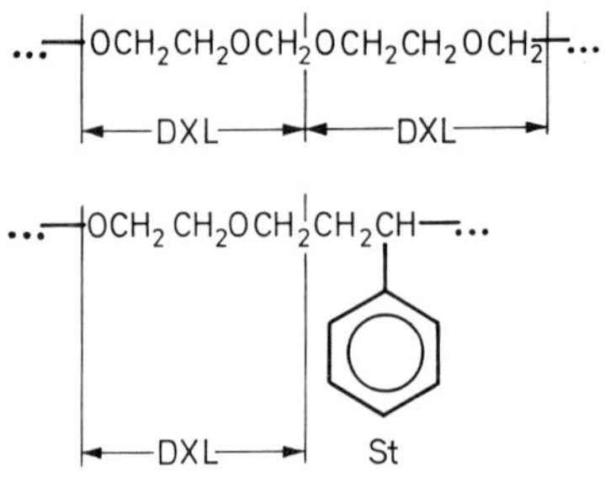

(7-47)

Thus, upon hydrolysis 1 mol of formaldehyde is liberated from the DXL-DXL dyad but not from the DXL-St dyad. It was shown [152] that the amount of formaldehyde formed upon hydrolysis is considerably lower than for a polymer composed entirely of DXL-DXL dyads although it is slightly higher than that calculated from DXL-St dyads for a completely random arrangement. Thus, it was concluded that there is a slight tendency to form St blocks, and unit distribution is not completely random.

More recently cyclic acetal-styrene systems were reinvestigated by the GPC technique [154,155]. Using double detection (UV and RI), Yamashita et al. showed that products of copolymerization of styrene with tri- and tetraethylene glycol formals have a unimodal molecular weight distribution, and that the maxima of both RI and UV traces coincide indicating that the products are true copolymers.

Similarly to styrene, isoprene (IP) also yields almost random copolymers with DXL [156]. Typical results of copolymerization are summarized in Table 7.21.

Table 7.21. Copolymerization of 1,3-dioxolane and isoprene $[DXL + IP] = 5 \cdot 10^{-2}$ mol, $[BF \cdot Et_2O] = 10^{-3}$ mol, solvent CH_2Cl_2 1 ml, 0 °C, 5 days [156]

DXL molar fraction in feed	Conversion %wt	DXL molar fraction in copolymer	$\overline{M}_n$ [a]
0.9	77.6	0.95	3.300
0.7	43.9	0.88	3.400
0.5	20.4	0.66	3.200
0.3	19.0	0.43	2.900
0	34.3	—	3.300

[a] Osmometry in benzene solution

7.3.8 Gel Permeation Chromatography of Polyoxymethylene

A major problem in gel permeation chromatography of polyoxymethylene is the lack of a suitable solvent in which polyoxymethylene can be dissolved at a relatively low temperature. In the first paper published on this subject, Japanese authors used a

DMF solution at 145 °C and obtained GPC curves both on an analytical and preparative scale [157]. The commercial polymer (Duracon®-Polyplastics Co. Ltd.) was separated by preparative GPC into seven to ten fractions with a quite narrow molecular weight distribution ($\overline{M}_w/\overline{M}_n = 1.3$–$1.4$). Decomposition was negligible and approximately 99% of polymer could be recovered.

Similar conditions were used by the Freiburg group [158]. It was shown that copolymers of TXN with ethylene oxide (EO) or propylene oxide (PO) have a bimodal molecular weight distribution. The main peak was obtained at an elution volume of

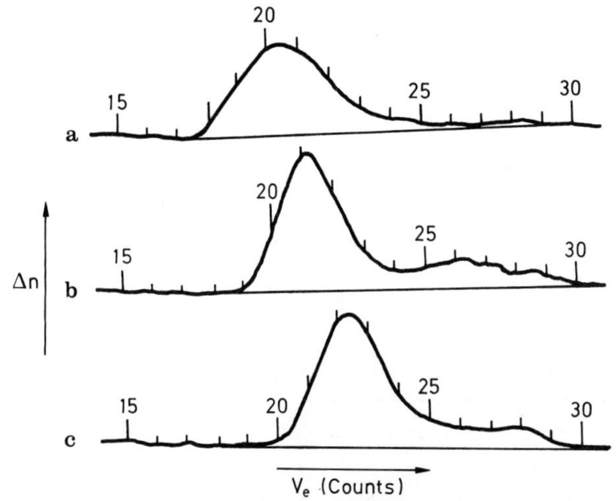

Fig. 7.11 a–c. GPC curves of trioxane-propylene oxide copolymers [158].
a 0.45% mol of oxyprolylene units;
b 2.7% mol of oxypropylene units;
c 4.7% mol of oxypropylene units

20–21 counts, which corresponds to a molecular weight in the range of 10^5–10^6 (polystyrene calibration). This peak was accompanied by a relatively small peak in the 25–30 counts region (molecular weight $10^3 - 3 \cdot 10^3$), the intensity of which increased with increasing EO or PO content. Fig. 7.11 shows representative GPC traces of such bimodal distribution indicating the presence of a low-molecular-weight "tail" composed presumably of cyclic oligomers in TXN copolymers obtained under various experimental conditions [158].

7.4 References

1. "Cyclic Monomers", Frisch, K. C. Ed., Wiley-Interscience, 1972, p. 144
2. Meerwein, H. in Houben-Weyl, Methoden der Organischen Chemie, Thieme, Stuttgart, 1965, Vol. 6/3 p. 204
3. "Analytical Chemistry of Polymers" Kline, G. M. Ed., Interscience, New York, 1959, Part I, pp. 17 and 295

4. "The Carbohydrates", Pigman, W., Horton, D. Eds. Academic Press, New York and London, 1972, Vol. I A, p. 391
5. Jedliński, Z. et al.: Macromolecules *9*, 622 (1976)
6. Krummenacher, B., Elias, H. G., Makromol. Chem. *150*, 271 (1971)
7. Plesch, P. H., Westermann, P. H.: Polymer, *10*, 105 (1965)
8. Laurent, P. A., Tarte, P., Rodrigues, B.: Bull. Soc. Chim. France *1959*, 946
9. Schulz, R. C. et al.: ACS Symposium Series, *59*, 77 (1977)
10. Okada, M., Kozawa, S., Yamashita, Y., Makromol. Chem. *127*, 66 (1967)
11. Hellermann, W., Schulz, R. C., Makromol. Chem. Rapid Commun. *2*, 585 (1981)
12. Okada, M., Mita, K., Sumitomo, H.: Makromol. Chem. *176*, 859 (1975)
13. Firat, Y., Plesch, P. H.: Makromol. Chem. *176*, 1179 (1175)
14. Okada, M., Mita, K., Sumitomo, H., Makromol. Chem. *177*, 895 (1976)
15. Firat, Y., Plesch, P. H.: J. Polym. Sci. Pol. Lett., *13*, 135 (1975)
16. Jedliński, Z., Solich, M., J. Polym. Sci., *A6*, 3182 (1968)
17. Sumitomo, H., Okada, M., Ito, H.: J. Polym. Sci. A, *9*, 3115 (1971)
18. Okada, M., Mita, K., Sumitomo, H., Makromol. Chem. *177*, 2055 (1976)
19. Okada, M., Yagi, K., Sumitomo, H.: Makromol. Chem. *163*, 225 (1973)
20. Albrecht, K. et al.: Makromol. Chem. *178*, 3191 (1977)
21. Rentsch, Ch., Schulz, R. C.: Makromol. Chem. *178*, 2535 (1977)
22. Kawakami, Y., Yamashita, Y.: Macromolecules, *10*, 837 (1977)
23. Andrews, J. M., Semlyen, J. A.: Polymer *13*, 141 (1972)
24. Rozenberg, B. A., Irzak, W. I., Enikolopyan, N. S.: "Interchain Exchange Reactions in Polymers" (in Russian), Chimia, Moscow, 1975 p. 34
25. Matyjaszewski, K., et al.: Makromol. Chem. *181*, 1469 (1980)
26. Hill, J. W., Carrothers, W. H.: J. Amer. Chem. Soc. *57*, 925 (1935)
27. Jones, F. R., Plesch, P. H.: Chem. Commun. *1969*, 1230, 1231
28. Kubisa, P., Penczek, S.: Makromol. Chem. *144*, 169 (1971)
29. Penczek, S., Kubisa, P.: Makromol. Chem. *165*, 121 (1973)
30. Chwiałkowska, W., Kubisa, P., Penczek, S.: Makromol. Chem. *183*, 753 (1982)
31. Kubisa, P., Penczek, S.: Makromol. Chem., *179*, 445 (1978)
32. Kubisa, P., Penczek, S.: Macromolecules, *10*, 1216 (1977)
33. Słomkowski, S., Penczek, S.: Chem. Commun. *1970*, 1347
34. Kabir-ud-Din, Plesch, P. H.: J. Chem. Soc. Perkin II, *1978*, 937
35. Szymański, R. et al.: Chem. Commun. *1976*, 33
36. Stolarczyk, A., Kubisa, P., Penczek, S.: J. Macromol. Sci. Chem. *11*, 2047 (1977)
37. Szymański, R., in preparation
38. Leader, G. R.: Anal. Chem. *42*, 16 (1970)
39. Walker, J. F.: "Formaldehyde", Reinhold Publ. Corp., New York 1964
40. Staudinger, H.: "Die Hochmolekularen Organischen Verbindungen" Springer-Verlag, Berlin, 1922
41. MacDonald, R. N. (to E. I. du Pont de Nemours Co.) US Pat. 2768994 (1956)
42. Walling, C. T., Brown, F., Bartz, K. W.: (to Celanese Corporation of America) US Pat. 3027352 (1962)
43. Polish Patents 57.935 and 57.963 (1969)
44. "Cyclic Monomers", Frisch, K. Ed., Wiley Interscience, New York, 1972, p. 132
45. Fejgin, J., Cieślak, J.: Polimery (Poland) *24*, 304 (1979)
46. Drösher, M., et al.: Makromol. Chem., *177*, 1695 (1976)
47. Bevington, J. C., May, H.: in "Encyclopedia of Polymer Science and Technology" Mark, H. F., Gaylord, N. G. Eds., Interscience, New York, London, Sydney, 1964, Vol. I. p. 609
48. Weissermel, K. et al., Angew. Chem. Internat. Ed. *6*, 526 (1967)
49. Booth, H. S., Martin, D. R.: "Boron Trifluororide and Its Derivatives" John Wiley, New York, 1949
50. Penczek, S. et al.: Makromol. Chem. *172*, 243 (1973)
51. Barker, S. J., Price, M. B.: "Polyacetals" Iliffe Books, London, 1970, p. 30
52. Penczek, S. et al.: Makromol. Chem. *116*, 203 (1968)
53. Jaacks, V., Adv. in Chem. Series, *91*, 371 (1969)
54. Jaacks, V., Kern, W.: Makromol. Chem. *62*, 1 (1963)

55. Kern, W.: Chem. Ztg., Chem. Apparatur *88*, 623 (1964)
56. Iguchi, M.: British Polym. J. *5*, 195 (1973)
57. Collins, G. L. et al.: J. Polym. Sci., Pol. Lett. *17*, 667 (1979)
58. Okamura, S., Higashimura, T., Miki, T.: Progress in Polymer Science, Japan, *3*, 97 (1972)
59. Burg, K., Schlaf, H., Cherdron, H.: Makromol. Chem. *145*, 247 (1971)
60. Perst, H.: "Oxonium Ions in Organic Chemistry", Verlag Chemie, Academic Press, 1971
61. Rozenberg, B. A. et al.: J. Polym. Sci. Pol. Chem. *11*, 1 (1973)
62. Kuzniecov, A. A. et al.: Dokl. Akad. Nauk SSSR, *197*, 1281 (1970)
63. Komarov, B. A. et al.: Vysokomol. Soed. *A17*, 691 (1975)
64. Staudinger, H., Signer, R.: Z. Kristall. *70*, 193 (1929)
65. Staudinger, H. et al.: J. Physik. Chem, *A126*, 425 (1927)
66. Leese, L., Bauber, M. W.: Polymer, *6*, 269 (1965)
67. Jaffe, M., Wunderlich, B.: Kolloid Z. Polymer *215–218*, 203 (1967)
68. Szymański, R., Penczek, S.: Makromol. Chem. *183*, 1587 (1982)
69. Arnett, E. M.: Progress Phys. Org. Chem. *1*, 1 (1963)
70. Hermann, H. D., Fischer, E., Weissermell, K.: Makromol. Chem. *90*, 1 (1966)
71. Grassie, N., Roche, R. S.: J. Polym. Sci. C *16*, 4207 (1968)
72. Brit. Pat. 941.647 (1963)
73. Mateva, R., Wegner, C., Lieser, G.: J. Polym. Sci., Pol. Lett. *11*, 369 (1973)
74. Lücke, A., Wegner, G., Polymer Prepr. *21*, 197 (1980)
75. Wegner, G. et al.: Makromol. Chem. *181*, 1763, (1980)
76. Iguchi, M., Murase, J., Watanabe, J., Watanabe, K.: Br. Pol. J. *6*, 61 (1974)
77. Wunderlich, B.: Adv. in Pol. Sci. *5*, 568 (1968)
78. Dröscher, S. et al.: Polymer *16*, 497 (1975)
79. Miki, T., Higashimura, T., Okamura, S.: J. Polym. Sci. *5*, 95 (1967)
80. Berlin, Al. Al., et al.: Dokl. Akad. Nauk SSSR, *184*, 1128 (1969)
81. Berlin, Al. Al., et al.: Vysokomol. Soed. *A12*, 443 (1970)
82. Bogdanova, K. A., et al.: Dokl. Akad. Nauk SSSR *197*, 618 (1971)
83. Bogdanova, K. A., et al.: Vysokomol. Soed. *A14*, 1976 (1972)
84. Bogdanova, K. A., et al.: Dokl. Akad. Nauk SSSR, *211*, 874 (1973)
85. Berlin, Al. Al. et al.: Vysokomol. Soed. *A17*, 643 (1975)
86. Bogdanova, K. A., et al.: Vysokomol. Soed., *A17*, 658 (1975)
87. Oreschnikova, T. F., Gruznov, A. G., Romanov, L. M.: Vysokomol. Soed. *A17*, 1927 (1975)
88. Bulai, A. H., et al.: Vysokomol. Soed. *A18*, 1372 (1976)
89. Oreschnikova, T. F. et al.: Vysokomol. Soed. *A19*, 24 (1977)
90. Berlin, Al. Al. et al.: Vysokomol. Soed. *A15*, 555 (1973)
91. Wilski, K.: Markomol. Chem.: *150*, 209 (1971)
92. Kern, W., Jaacks, V.: J. Polym. Sci. *48*, 399 (1960)
93. Miki, T., Higashimura, T., Okamura, S.: J. Polym. Sci. *A5*, 2997 (1967)
94. Ivanova, L. J. et al.: Vysokomol. Soed. *A17*, 1229 (1975)
95. Dimonie, M., Dragan, G., Negulescu, I. I.: 28th IUPAC Macromołecular Symposium, Amherst, 1982, p. 156
96. Berlin, Al. Al. et al.: Dokl. Akad. Nauk SSSR *184*, 1128 (1969)
97. Berlin, Al. Al. et al.: Vysokomol. Soed. *A15*, 554 (1973)
98. Vorobieva, G. A., et al.: Dokl. Akad. Nauk SSSR *214*, 273 (1974)
99. Kučera, M., Spousta, E.: Makromol. Chem. *82*, 213 (1965)
100. Baader, H., Jaacks, V., Kern, W., Makromol. Chem. *82*, 213 (1965)
101. Jaacks, V.: Makromol. Chem. *99*, 300 (1966)
102. Penczek, S.: Makromol. Chem. *175*, 1217 (1974)
103. Kern, W. et al.: Pure & Appl. Chem. *12*, 372 (1966)
104. Jaacks, V. et al., Makromol. Chem. *115*, 290 (1968)
105. Meerwein, H. et al.: Liebiegs Ann. Chem. *635*, 1 (1960)
106. Weissermel, K. et al.: Kunstoffe *54*, 410 (1064)
107. Leece, L., Baumber, M.: Polymer *5*, 380 (1964)
108. Morelli, F., Tartarelli, R., Masetti, G., Eur. Pol. J. *4*, 555 (1968)
109. Onyon, P., Taylor, K.: Eur. Pol. J. *1*, 133 (1965)
110. Smirnov, Ju. N. et al.: Vysokomol. Soed. *16*, 288 (1974)

111. Kohlschütter: Annalen *482*, 75 (1930)
112. Mesrobian, R. B.: J. Chem. Phys. *22*, 565 (1954)
113. Okamura, S., Hayashi, K., Nokamura, N.: Isotopes and Radiation, *3*, 416 (1960)
114. Chapiro, A.: in Encyclopedia of Polymer Science and Technology, Mark, H., Gaylord, N. S. Eds., Vol. 11 p. 703
115. Carazzolo, G., Leghissa, S., Mammi, M.: Makromol. Chem. *60*, 171 (1963)
116. Wegner, G., Fischer, E. W., Munoz-Escalona, A.: Makromol. Chem. *Suppl. 1*, 521 (1975)
117. Wunderlich, B.: "Macromolecular Physics", Academic Press, New York, 1973, Vol. I
118. Komaki, A., Matsumoto, T.: J. Polymer Sci. *B* , 671 (1963)
119. Sagu, M. L. et al.: Angew. Makromol. Chem. *62*, 45 (1977)
120. Colson, J. P., Reneker, D. H.: J. Appl. Phys. *44*, 4293 (1973)
121. Eastmond, G. C.: in "Progress in Polymer Science", Jenkins, A. D., Ed., Pergamon Press, Oxford, 1970, Vol. II, p. 3
122. Chatani, Y.: in "Progress in Polymer Science Japan" Otsu, t., Takayanagi, M., Eds., A Halsted Press Book, Tokyo, John Wiley, New York, 1974, Vol. VII, p. 149
123. Okamura, S., Hayashi, K.: Makromol. Chem. *47*, 230 (1961)
124. Bassett, D.: Nature *215*, 731 (1967)
125. Renecker, D. H.: J. Appl. Phys. *41*, 4296 (1970)
126. Voigt-Martin, J., Makromol. Chem. *175*, 2669 (1974)
127. Tupikov, V. I., Pshezhetskii, S. Y.: Zh. Fiz. Khimii *38*, 2430 (1964)
128. Marans, N. S., Wessels, F. A.: J. Appl. Polym. Sci. *19*, 3681 (1965)
129. Kiss, G. et al.: J. Appl. Polym. Sci. *26*, 2485 (1981)
130. Jamison, S. E., Noether, H. D., J. Polym. Sci., *B1*, 51 (1963)
131. Okamura, S., Kobayashi, E., Higashimura, T.: Makromol. Chem. *95*, 20 (1966)
132. Okamura, S., Kobayashi, E., Higashimura, T.: Makromol. Chem. *88*, 1 (1965)
133. Rokova, G. V., Romanov, L. M., Enikolopyan, N. S.: Vysokomol. Soed. *6*, 2184 (1964)
134. Fleischer, D., Schulz, R. C.: Makromol. Chem. *176*, 677 (1975)
135. Opitz, G.: Plaste und Kautschuk, *22*, 951 (1975)
136. Kučera, M., Pichler, J.: Collect. Czech. Chem. Commun. *30*, 991 (1965)
137. Kučera, M., Pichler, J.: Polymer *5*, 371 (1964)
138. Fejgin, J., Tomaszewicz, M., Cieslak, J.: Polimery (Poland) *21*, 298 (1976)
139. Inoue, M.: J. Appl. Polym. Sci. *8*, 2225 (1964)
140. USP 3 293 221 (1966)
141. USP 3 686 142 (1968)
142. Wissbrun, K. F.: Makromol. Chem. *118*, 211 (1968)
143. Hermann, H. D., Burg, K. H.: "Angew. Makromol. Chem. *15*, 219 (1971)
144. DeMejo, L., MacKnight, W. J., Vogl, O.: Polymer *19*, 956 (1978)
145. DeMejo, L., MacKnight, W. J., Vogl. O.: Polymer J. *11*, 15 (1979)
146. Fejgin, J., Penczek, S., Tomaszewicz, M.: Polimery (Poland) *21*, 11 (1976)
147. Rösinger, S., Hermann, H., Weissermel, K.: J. Polym. Sci. *5*, 183 (1967)
148. Iwabuchi, S., Jaacks, V., Kern, W.: Makromol. Chem. *177*, 1929 (1976)
149. Höhr, L., Cherdron, H., Kern, W.: Makromol. Chem. *52*, 59 (1962)
150. Higashimura, T. et al.: J. Polym. Sci. *5*, 1927 (1967)
151. Minoura, Y., Mitoh, M., Mabuchi, Y.: Makromol. Chem. *119*, 86 (1968)
152. Okada, M., Yamashita, Y., Ischii, Y.: Makromol. Chem. *94*, 181 (1966)
153. Yamashita, Y., Okada, M., Hirota, M.: Makromol. Chem. *122*, 284 (1964)
154. Kawakami, Y., Yamashita, Y.: Polymer J. *9*, 227 (1977)
155. Kawakami, Y., Suzuki, J., Yamashita, Y.: Polymer J. *9*, 519 (1977)
156. Okada, M. et al.: Polymer J. *1*, 1 (1970)
157. Ischigaki, I. et al.: J. Appl. Polym. Sci. *18*, 1927 (1974)
158. Lücke, A.: Ph. D. Thesis, Freiburg, 1979, p. 137

8 Bicyclic Monomers

8.1 Introduction

During the last two decades, there has been a growing interest in the polymerization of bicyclic monomers containing one or more oxygen atoms, i.e. monomers belonging to one of the following groups:

Bicyclic ethers	Bicyclic acetals	Bicyclic orthoesters	Spiro-orthoesters	Spiroortho-carbonates

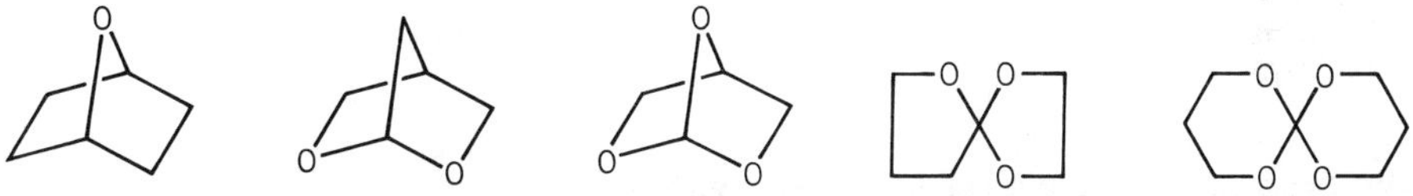

7-oxabicyclo |2.2.1|heptane	2,6-dioxabicyclo |2.2.1| heptane	2,6,7-trioxabicyclo |2.2.1| heptane	1,4,6-trioxaspiro |4.4|nonane	1,5,7,11-tetraoxaspiro |5.5| undecane

(8-1)

Another group of monomers, which also belongs to this category, are anhydro sugars, i.e. bicyclic compounds formed by intermolecular condensation of two hydroxyl functions of sugars. Thus, for example 1,6-anhydro glucopyranose has a of 6,8-dioxabicyclo|3.2.1|octane skeleton.

1,6-anhydro-β-D-glucopyranose   6,8-dioxabicyclo |3.2.1| octane

Interest in the polymerization of bicyclic compounds stems from possibilities offered in synthetic polymer chemistry by this group of monomers. Thus, polymerization of anhydro sugars, i.e. substituted bicyclic acetals, leads to synthetic biopolymers-polysaccharides or their analogs. Polymerization of certain bicyclic monomers provides systems expanding on conversion from monomer into polymer.

Various aspects of polymerization of bicyclic monomers have been recently reviewed. Sumitomo and Okada summarized research on the polymerization of bicyclic acetals, oxalactones, oxolactams and related heterobicyclic compounds [1]. Yokoyama and Hall gave an authoritative review on the ring-opening polymerization of atom-bridged and bond-bridged bicyclic ethers, acetals and orthoesters [2]. Bailey reviewed polymerizations accompanied by volume expansion [3,4]. Progress in the polymerization of anhydro sugars was summarized by Schuerch [5,6].

In this chapter we will cover two subjects in detail. First, the cationic polymerization of anhydro sugars and their models. Natural polysaccharides are used at a large scale mainly in the textile and food industry and in medicine. Although synthetic polysaccharides cannot compete with inexpensive natural products (due to the complex monomer synthesis), they provide useful models for studying the relationships between structure and properties, which is especially important in medical applications. This is because synthetic methods give polymers having desired uniform, regular structures, mimicking the structures of polysaccharides isolated from natural sources.

Another topic is the polymerization with expansion in volume. Shrinkage, which accompanies the polymerization of typical monomers, creates serious problems in many applications, including precision castings, high-strength adhesives or dental fillings. The observation that some spiro- and bicyclic orthoesters undergo polymerization with almost zero shrinkage or slight expansion in volume, may be of practical importance.

8.2 Polysaccharides and Their Models

8.2.1 Polymerization of Anhydro Sugars and Related Compounds

Polysaccharides, high-molecular-weight polymers built up of sugar units, have found numerous commercial applications. Such materials as cellulose, chitin, dextran, glycogen, starch or a variety of so-called gums, are widely used as food additives, paper additives, plastic textile finishes, adhesives or coatings. Polysaccharides are also used in medicine as emulsifiers, for coating, tableting and viscosity control aids, as polysaccharide drug complexes and for a number of specific therapeutic uses [6].

On a commercial scale, only polysaccharides from natural sources such as land plants and seaweeds are used. Medical applications, however, stimulate research on synthetic polysaccharides, which could be used as well defined models for studying the biological functions of different polysaccharides.

Polysaccharides arise from monomeric sugars by elimination of water, i.e., either by polycondensation of sugars or ring-opening polymerization of anhydro sugars:

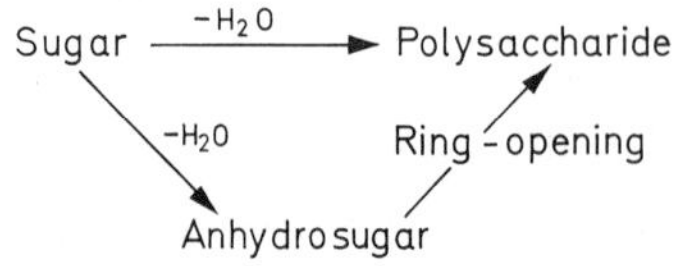

(8-3)

Mechanistic and stereoelectronic aspects of anhydro sugar formation are discussed in Ref. [7].

8.2.2 Structures of Natural Polysaccharides

Polymerization of anhydro sugars is aimed at the synthesis of models of naturally occurring polysaccharides and the preparation of polysaccharides resembling natural products. We briefly summarize the chemical structures of major natural polysaccharides[1].

From the multitude of stereoisomeric monosaccharides only a few are found in natural polysaccharides. Among them are hexoses[2]:

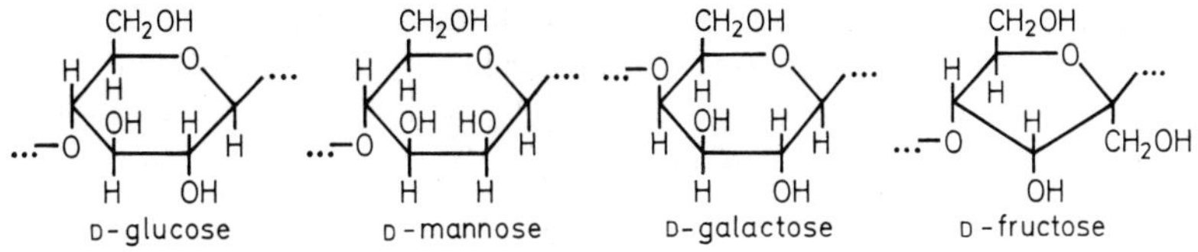

(8-5)

and pentoses:

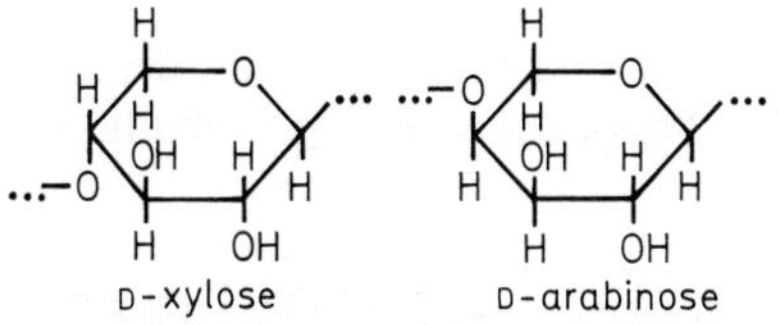

(8-6)

8.2.1.1 Anhydro Sugars

The anhydro sugars (or sugar anhydrides) are monosaccharide derivatives which contain, in addition to the normal 5-membered furanose ring or 6-membered pyranose ring, another oxygen-containing ring, formed through the elimination of water from any two hydroxyl groups. The newly formed ring may arise by linking adjacent hydroxyl groups (3-membered ring) or hydroxyl groups separated by one or more CHOH groups. This is illustrated by a scheme showing the structures of anhydro sugars which may be obtained from D-galactopyranose:

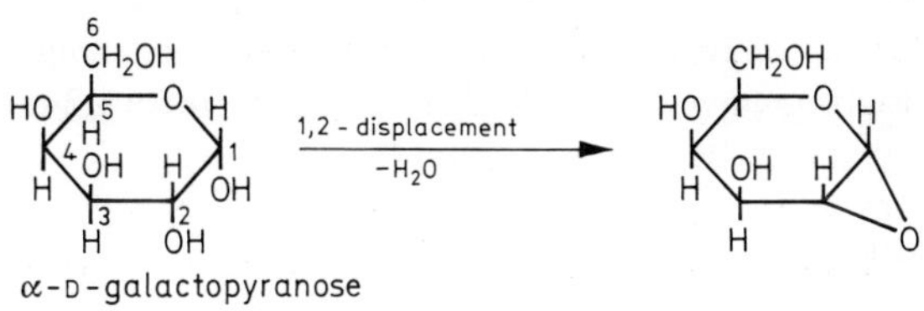

1 Throughout this chapter, Haworth formulas are generally used. In some cases, "conformational formulas" are used to indicate particular conformations

2 Hexose-monosaccharide containing six carbon atoms
Pentose-monosaccharide containing five carbon atoms
Pyranose-monosaccharide containing five or six carbon atoms in form of the 6-membered ring.
Furanose-monosaccharide containing five or six carbon atoms in form of the 5-membered ring

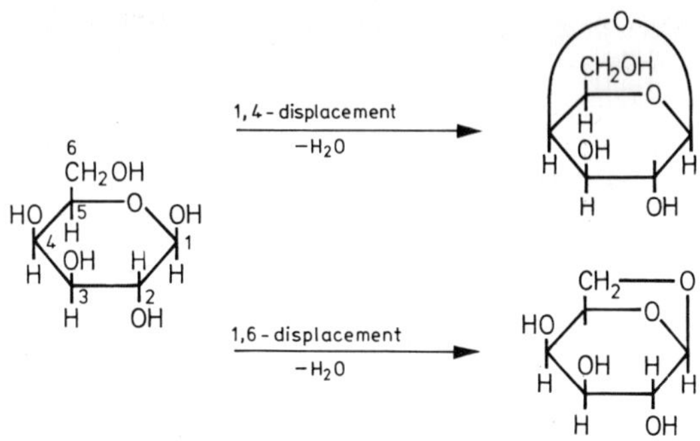

β-D-galactopyranose (8-4)

In most natural polysaccharides, monosaccharide units are linked by α- or β-(1,4)-linkages, and the linkage (α- or β-) is repeated uniformly through the chain. For example, in cellulose the monosaccharide units are linked by β-(1,4)linkages:

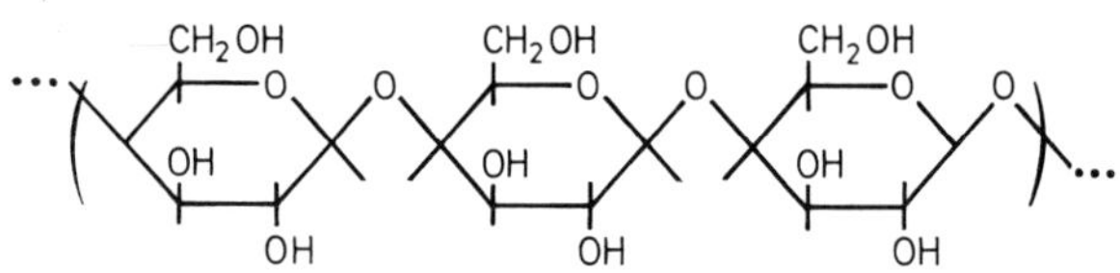

(8-7)

1,6 linkages formed by ring-opening polymerization of 1,6-anhydro sugars are found in natural polysaccharides such as polytran, gum arabic, gum ghatti and guar gum.

8.2.3 Polysaccharides by Polymerization of Anhydro Sugars

Cationic ring-opening polymerization of anhydro sugars is possible because of their relatively high basicity and because their steric and/or conformational strain is reduced by opening of the anhydride ring. As shown in Scheme (8-4) anhydro sugars are composed of two fused rings of different size, depending on which hydroxyl groups are involved in the formation of the anhydride ring.

As shown in this section, early work usually gave only low-molecular-weight oligomers, whereas more recently high polymers were obtained.

8.2.3.1 Anhydro Sugars Containing 3-Membered Rings

1,2-displacement leads to the formation of 3-membered oxirane rings. Only a few anhydro sugars in this group have been polymerized, among them 1,2-anhydro-3,4,6-tri-O-acetyl-D-glycopyranose (Brigl's anhydride) [8].

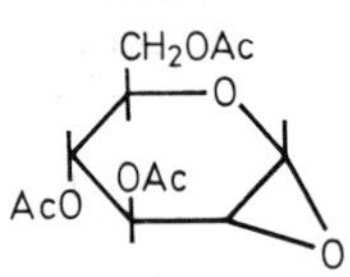

(8-8)

Polymerization of (8-8) with Lewis acid initiators involves opening of the oxirane ring and leads to low-molecular-weight polymers; in addition crosslinked gel is formed due to the reaction of cationic species with the ester functions. For example, the polymerization of (8-8) in CH_2Cl_2 soln. (1 g monomer in 3 ml CH_2Cl_2) in the presence of PF_5 or BF_3 at −78 °C gives in 100 h 20–90% oligomers with $\overline{DP}_n = 2$ to 4. Conversion increases with increasing initiator concentration. At the same time the proportion of the insoluble fraction decreases; with 20 mol% of PF_5, 92% soluble polymer was reported.

Anhydro sugars containing an oxirane ring as a side-chain to the sugar moiety have also been polymerized by the cationic mechanism [9]:

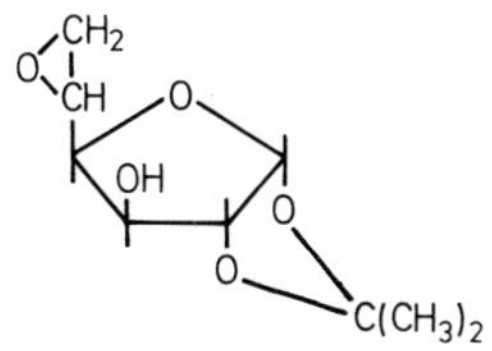

5,6-anhydro-1,2-O-isopropylidene-α-D-glucofuranose

(8-9)

Polymerization of (8-9) proceeds by opening of the oxirane ring and leads, to a structure not occuring in nature. In this structure the sugar moiety is not a part of the main chain, but is a pendant group attached to the polyoxyethylene backbone.

8.2.3.2 Anhydro Sugars Containing 4-Membered Rings

Displacement involving hydroxyl groups separated by three carbon atoms leads to the oxetane ring. Recently the anhydro sugar belonging to this class was polymerized by a cationic mechanism [10].

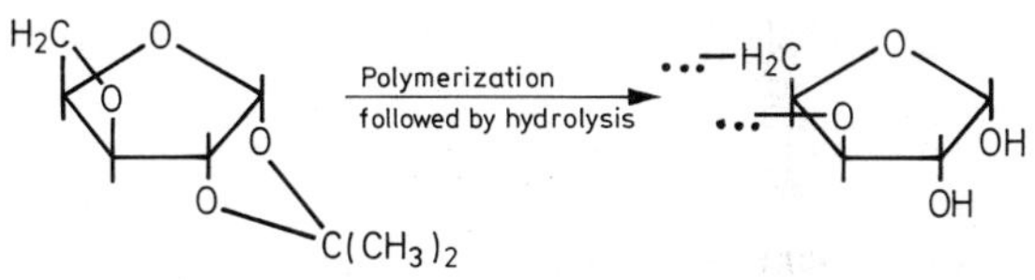

(8-10)

3,5-Anhydro-1,2-O-isopropylidene-α-D-xylofuranose was polymerized with PF_5 or $BF_3 \cdot OEt_2$ initiators between 0 and −78 °C and 14–60% conversion to polymer with $\overline{M}_n = 7000$–11000 ($\overline{DP}_n = 40$–65) was reported. The protecting group can be removed by acid hydrolysis. According to ^1H- and ^{13}C-NMR spectra, the polymer has a regular structure, which indicates that the polymerization proceeds exclusively by breaking the C-5-O bond (the other possibility would be the breaking of the C-3-O bond which should lead to inversion on C-3). In the ^{13}C-NMR spectrum only five signals were observed indicating a simple head-to-tail structure.

Polymerization of anhydro sugars containing 3- or 4-membered rings proceeds exclusively through opening of the oxirane or oxetane ring due to the relatively large ring strains.

8.2.3.3 Anhydro Sugars Containing 5-Membered Rings

1,4-displacement and 1,6-displacement leads to 5-membered rings:

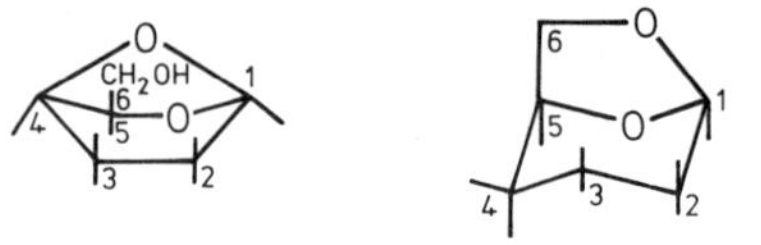

(8-11)

Several 1,4-anhydro sugars have been polymerized cationically, for example:

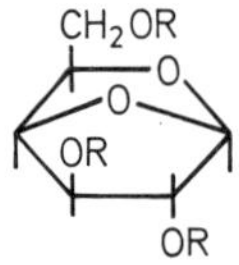

(8-12)

1,4-anhydro-2,3,6-tri-O-methyl-β-D-galactopyranose [11)]

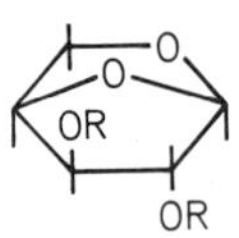

(8-13)

1,4-anhydro-2,3-di-O-methyl-α-L-arabinopyranose [11)]

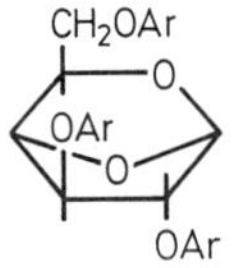

(8-14)

1,4-anhydro-2,3,6-tri-O-benzyl-α-D-glucopyranose [12,13)]

1,4-Anhydro sugars can be polymerized to relatively high-molecular-weight polysaccharides.

Polymerizations carried out in CH_2Cl_2 soln. (0.5 g monomer in 5 ml solvent) with 2.5 mol % PF_5 initiator at -78 °C gave after 100 h 85 % of polymers of galactose (8-12) and arabinose (8-13) with $\overline{DP}_n = 90$. Higher $\overline{DP}_n$ values (up to 250, $\overline{M}_n = 40000$) were reported for the polymerization of the glucose derivative (8-14) under essentially the same conditions [12,13)].

The polymerization of 1,6-anhydro sugars have been studied most thoroughly. Many 1,6-anhydro glucopyranoses derivatives polymerize easily in the presence of Lewis acids; however, some 1,6-anhydro glucofuranoses may adopt strainless conformations which reduces their tendency for polymerization. Thus, monomer (8-15) polymerizes very reluctantly [14)] (Ar = benzyl):

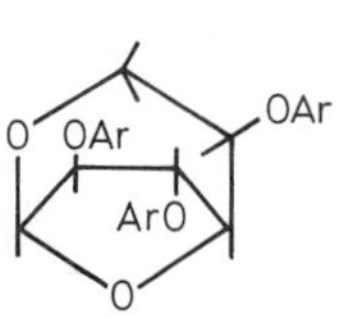

(8-15)

1,6-Anhydro-2,3,5-tri-O-benzyl-α-D-galactofuranose polymerized in CH_2Cl_2 solvent in the presence of 13% mol PF_5 at —78 °C gives only a trace of polymer after 48 h while at 0 °C with 22.7%mol PF_5 after 125 h polymer with $\overline{M}_n = 1130$ was obtained in 57% yield. $BF_3 \cdot OEt_2$ failed to initiate the polymerization at 0 °C.

1,6-Anhydro glucopyranoses usually cannot adopt strainless conformations and can be polymerized with reasonable rates to higher-molecular-weight polysaccharides. Thus, the polymerization of:

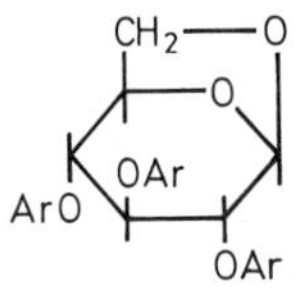

(8-16)

1,6-anhydro-2,3,4-tri-O-benzyl-β-D-glucopyranose [15] in CH_2Cl_2 solution (33%vol.) with 1%mol PF_5 initiator at —60 °C gives in 2 h 92% polymer with $\overline{DP}_n = 900$.

Various other 1,6-anhydro sugars have been polymerized to relatively high-molecular-weight polysaccharides. All of them contain fused 1,3-dioxolane (O,1,O,5,6), oxane (tetrahydropyran (O,1,2,3,4,5) and oxepane (O,1,2,3,4,5,6) rings and differ only by the configuration on carbons 2,3 and 4:

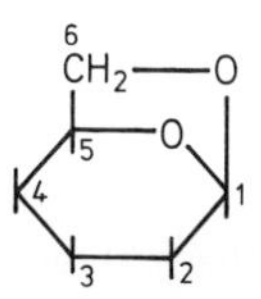

(8-17)

Basic characteristics of 1,6-anhydro sugar polymerization are given in Table 8.1.

Table 8.1. Polymerization of 1,6-anhydro sugars

Monomer	Initiator	Yield %	$\overline{DP}_n$	$\overline{DP}_n$ of free polysaccharide	Ref.
1,6-anhydroglucose methyl ether	$ClSO_3H$, BF_3, PF_5	90	200		16,17,18,19)
1,6-anhydroglucose benzyl ether	PF_5	95	930	260	20,21,22,23)
1,6-anhydromannose benzyl ether	PF_5	90	2200	350	25,26)
1,6-anhydrogalactose benzyl ether	PF_5	80	400	150	27,28,29)
1,6-anhydromaltose benzyl ether	PF_5	75	30		30)
1,6-anhydrocellobiose benzyl ether	PF_5	70	14		31)

Polymerization of the parent monomer of this class: 6,8-dioxabicyclo|3,2,1|octane is treated in Section 8.3.1.1. [3]

8.2.4 Stereochemistry of Anhydro Sugar Polymerization

Synthetic polysaccharides are prepared mainly to obtain models which may be helpful in establishing the biological functions of natural polysaccharides. It is desirable to find systems in which the polymerization of anhydro sugars proceeds stereospecifically, i.e. to prepare uniform and well defined structures.

In the polymerization of anhydro sugars (substituted bicyclic acetals) two possible sources of inhomogenities exist. The anhydro sugar molecule comprises two fused rings and if the difference in the strain energies is insufficient, both rings may open under polymerization conditions. For example, polymers prepared from 1,4-anhydro glucopyranoses (i.e. compounds having a 6,8-dioxabicyclo|2.2.1|heptane skeleton) may contain both 6-membered pyranose and 5-membered furanose rings in the main chain.

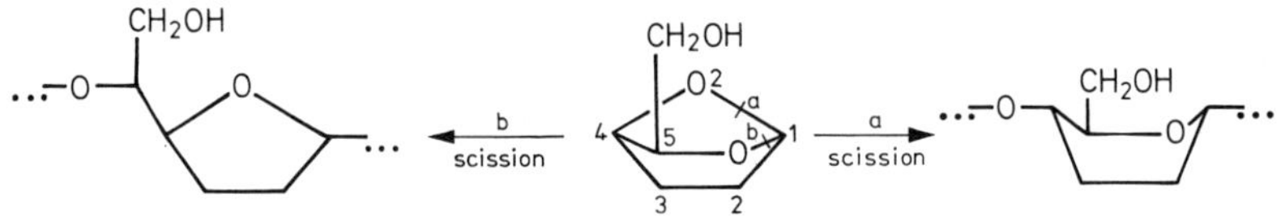

(8-18)

This type of lack of specificity is thus related to ring strain.

The other source of inhomogenity, related to the mechanism of polymerization, is the possibility of inversion on C-1 during monomer incorporation into the growing chain:

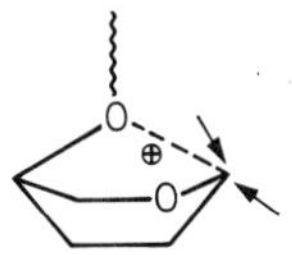

(8-19)

If this reaction proceeds exclusively with inversion the resulting polymer would contain α- or β- linkages. If, however, racemization occurs, the polymer would contain both α- and β-linkages.[4]

NMR analysis and identification of hydrolysis products, have shown that the polymerization of 1,2-anhydro sugars proceeds exclusively by opening of the oxirane ring and that the polymer contains exclusively pyranose rings [6].

---

3 According to the nomenclature used in carbohydrate chemistry only carbon atoms are numbered [as in (8–17)]. If (8–17) is treated as a bicyclic acetal, all atoms are numbered, thus number 6 is assigned to the oxygen atom in the 6-membered ring, number 7 to the carbon atom denoted as 6 in (8–17) and number 8 is assigned to the second oxygen atom

4 Troughout this chapter Haworth convention is mainly used and we follow the rule given in Ref. [7] according to which in Haworth formulas the CH_2OH group is directed upward for the D-series and downward for the L-series, thus in the D-series the OH (or -OR) group at C-a is directed downward for the α-anomer and upward for the β-anomer. The reverse is true for the L-series

Polymers obtained from 1,2-anhydro sugars contain, however, varying proportions of α- and β-linkages whose content is determined by the specific rotation of the polymer (provided that the specific rotation of the corresponding glycosides, being models of α- and β-linkages, are known).

The observed racemization on C-1 indicates that the cationic center in the polymerization of 1,2-anhydro sugars has carbenium- rather than oxonium-ion character, and that the attack of the incoming monomer is possible from either side of the carbenium-ion plane, as shown in the scheme below:

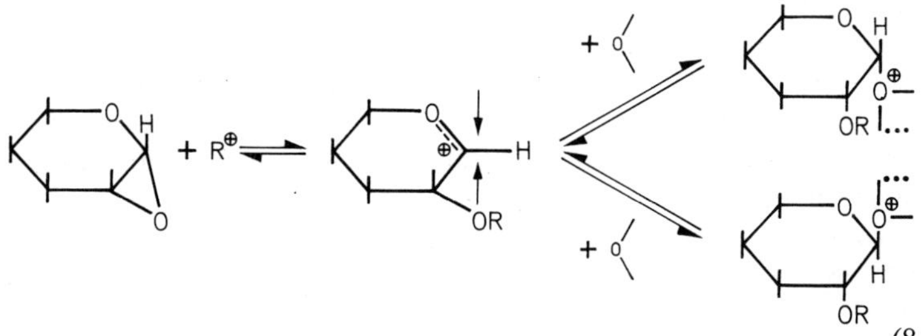

(8-20)

The unimolecular opening of the C-1-O-bond is facilitated by the stabilization of the positive charge on C-1 by the oxygen atom in the ring α-position to the cationic center. A more thorough discussion of the role of carbenium and oxonium ions in cationic polymerization is given in Vol. 1, Adv. Polymer Sci. **37**, 1980 Section 4.1.4.

Thus, the polymerization of 1,2-anhydro sugars gives polymers, which, although composed of the same repeating units, are not regular because these repeating units are linked by both α- and β-linkages [6].

1,4- and 1,6-anhydro sugars are composed of fused 5- and 6-membered rings:

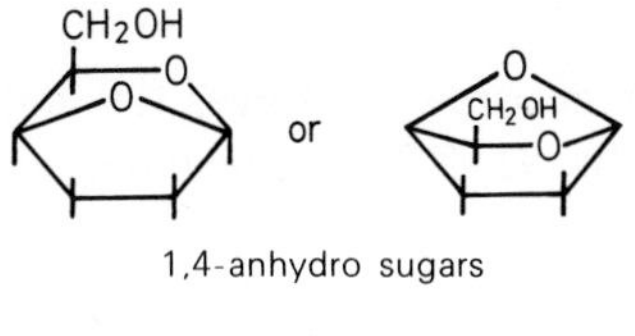

(8-21)

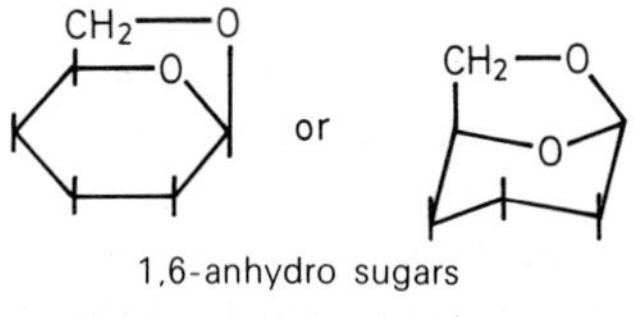

(8-22)

Polysaccharides prepared from 1,4-anhydro sugars contain both pyranose (6-membered) and furanose (5-membered) rings in the main chain [12]. The presence of both structures was established by hydrolysis which showed that either of the chemically equivalent, C-1-O-bonds may be broken during polymerization:

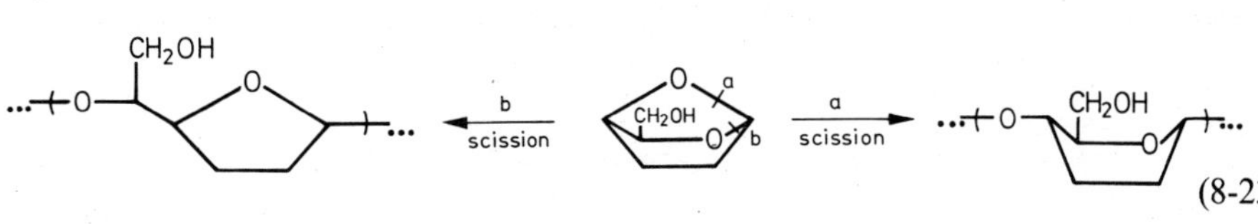

(8-23)

The proportion of furanose and pyranose rings may be to some extent altered by changing the reaction conditions but complete specificity has not yet been achieved. It has been reported, however, that for particular monomers, e.g. for:

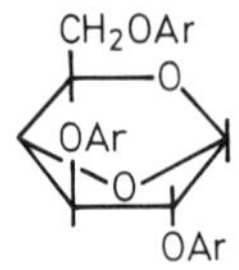

(8-24)

1,4-anhydro-2,3,6-tri-O-benzyl-α-D-glucopyranose, conditions may be found under which the corresponding polysaccharides contain only a few or no furanose rings [13].

Also with respect to the α- and β-linkage content, the polymerization of 1,4-anhydro sugars is not stereospecific, with the exception of (8-24), for which a predominant formation of β-linkages was reported [13].

In contrast to 1,4-anhydrides, the polymerization of 1,6-anhydro sugars with fused 5- and 6-membered rings, gives polymers containing exclusively pyranose rings in the main chain.

Analysis of the strain energies in bicyclic acetals has recently been published by Hall [2] and in the subsequent discussion we will follow his reasoning.

In 1,4-anhydro sugars the 1 and 4 positions are linked by oxygen bridges, thus, as shown in Scheme (8-21), the pyranose ring may exist in only one, boat conformation. This may introduce considerable strain into the otherwise essentially strainless 6-membered pyranose ring. The enthalpy of boat conformation of cyclohexane is higher than that of the chair conformer by 27 kJ/mol (6.4 kcal/mol), while the strain energies of tetrahydropyran (6-membered ring) and 1,3-dioxolane (5-membered ring) differ by only 22 kJ/mol (5 kcal/mol) [2]. Thus, the strain energy of the 6-membered ring in an unfavourable conformation may be even higher than that of the 5-membered ring.

The ring strains of hydrocarbons having the same skeleton as 1,4- and 1,6-anhydro sugars namely: bicyclo-|2.2.1|heptane and bicyclo|3.2.1|octane differ considerably [2]:

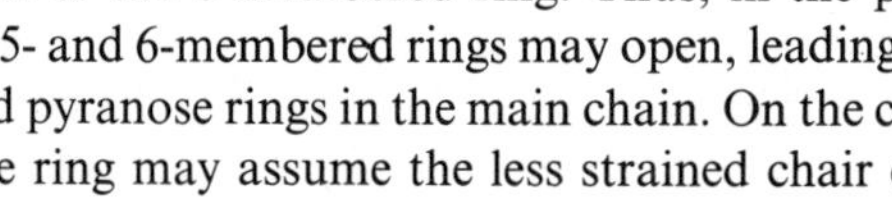

(8-25)

71.9 kJ/mol (17.2 kcal/mol) 51.8 kJ/mol (12.4 kcal/mol)

Apparently, the excess strain in bicyclo |2.2.1|heptane is due to the unfavourable conformation of the 6-membered ring. Thus, in the polymerization of 1,4-anhydro sugars both 5- and 6-membered rings may open, leading to polysaccharides containing furanose and pyranose rings in the main chain. On the contrary, in 1,6-anhydro sugars the pyranose ring may assume the less strained chair conformation, which prevents the opening of the 6-membered ring and favours the exclusive opening of the 5-membered ring:

(8-26)

Thus, polymerization of 1,6-anhydro glucopyranoses proceeds exclusively by the opening of the 5-membered ring. As discussed in Vol. 1, Adv. Polymer Sci. **37** (1980) Sect. 4.1.4, in the polymerization of cyclic acetals the predominant species are the tertiary oxonium ions and propagation proceeds by the S_N2 mechanism. Thus, in contrast to the polymerization of 1,2-anhydro sugars in which the contribution of carbenium-ions is significant and racemization on C-1 occurs, the polymerization of 1,6-anhydro sugars proceeds exclusively with inversion of configuration at C-1. For this reason α-1,6-anhydro monomers give β-1,6-linked polysaccharides and vice versa:

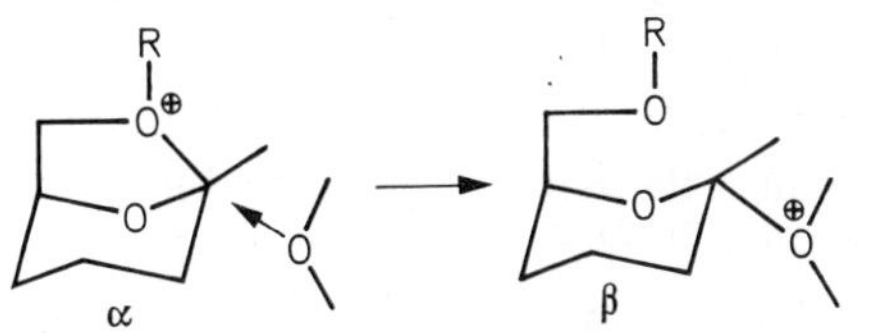

(8-27)

This may be illustrated by using the example of the polymerization of 1,6-anhydro-2,3,4-tri-O-benzyl-β-D-glucopyranose:

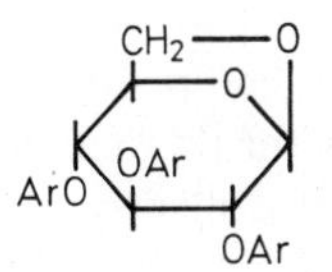

(8-28)

This anhydro sugar, polymerized at −60 °C in CH_2Cl_2 solvent with 0.5–1.0 mol% PF_5 initiator, gives high-molecular-weight polymer ($\overline{M}_n$ up to $7 \cdot 10^5$) with a >50% yield, which after debenzylation shows a specific rotation between +181° and +197° [29].

This polymer is identical with the corresponding naturally occurring polysaccharide (dextran), for which a highly specific α-structure has been established. Natural dextrans are known to have a $|\alpha|_D^{25}$ from +196° to +199° [29].

A comparison of the specific rotation values indicates that the polymer of (8-28) has an α-stereospecificity almost as high as natural dextran [29].

Thus, the polymerization of 1,6-anhydro sugars may be represented by the scheme:

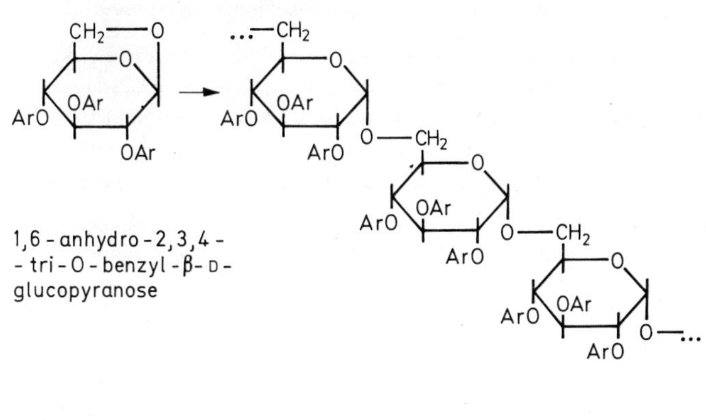

(8-29)

8.2.5 Mechanistic Aspects of Anhydro Sugar Polymerization Related to Polymer Structure and Synthetic Applications

Anhydro sugars contain free hydroxyl functions which have to be blocked prior to cationic polymerization. Otherwise, they would lower the degree of polymerization by participating in transfer reactions. Ester functions in the monomer may also interfere with the cationic centers. Thus free hydroxyl groups are usually transformed into ether groups. Benzyl ether blocking substituents are preferred, although other aliphatic or aromatic ether blocking groups may also be used [5,6].

The recovery of the deblocked polysaccharide may be most conveniently achieved by sodium in liquid ammonia. Usually, deblocking is accompanied by chain scission, but this is negligible under proper conditions. For example, debenzylation of a polymer with $\overline{DP}_n$ close to 900, which requires the removing of 2700 benzyl ether functions, causes a three- to four-fold decrease in $\overline{DP}_n$ [20,21]. Sterically pure linear polysaccharides with a number average degree of polymerization of 150–300 can be prepared by cationic polymerization of ether derivatives of 1,6-anhydro sugars followed by deblocking.

Research in this field was directed mainly towards synthetic aspects and information concerning the mechanism and kinetics of the polymerization are scarce.

Lewis acids were usually used as initiators, PF_5 being the most common choice. Recently, ^{31}P- and ^{19}F-NMR was used to study kinetic features of anhydro sugar polymerization initiated with PF_5 [29,30]. PF_5 generated by the thermal decomposition of p-$ClC_6H_4N_2^+PF_6^-$ was introduced into a solution of anhydro sugar (8-30) in CH_2Cl_2,

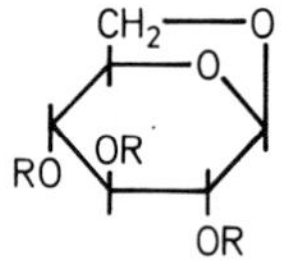

(8-30)

1,6-anhydro-2,3,4-tri-O-benzyl-β-D-glucopyranose

and ^{31}P- and ^{19}F-NMR spectra were recorded after different reaction times. PF_6^-, PF_4O- and POF_3 were identified on the basis of characteristic chemical shifts and

Table 8.2. Chemical shifts and coupling constants determined by ^{31}P- and ^{19}F-NMR for species existing in the system: 1,6-anhydro-2,3,4-tri-O-benzyl-β-D-glucopyranose-PF_5 [29]

Species	^{31}P-NMR spectra		^{19}F-NMR spectra	
	Chem. shift ppm	Coupling const. Hz	Chem. shift ppm	Coupling const. Hz
PF_4O^-	81.1 (quintet)	886.3	−11.7 (doublet)	856
PF_6^-	146.8 (septet)	712.9	−5.7 (doublet)	696
POF_3	37.6 (quartet)	1071.8	9.2 (doublet)	1044
$PF_5 \cdot O\langle$ (monomer)	134.8 (sextet)	817.9	−13.9 (doublet)	808

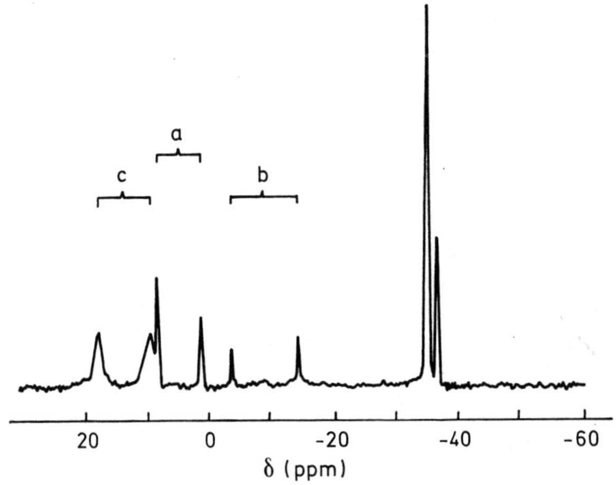

Fig. 8.1. ^{19}F-NMR spectrum of the reaction mixture of 1,6-anhydro-2,3,4-tri-O-benzyl-β-D-glucopyranose-PF_5 (10:1 molar ratio) at −80 °C. Accumulated from 44 to 89 min. after initiation [30].
a — $PF_6^{\ominus}$; b — POF_3; c — PF_5-monomer complex

coupling constants. The ^{19}F-NMR spectrum is shown in Fig. 8.1. At low temperatures (−80 °C) an additional signal was observed. This signal, which disappeared as the temperature increased, was attributed to the $PF_5 \cdot O =$ complex. Chemical shifts and coupling constants are given in Table 8.2.

The intensity of $PF_6^{\ominus}$ signals was constant troughout the polymerization, whereas the intensity of the signals of PF_4O- decreased with time. On the basis of these observations the following reaction scheme was proposed.

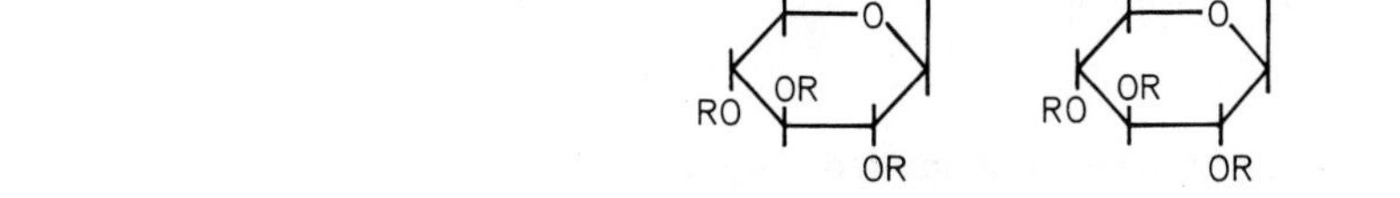

(8-31)

The intensity of the F_4PO-signal decreases with time, indicating the limited stability of this group. It apparently decomposes, possibly to F_3PO and FCH_2- (although only the former species was identified). The intensity of the $PF_6^{\ominus}$ signal does not change, which implies that the concentration of the active species is constant and the polymerization is a living one. This conclusion was confirmed by molecular weight measurements: the $\overline{DP}_n$ measured by GPC was nearly equal to the value calculated from the [monomer]/[$PF_6^{\ominus}$] ratio assuming that the concentration of the growing species is equal to the concentration of PF_6^- measured by ^{19}F-NMR spectroscopy.

Table 8.3. Calculated and measured $\overline{DP}_n$ of 2,3,4-tris-O-benzyl-α-D-glucopyran [29]

No.	$[Monomer]_0/[PF_5]_0$ [a]	Polymerization		$\overline{DP}_n$ [b] calcd.	$\overline{DP}_n$ meas.
		Temp. °C	time, hrs		
1	10	−60	1.7	40.2	49.1
2	10	−60	2.6	30.5	45.9
3	10	−60	3.1	36.2	36.2
4	20	−60	2.4	96.9	76.3
5	10	−40	20	37	43
6	20	−40	20 + 24	74	74

[a] Calculated from the concentration of $pClC_6H_4N^{\oplus}PF_6^{\ominus}$;
[b] Calculated as $[monomer]/[PF_6^{\ominus}]$, $[PF_6^{\ominus}]$ measured by ^{19}F-NMR

Results shown in Table 8.3 indicate that the concentration of the active species is indeed equal to the concentration of $PF_6^{\ominus}$.

Peculiarly there is no relationship between the concentration of the growing species and the amount of $p\text{-}ClC_6H_4N_2^{\oplus}PF_6^{\ominus}$ used. This is probably due to the irreproducibility of the thermal decomposition of the diazonium salt. In experiments 1-3 (cf. Table 8.3) the amount of liberated PF_5 was apparently different, although the amount of $pClC_6H_4N_2^+PF_6^-$ was kept constant. This observation weakens the authors [29] earlier statement concerning the effect of initiator concentration (expressed in terms of precursor concentration) on the polymerization kinetics and molecular weights.

$\overline{DP}_n$ does not increase gradually with conversion, but passes through a maximum and then reaches a constant value. The maximum on the $\overline{DP}_n$ curve occurs when the conversion is already high, and according to GPC measurements the decrease of $\overline{DP}_n$ is due to the formation of low-molecular-weight polymer at this stage [30].

According to the authors [30], the high viscosity of the mixture (polymers with $\overline{DP}_n$ up to 1500 are formed) prevents the formation of longer chains due to restricted monomer diffusion to the growing species. In our opinion it is more likely that, due to slow initiation, some initiating species still exist at this stage of reaction and the observed decrease of $\overline{DP}_n$ is due to a molecular weight redistribution of already existing macromolecules. Indeed, as it was shown by the same authors, addition of PF_5 to a solution of preformed polysaccharide caused a considerable decrease of $\overline{DP}_n$ (from 870 to 440 in 10 h) [29].

8.2.6 Copolymerization of Anhydro Sugars

The copolymerization of anhydro sugars was studied for the following monomers:

(8-32)

1,6-anhydro-2,3,4-tri-O-methylbenzyl-β-D-glucopyranose (TXGL)

1,6-anhydro-2,3,4-tri-O-benzyl-β-D-galactopyranose (TBGA)

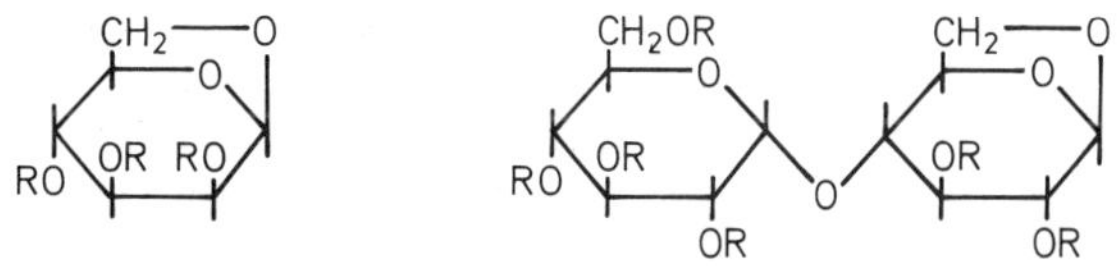

1,6-anhydro-2,3,4-tri-O-benzyl-β-D-mannopyranose (TBBM)

1,6-anhydro-2,3-di-O-benzyl-4-O-(2,3,4,6-tetra-O-benzyl-α-D-glucopyranosyl)-β-D-glucopyranose (MBMA)

(8-33)

Reactivity ratios of copolymerization of TXGL with TBGA, TBMM and MBMA have been determined [31]. Some of these results are summarized in Table 8.4.

Table 8.4. Reactivity ratios in copolymerization of various anhydro sugars [31]

Monomer		Reactivity ratios	
A	B	r_A	r_B
TXGL	TBGL	1.21	1.22
TXGL	TBBM	0.77	8.31[a]
		0.95	9.58
TXGL	TBGA	1.38	0.27[a]
		1.41	0.31
TXGL	MBMA	1.51	0.55[a]
		1.84	0.30

[a] From two sets of experiments
TBGL — tribenzyl analogue of TXGL

Comparison of the rate constants of homo- and cross-propagations led initially to the conclusion that the reactivity of anhydro sugars is governed by the energy required for the change of conformation of the entering monomer when it becomes incorporated into the growing center:

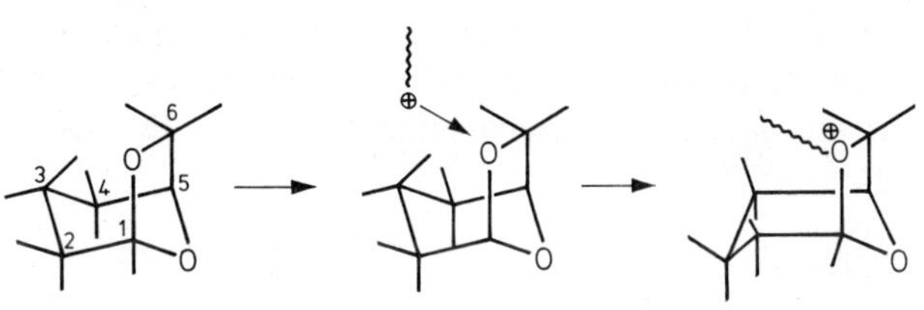

(8-34)

Such a change of conformation would convert the chair conformation, in which the substituent on C-3 provides a considerable steric hindrance for approaching the cationic center, to the boat-like conformation, allowing more space for the approach of both cation and counterion. On the basis of more detailed analysis [32], Schuerch concluded that reactivity differences cannot be due to conformational changes but are rather due to the inherent monomer reactivity.

8.3 Polysaccharide Analogs Prepared by Cationic Ring-opening Polymerization

8.3.1 Bicyclic Acetals

8.3.1.1 6,8-Dioxabicyclo|3.3.1|octane

Polymerizability of anhydro sugars is governed both by properties of the fused ring system and by spacial arrangement of substituents (blocked hydroxyl groups). Studies of model monomers containing anhydro sugar skeletons devoid of substituents allow the separation of these two factors. The most extensively studied monomer of this class, 6,8-dioxabicyclo|3.2.1|octane, has a skeleton of 1,6-anhydro glucopyranose, but is devoid of hydroxyl groups present in the sugar:

6,8-dioxabicyclo |3.2.1| octane (DBO) (heterocyclic nomenclature)

1,6-anhydro-2,3,4-trideoxy-β-D-glucopyranose (carbohydrate nomenclature)

1,6-anhydro-β-D-glucopyranose

(8-35)

DBO can be conveniently prepared from acrolein dimer [33]:

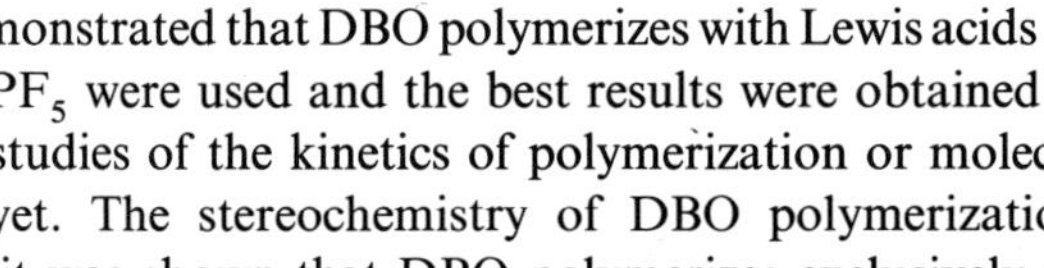

(8-36)

The cationic polymerization of DBO was studied by three independent groups [34, 35, 36].

It was demonstrated that DBO polymerizes with Lewis acids as initiators. $BF_3 \cdot OEt_2$, $SbCl_5$ and PF_5 were used and the best results were obtained with PF_5; however, no systematic studies of the kinetics of polymerization or molecular weights have been published yet. The stereochemistry of DBO polymerization was investigated in detail, and it was shown that DBO polymerizes exclusively through opening of the 5-membered ring:

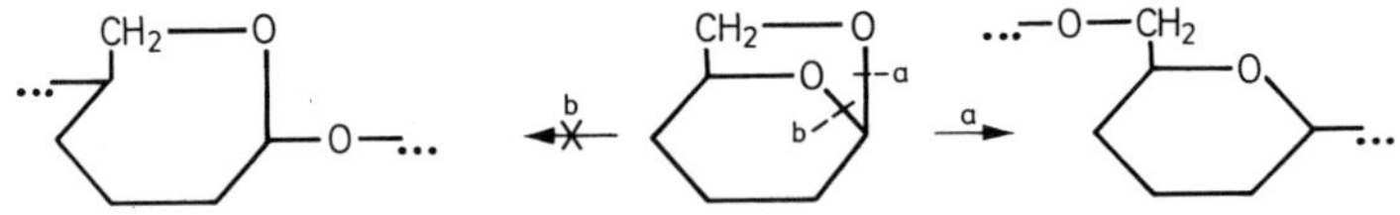

(8-37)

Polymerization of DBO may lead to α- and β-stereoisomers, due to the presence of the assymetric C-1 atom:

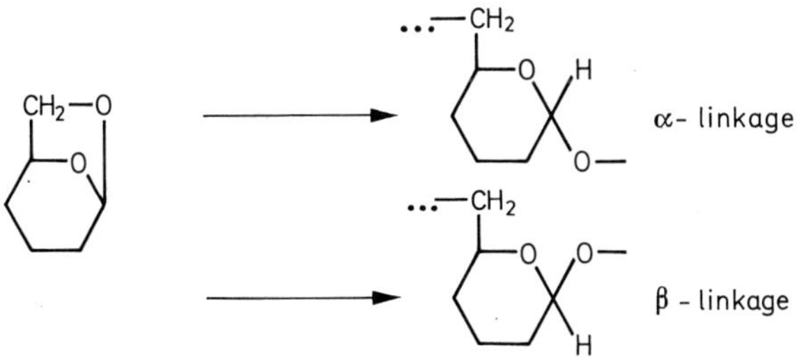

(8-38)

In the α- anomer the hydrogen atom on carbon C-1 occupies the equatorial position, while in the β-anomer it is located in the axial position:

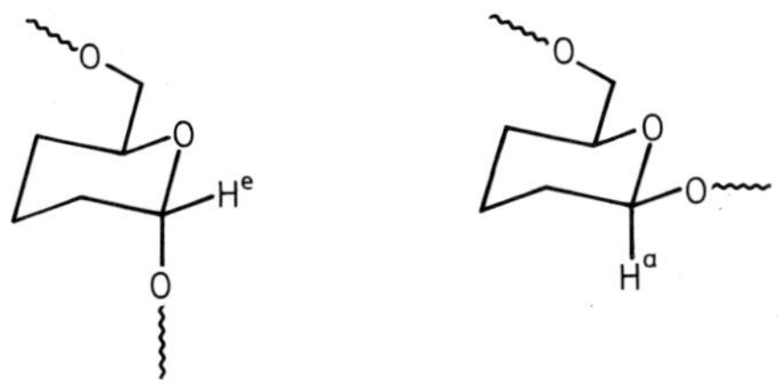

(8-39)

Equatorial hydrogens in cyclohexane and pyranose rings generally give signals at lower fields than their axial counterparts [37]. Thus, stereoregularity of polymers can be evaluated by ^{1}H-NMR spectroscopy, by comparing the intensities at 4.85 ppm (equatorial) and 4.40 ppm (axial) signals. This is often expressed in terms of molar percent of equatorial hydrogens, denoted EQ-H %. A typical ^{1}H-NMR spectrum is shown in Fig. 8.2. The EQ-H % for polyDBO depends on polymerization conditions [38]. At −80 °C, EQ-H % is close to 100 %, while a much lower EQ-H % index was measured for polymer prepared at 40 °C. This index changes troughout the polymerization, decreasing gradually with an increasing conversion and still conside-

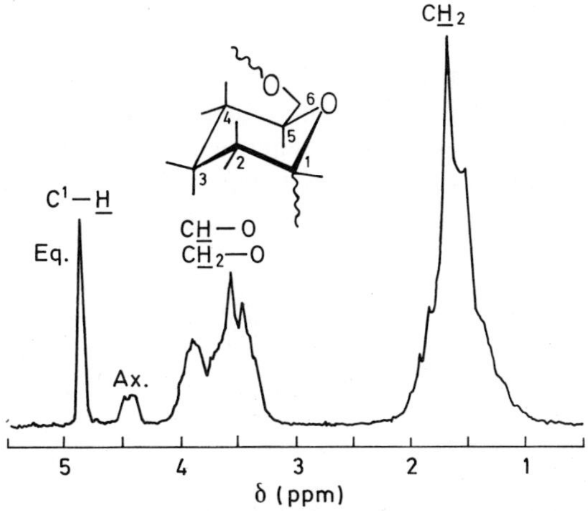

Fig. 8.2. 100 MHz ^{1}H-NMR spectra of poly-6,8-dioxabicyclo|3.2.1|octane at 0 °C. 5 % soln. in $CDCl_3$, room temperature [38]

Table 8.5. Stereochemistry of polymerization of 6,8-dioxabicyclo-|3,2,1|octane (DBO), $[DBO]_0 = 2.0\ mol \cdot l^{-1}$, $[BF_3 \cdot OEt_2] = 2 \times 10^{-3}\ mol \cdot l^{-1}$, CH_2Cl_2, —40 °C [38)]

Time, days	Conversion, %	EQ-H, %	$\overline{M}_n$
1	29.9	91	2900
2	94.6	84	5400
7	95.7	77	5600
9	95.7	76	4300
12	95.7	75	4500

rably decreasing when the reaction mixture is kept for a prolonged time after completion of polymerization. This is illustrated in Table 8.5.

The EQ-H % index indicates that polymer prepared at —40 °C and isolated at low conversions contains 91 % of α-linked units (polymer prepared at —80 °C contains almost exclusively α-linkages independent on conversion), whereas at 94.6 % conversion, polymer contains 84 % of α- and 16 % of β-linkages. The proportion of β-linkages increases further although both conversion and molecular weight do not change appreciably.

These results indicate that at —80 °C one direction of ring opening is clearly preferred and that the resulting polymer is regular. At higher temperatures this preference decreases and the polymer contains both α- and β-linkages. Several possibilities responsible for such a behaviour have been discussed. The observed dependence may be an artefact caused by the flipping of the tetrahydropyran ring, which could convert an equatorial proton into an axial one without affecting the configuration on C-1:

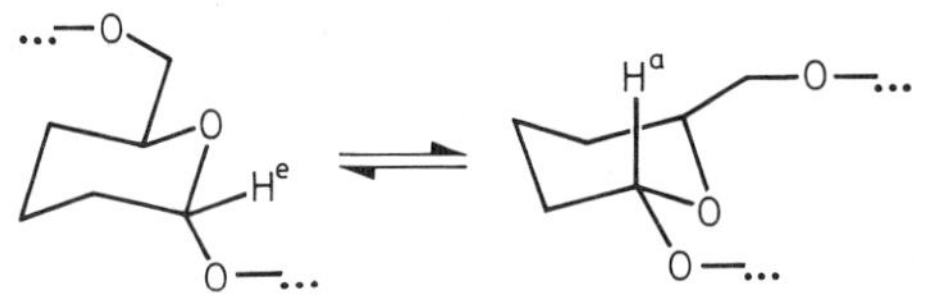

(8-40)

The equilibrium constant of the interconversion (8-40) should depend on temperature. However, the EQ-H % index of poly-DBO, determined at different temperatures (room temp. to 60 °C) is constant, indicating that flipping of the THP ring does not occur within the temperature range studied. There is also a possibility that flipping occurs on the penultimate unit of the growing cation.

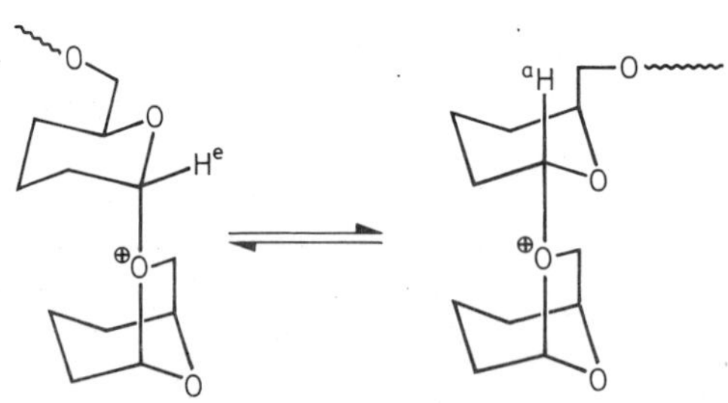

(8-41)

In order to eliminate this possibility, the configuration on other assymetric carbons should be known, this, however, cannot be determined by NMR spectroscopy, because the corresponding signals and those of the methylene groups adjacent to oxygen atoms overlap (at least at 60 MHz).

Assuming that the EQ-H % index corresponds to the proportion of α- or β-linkages in the resulting polymer and is not due to conformational changes as in Schemes (8-40) or (8-41), one has to find a reason for a different course of reaction at different temperatures and conversions. One likely explanation is due to Schuerch [20] who assumed that in the polymerization of anhydro sugars two types of growing species, namely tertiary oxonium ions and carboxonium ions, are in equilibrium and both may propagate. This view is in good agreement with conclusions derived from model studies of 1,3-dioxolane polymerization (Adv. Polymer Sci. 37). Propagation on the tertiary oxonium ion is a S_N2-type reaction and proceeds exclusively with inversion of configuration.

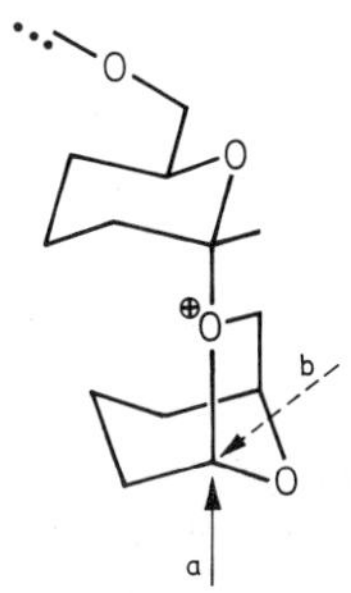

(8-42)

DBO approaches the terminal unit from the opposite side of the positively charged oxygen (a) because approach from the other side (b) is sterically hindered. If cyclic structure (8-42) were in equilibrium with the product of the unimolecular ring opening, i.e. carbenium ion, then the situation should be different as the carboxonium ion is nearly flat.

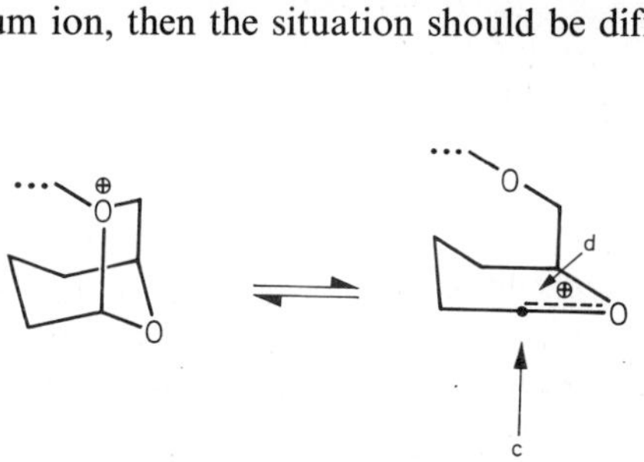

(8-43)

In this case the approach of DBO would be almost equally possible from both sides (c or d) and propagation would lead to polymer containing both α- and β-linkages, as observed for the polymerization of anhydro sugars containing oxirane rings (cf. Sect. 8.2.3.1). Model studies show that equilibria between oxonium and carboxonium ions are shifted to the direction of oxonium ions at lower temperatures [39], thus the formation of stereoregular polymer at low temperatures and irregular polymer at higher temperatures is in line with our present knowledge of the oxonium/carboxonium equilibria.

There is, however, an additional experimental fact which cannot be easily explained by the oxonium-carboxonium equilibrium. According to the data of Table 8.5 the EQ-H% index, which is a measure of steroregularity, decreases considerably when the equilibrated system is kept for a prolonged period of time. These changes in proportions of α- and β-linkages can be explained by assuming that additional reactions proceeding in the equilibrated but still living system, lead to an increase of the proportion of β-linked units. It is suggested that secondary reactions, in which tertiary oxonium ions react with oxygen atoms in a polymer chain, lead to randomization. These processes are enhanced in the systems discussed here, because chain transfer to polymer is favoured due to similar basicities of monomer ($pK_b = 6.94$) and polymer ($pK_B = 6.97$) [38]. In DBO polymerization, chain transfer to polymer leads to the following structure:

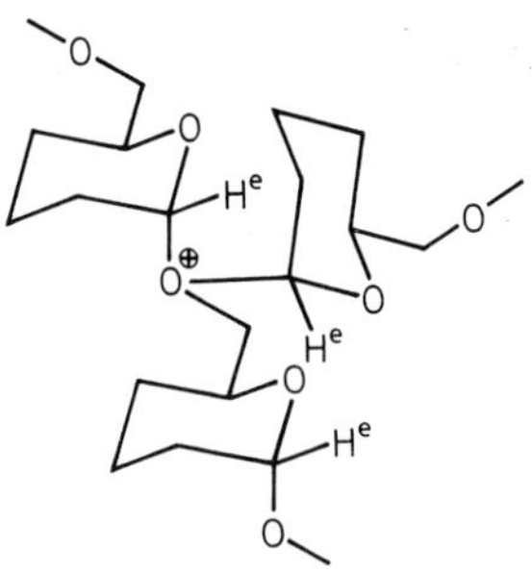

(8-43)

Attack of the next DBO molecule occurs from the opposite side of the positively charged oxygen atom (as in propagation) and leads to inversion at carbon C-1; thus each reaction step changes of one α-linkage into a β-linkage or *vice-versa*. These processes repeated many times would eventually lead to complete randomization.

More recently Sumitomo and Okada studied the polymerization of 4-bromo-6,8-dioxabicyclo|3.2.1|octane with a substituent at the axial or equatorial position [40]:

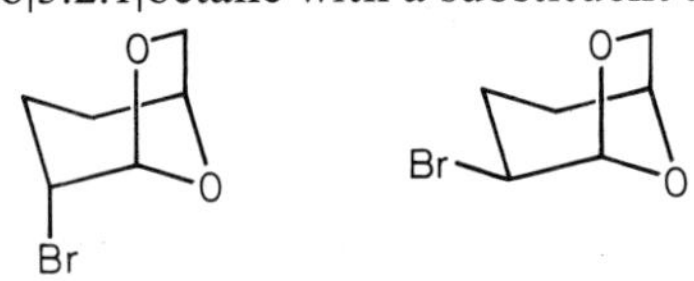

4(a) Br-DBO 4(e) Br-DBO (8-44)

According to these authors 4(e) DBO even at —30 °C gives polymer composed almost exclusively (>97%) of α-linked units. The (a) isomer gives under these conditions polymers containing considerable amounts of β-linked units and even at —78 °C mixtures of 42% (a) and 58% (e) isomers yield polymer with 22% β-linked units.

This was explained by the participation of bromonium ions facilitating the attack leading to the β-form for (a) isomer:

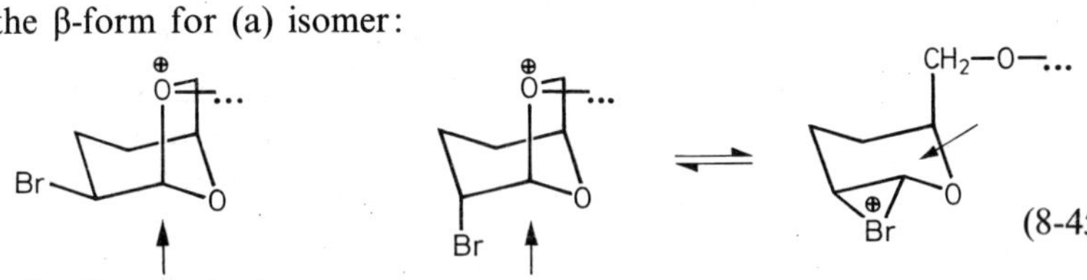

(8-45)

Direction of attack leading to exclusive formation of the α-form

Formation of α-linkage

Formation of β-linkage

The understanding of the stereochemistry of bicyclic acetal polymerization has recently been advanced by the synthesis of optically active DBO enantiomers by the following procedure [41,42]:

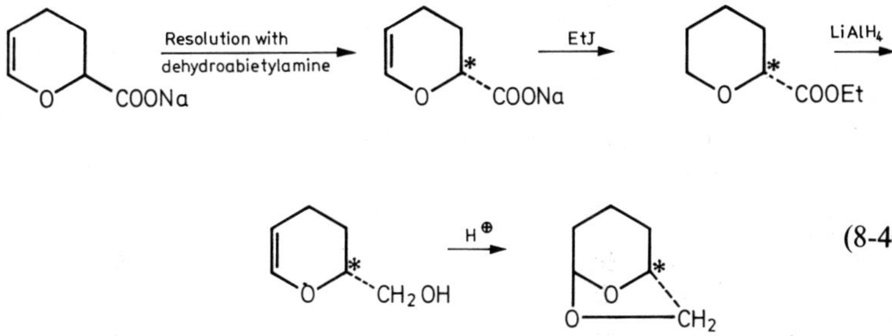

(8-46)

The ^{13}C-NMR spectrum of polymer prepared from racemic DBO exhibits six signals each separated into two lines. These signals were assigned to the following carbon atoms (chemical shifts given in ppm δ)

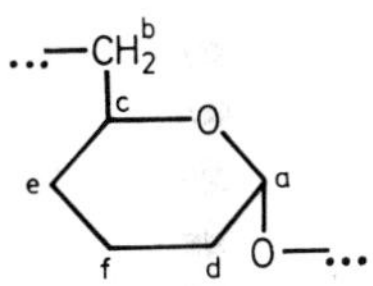

a : 97.27 , 96.49
b : 69.68 , 69.05
c : 68.23 , 67.39
d : 29.43
e : 27.87 , 27.34
f : 17.88

(8-47)

The corresponding ^{13}C-NMR spectrum is shown in Fig. 8.3. One set of lines (more abundant) is identical with the spectrum of the polymer obtained from the optically pure enantiomer. Thus it was concluded that in the polymer prepared from racemic DBO two sets of signals corresponding to isotatic and syndiotactic dyads are present.

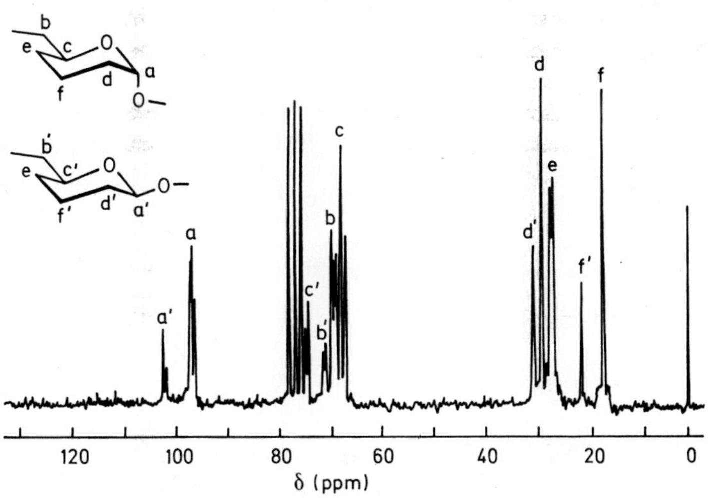

Fig. 8.3. 25 MHz ^{13}C-NMR spectrum of racemic poly(tetrahydropyran-2,6-diyloxymethylene) prepared in CH_2Cl_2 at −78 °C. 7% wt soln in $CDCl_3$ [42]

Table 8.6. Cationic polymerization of racemic 6,8-dioxabicyclo|3,2,1|octane (DBO), DBO = 1.27 — 1.44 g, solvent = 4.0 ml, initiator ($BF_3 \cdot OEt_2$) = 2 mol %, time = 24 hrs [42)]

Solvent	Temp.,	Conversion	M_n	Isotatic diads	α-Form %	
	°C	%		%	^{13}C-NMR	^{1}H-NMR
CH_2Cl_2	0	54	1400	48	73	74
CH_2Cl_2	—30	22	1900	52	77	83
CH_2Cl_2	—60	81	26000	57	87	88
CH_2Cl_2	—78	95	55000	75	100	100
CH_2Cl_2	—90	92	129000	84	100	100
$C_6H_5CH_3$	—90	38	41000	80	100	100
$C_3H_7NO_2$	—90	88	15000	80	100	100

Comparisons of the intensity of signals led to the conclusion that the polymer is richer in isotactic dyads (Table 8.6).

To confirm that indeed in this system some stereoselection occurs, polymerization of racemic DBO enriched with one enantiomer was studied [42)]. If selection occurs through the growing chain control mechanism, polymer should, at moderate conversions, be further enriched with enantiomer being in excess. This is due to the fact that, if for example L monomer is in two-fold excess, both the concentration of L-growing species and L monomer is twice as high as D-growing species and D monomer. This gives for the second order kinetics a 4-fold higher rate of incorporation of L monomer.

Such enrichment was indeed observed in DBO polymerization, and it was concluded that selection is due to the steric repulsion between the equatorial proton on C-3 of the growing center and the equatorial proton on C-4 of the incoming monomer of opposite chirality. This steric repulsion is minimised if both terminal units and incoming monomers have the same chirality.

8.3.1.2 Dioxabicyclo|3.2.1|oct-3-ene

Polymerization of saturated bicyclic acetals leads to polysaccharide analogs devoid of hydroxyl groups. To obtain models resembling more closely natural polysaccharides, bicyclic acetal containing 1,6-anhydro sugar skeleton and double bond, namely 6,8-dioxabicyclo|3.2.1|oct-3-ene was polymerized [43)]:

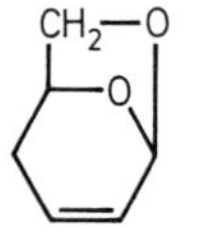

(8-48)

6,8 - dioxabicyclo |3. 2.1| oct - 3 - ene (DBOE)

DBOE (1.5 mol · l^{-1}) polymerizes readily in the presence of $BF_3 \cdot OEt_2$ (10^{-2} mol · l^{-1}). At —78 °C up to a 97% conversion to polymer with $\overline{M}_n$ in the range of 2000–6000 was attained within 24 hrs. Typical results are shown in Table 8.7. The resulting polymer contained a variable amount of CH_2Cl_2 insoluble fraction which may be crosslinked;

Table 8.7. Cationic polymerization of 6,8-dioxabicyclo|3,2,1|oct-3-ene in CH_2Cl_2 [43)]

DBOE $mol \cdot l^{-1}$ [a]	$BF_3 \cdot OEt_2 \cdot 10^2$ $mol \cdot l^{-1}$ [a]	Temp. °C	Time h	Yield %	Soluble[b] fraction %	$\overline{M}_n$
1.4	1.4	−78	24	77	97	6400
1.1	1.1	−78	24	64	100	6000
3.3	1.0	−78	24	87	70	6200
1.4	0.7	−78	24	74	98	—
1.4	0.7	−60	24	77	81	—
1.4	1.4	−40	24	72	36	—
0.7	1.4	0	19	97	0	—

[a] Calculated on the basis of original data from Ref. [43)];
[b] Polymer soluble in CH_2Cl_2, C_6H_6, CCl_4, $CHCl_3$, 1,4-dioxane, THF and DMF

the proportion of this fraction increased with increasing temperature and initial DBOE concentration. Polymerizations carried out at −78 °C gave essentially 100% soluble polymer having the following structure (according to NMR and IR spectra):

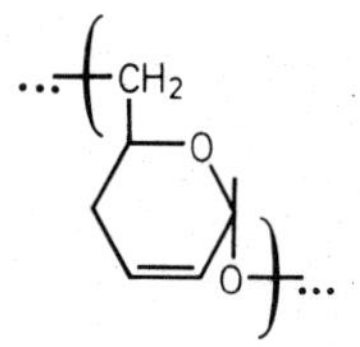

(8-49)

The following assignment for the ^1H- and ^{13}C-NMR spectra were given:

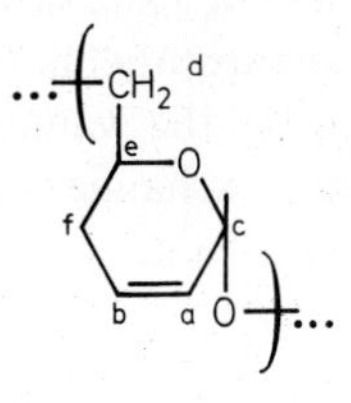

	^1H ppm δ		^{13}C ppm δ
a :	5.5 − 6.0	(m)	130
b :			125
c :	5.0	(s)	94.4
d :	3.3 − 4.0	(m)	70
e :			65
f :	2.0	(s)	25

(8-50)

Hydrogenation of polyDBOE gave polyDBO. NMR spectra of hydrogenated polyDBOE were indentical with those of stereoregular polyDBO, which indicates that polymerization of DBOE also proceeds stereospecifically.

PolyDBOE can be converted into its dihydroxyl derivative, in which the hydroxyl groups are either in cis or trans position. Thus, epoxidation of polyDBOE with m-chloroperbenzoic acid, followed by hydrolysis, leads to poly(trans-3,4-dihydroxy-tetrahydropyran-6.2-diyloxymethylene):

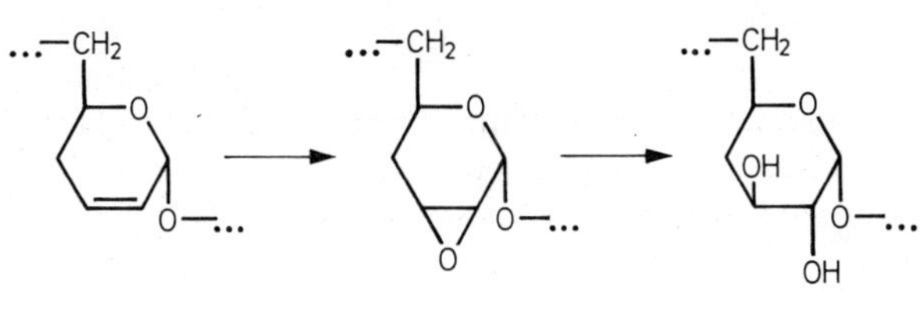

(8-51)

Oxidation of polyDBOE with osmium tetraoxide leads to the cis-isomer:

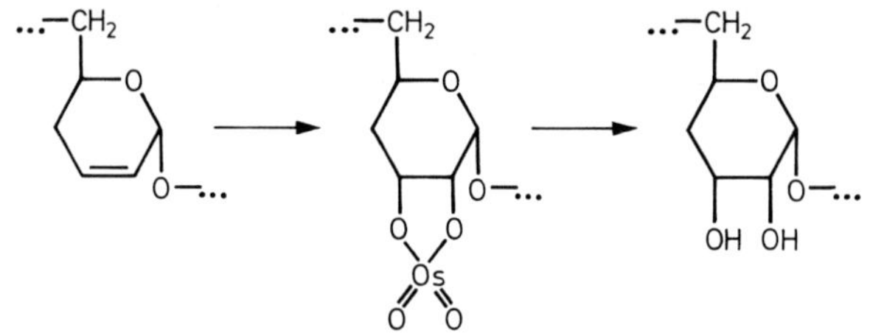

(8-52)

^{1}H-NMR spectra of the trans-isomer consist of three signals of acetal protons, as shown in Fig. 8.4. This was attributed to the presence of different isomers (and conformers). Thus, for the α-form the following structures are possible:

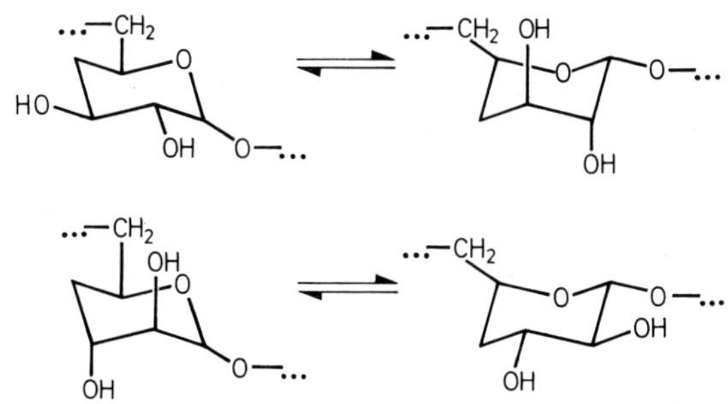

(8-53)

Additional four structures may exist for β-linked units, the proportion of β-linked units was, however, <10%. Conformational free energies for each structure were estimated from non-bonding interaction energies between substituents of the tetrahydropyran ring and anomeric interaction energies, giving the values shown in Table 8.8. Within each pair, both conformers undergo fast exchange, consequently only one NMR signal is observed for each pair at 25 °C.

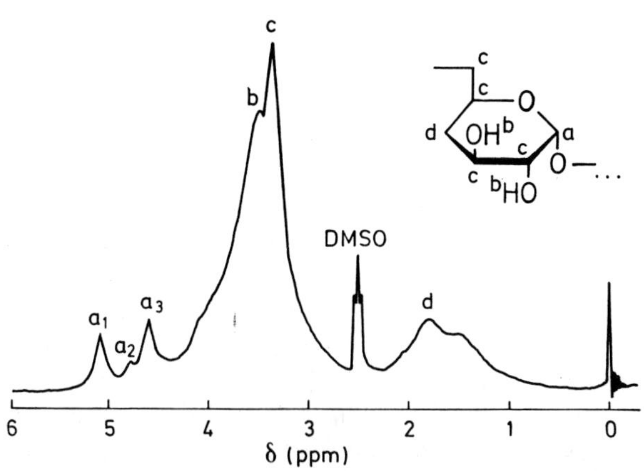

Fig. 8.4. 100 MHz ^{1}H-NMR spectrum of poly(trans-3,4-dihydroxytetrahydropyran-2,6-diyloxymethylene). 5% soln in DMSO-d_6, room temperature [43)]

Table 8.8. Conformational free energies for structural units of poly(6,8-dioxabicyclo|3,2,1|oct-3-ene) (polyDBO)[43)]

Repeating unit	Free energy		K_e (25 °C)	Proportion
	kJ/mol	(kcal/mol)		
I	6.7	(1.6)	$2.9 \cdot 10^{-3}$	~100
II	23.0	(5.5)		~0
III	11.9	(2.85)	$7.1 \cdot 10^{-1}$	~58
IV	12.8	(3.05)		~42

8.4 Cationic Ring-Opening Polymerization with Expansion in Volume

8.4.1 Volume Change in Polymerization

Polymerization of typical monomers proceeds with a negative change of volume (shrinkage). The molar contraction during bulk polymerization varies from a few to several tens percent, depending on monomer structure. Some pertinent data are given in Table 8.9[4)].

Table 8.9. Shrinkage accompanying polymerization calculated from bulk densities of monomer and polymer[4)]

Monomer	Shrinkage %	Monomer	Shrinkage %
Ethylene	66.0	Ethylene oxide	23
Propylene	39.0	Propylene oxide	17
Butadiene	36.0	Tetrahydrofuran	10
Vinyl chloride	34.4	Styrene oxide	9
Acrylonitrile	31.0	Cyclooctadiene	3
Methyl methacrylate	21.2	Octamethylcyclotetrasiloxane (D_4)	2
Styrene	14.5		
N-vinyl carbazole	7.5		

Shrinkage during polymerization stems from the conversion of Van der Waals' distances between monomer molecules to distances of chemical bonds between corresponding units in the polymer chain. Of lesser importance are such factors as free volumes in amorphous polymers and the ability of monomer and/or polymer to pack densely in the crystalline phase.

Two conclusions may be drawn from the data of Table 8.9. Shrinkage is considerably lower for ring-opening polymerization than for vinyl polymerization (e.g. ethylene oxide vs. ethylene, or styrene oxide vs. styrene). Moreover, the larger the monomer molecule, the lower the shrinkage. This can easily be rationalized on the basis of the explanation given above. During polymerization of ethylene oxide two monomer molecules, separated by a Van der Waals distance, are linked by a chemical bond but at the same time the oxygen atom of the oxonium ion [underlined in Scheme (8-54)], originally separated by covalent-bond distance from the carbon atom (underlined), moves to a larger distance which reduces the net shrinkage.

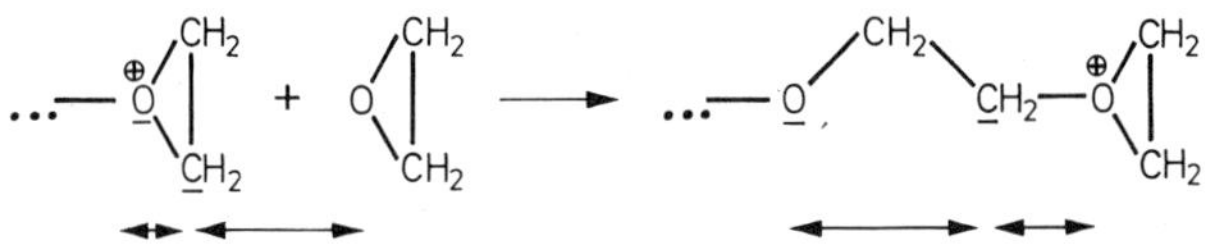
(8-54)

The smaller molar contraction for larger molecules is due to the larger proportion of atoms linked by covalent bonds in the monomer. Thus, the proportion of the newly formed to the originally present covalent bonds is lower for larger molecules, which results in less shrinkage.

The extent of shrinkage occuring during polymerization influences the properties of curable polymers (casting, coating) because contraction accompanying curing may result in the formation of microcracks and voids in the final product. Shrinkage may be to some extent reduced by mixing monomer or prepolymer with a neutral filler. Thus, introducing 50% vol. of a filler should reduce shrinkage by a factor of 2. Epoxide resins and fillers have been developed for which shrinkage upon curing is considerably reduced. Some examples of such systems are given in Table 8.10 [44].

Table 8.10. Effect of fillers on shrinkage in the polymerization of bisphenol A diglycidyl ether cured with primary amines [44]

Filler (20% wt.)	Shrinkage %
None	0.91
Asbestos	0.48
Silica	0.77
Flint powder	0.41
Limestone	0.47

In such systems, however, the filler will not permit shrinkage in overall dimensions and considerable stresses may build up in the composite.

8.4.2 Examples of Polymerization with Expansion in Volume

In 1972, Bailey and Sun reported that the cationic polymerzation of 1,4,6-trioxaspiro-|4.4|nonane:

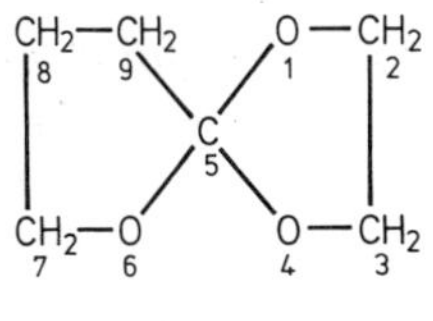
(8-55)

proceeds with a slight expansion in volume (0.1 %) [45]. This was rationalized in terms of changing distances between the groups involved in polymerization, as discussed in the previous section:

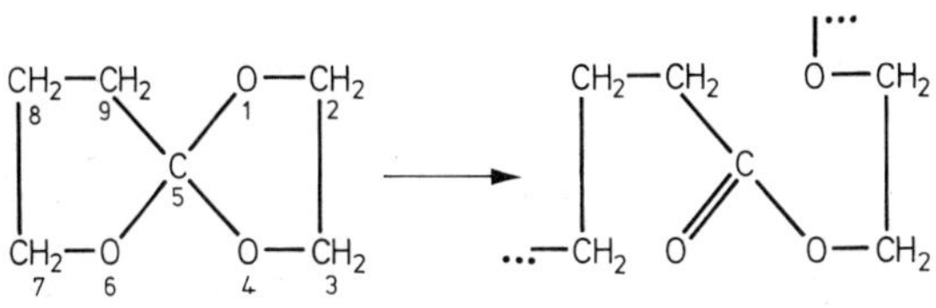

(8-56)

Thus, O-1 moves closer to the growing center (covalent bond) and O-6 closer to C-5 (carbonyl double bond). At the same time the distances between O-1 and C-5 and between O-6 and C-7 increase, due to the breaking of single bonds. By this process a compact, dense monomer molecule is transformed into a rather loose polymer segment.

A number of spiro orthoesters have been synthetized and the volume changes during the cationic ring-opening polymerization have been studied [46].

More recently, Hall reported that in the polymerization of some bicyclic monomers, opening of one ring leads to contraction whereas opening of the second ring is accompanied by expansion [47]. In the following sections experimental details concerning some typical systems studied to date are are briefly summarized.

8.4.2.1 Synthesis and Polymerization of Fused Bicyclic Orthoesters

The polymerization of a bicyclic monomer, in which only one or both rings are opened has recently been reported by Hall [47]. This monomer 2,6,7-trioxabicyclo|2.2.1|-heptane (TBH) may be prepared in 73 % yield by the reaction of glycerol and triethylorthoformate:

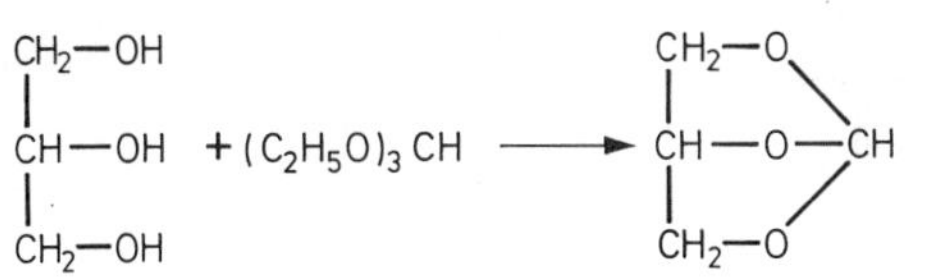

b.p. 80°C / 37 mm Hg

b.p. 20°C / 0.6 mm Hg

(8-57)

Polymerization of TBH in CH_2Cl_2 solution (1 mol · l^{-1}) at —78 °C with $BF_3 \cdot OEt_2$ or PF_5 gives in 4 hrs nearly 100 % conversion; the molecular weight of the polymer is in the 100–8000 range (by VPO in tetrahydrofuran solution at 37 °C). The ^{1}H-NMR spectrum of a typical polymer exhibits a sharp singlet at δ 5.81, a complex multiplet spectrum of a typical polymer exhibits a sharp singlet at δ 5.81 and a complex multiplet in the range of δ 3.4–4.6 and two small singlets appear at δ 5.68 and δ 5.24. By comparing the observed chemical shifts with chemical shifts of protons in model orthoformates, the following assigments were made:

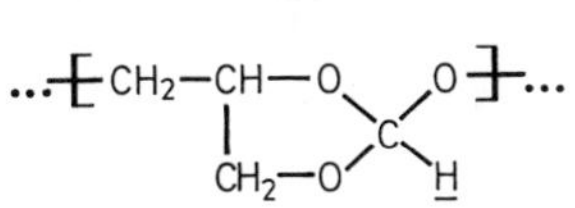

δ = 5.81

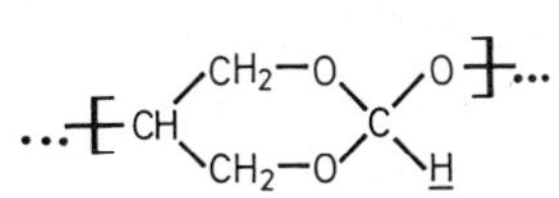

δ = 5.24

(8-58)
(8-59)

The intensity of the signals due to the orthoformate protons indicates that the polymer contains almost exclusively (90–100%) five-membered rings:

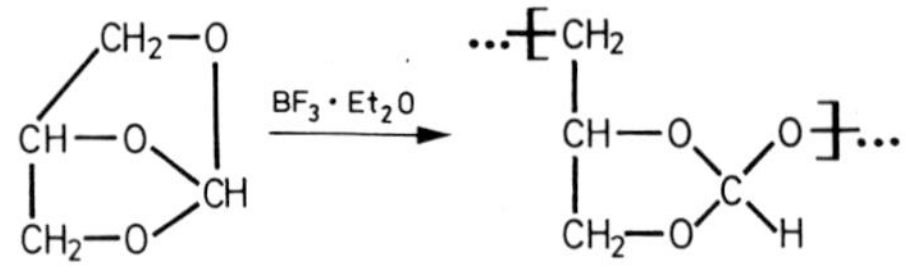

(8-60)

The polymerization is accompanied by approximately 6% shrinkage (calculated from densities of monomer and polymer at 25 °C). Upon heating the polymer solution in the presence of trifluoromethanesulfonic acid the substituted 1,3-dioxolane ring opens and linear polymer is formed. This is accompanied by a 9.7% expansion in volume.

Bulk polymerization of 2,6,7-trioxabicyclo|2.2.1|heptane with 1 mol% CF_3SO_3H initiator at 80 °C directly gives linear polymer with a quantiative yield (after 1 h). The structure of this polymer was determined from ^{1}H-NMR spectra, which showed two signals at δ 8.1 and δ 5.2 instead of the δ 5.8 signal corresponding to the 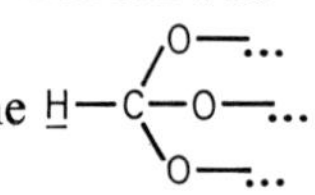 group. The following assigments are made:

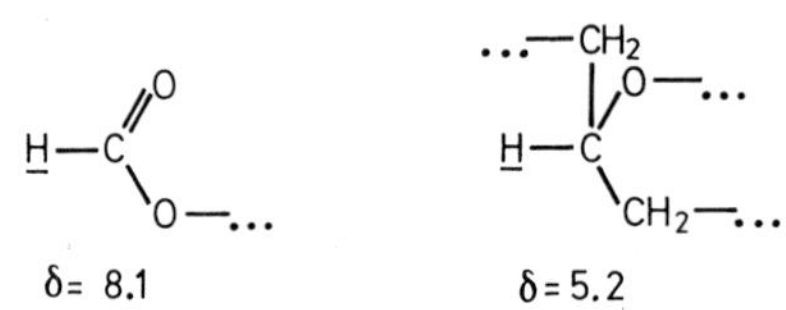

(8-61)

Comparison of the intensities of the corresponding signals indicated that the linear polymer contains two types of units:

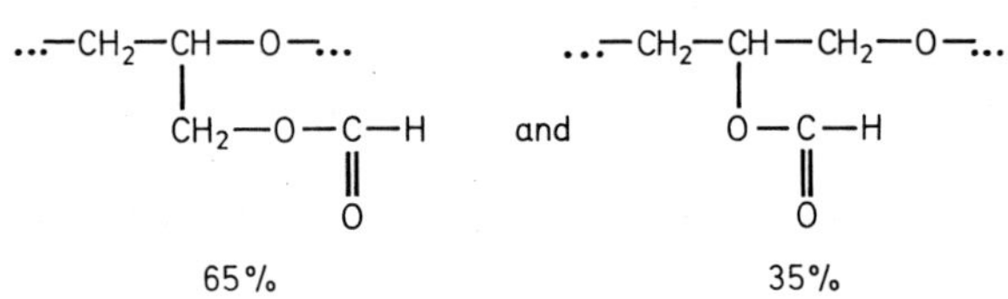

(8-62)

Thus, the complete reaction scheme may be visualised as follows:

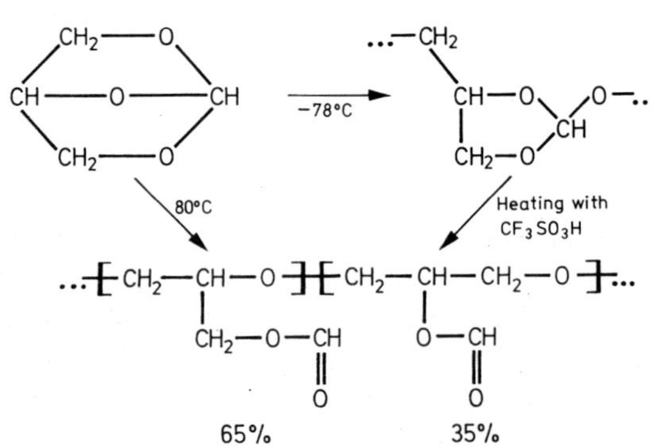

(8-63)

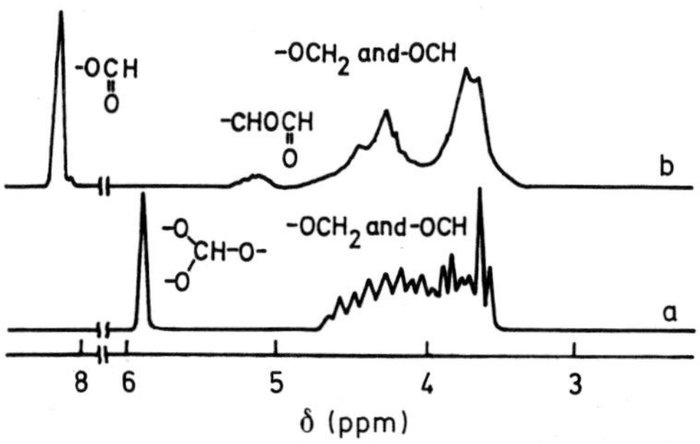

Fig. 8.5. ^{1}H-NMR spectra of poly-2,6,7-trioxabicyclo|2.2.1|heptane in $CDCl_3$. a — polymer obtained at — 78 °C; b — polymer obtained at 80 °C or by heating polymer a with $HOSO_2CF_3$ [47)]

Table 8.11. Volume changes during polymerization and/or isomerization of TBH calculated from bulk densities of monomer and polymer. (+ expansion, — shrinkage) [47)]

Temp., °C	Changes of volume, %[a]		
	M → CP	M → LP	CP → LP
25	−6.2	+4.0	+9.7
50	−7.4	+2.5	+9.8
80	−7.1	+1.9	+8.2
100	−8.4	−0.3	+8.1
120	−9.8	−1.4	+8.6

[a] M — monomer; CP — cyclic polymer; LP — linear polymer

The corresponding ^{1}H-NMR spectra are shown in Fig. 8.5. The volume changes during polymerizations, as calculated from densities of monomer and polymers, are given in the Table 8.11 [47)].

The same phenomenon, i.e. opening of one ring under mild conditions and opening of both rings (with volume expansion) under more drastic conditions, was observed in the polymerization of 2,6,7-trioxabicyclo|2.2.2.|octane [48)]. This monomer was obtained by reaction of the acyclic triol and orthoester:

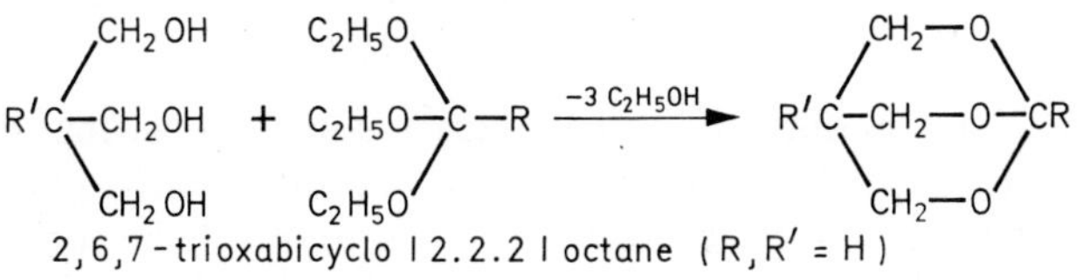

2,6,7-trioxabicyclo |2.2.2| octane (R, R' = H) (8-64)

2,6,7-Trioxabicyclo|2.2.2|octane polymerized at 0 °C in CH_2Cl_2 with $BF_3 \cdot O(C_2H_5)_2$ initiator to polymer insoluble in common organic solvents. Soluble polymer, with $[\eta]_{inh} = 0.10$ dl/g in $(CH_3)_2SO$ solution, was obtained, however, at higher (5–10 mol %) initiator concentrations. Polymer prepared at low initiator concentration is insoluble in the reaction medium due to high crystallinity and high molecular weight and not to cross-linking [48)].

According to ^{1}H- and ^{13}C-NMR spectra, the repeating unit of the polymer results from the opening of one ring:

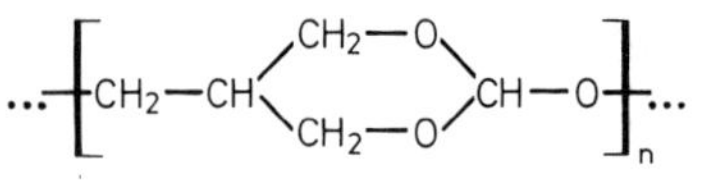

(8-65)

The presence of only one signal of the orthoester carbon in the ^{13}C-NMR spectrum (at 109.04 ppm), the simple ^{1}H-NMR spectrum (singlet at δ 5.28 for the orthoester group) and the absence of the formate group indicate high polymer stereoregularity and the essential absence of units resulting from opening of both rings:

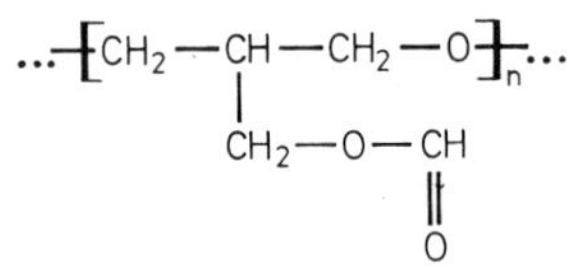

(8-66)

If, hovewer, the same monomer is polymerized under more drastic conditions [in bulk with $BF_3 \cdot O(C_2H_5)_2$ at 5 °C or 70 °C], both rings open and a substituted polyether is obtained. The formation of this thermodynamic product is accompanied by volume expansion [47, 49, 50].

According to these results expansion in volume during polymerization occurs only when both fused rings open upon polymerization.

8.4.2.2 Synthesis and Polymerization of Spiroorthoesters

8.4.2.2.1 1,4,6-Trioxaspiro|4.4|nonane

1,4,6-Trioxaspiro|4.4|nonane is a spiroorthoester that contains a tetrahydrofuran and a 1,3-dioxolane ring. The monomer can be prepared by reaction of γ-butyrolactone with ethylene oxide with 33% yield [51].

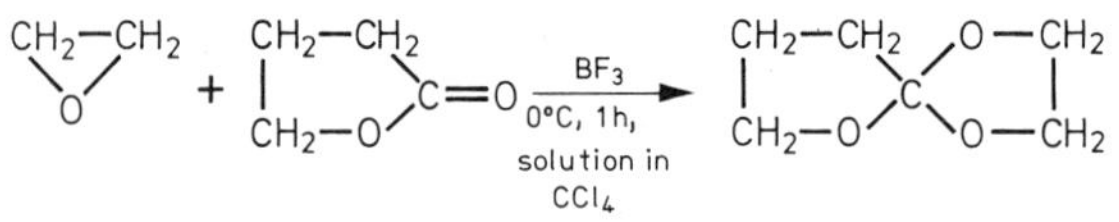

(8-67)

The density of the product (d_4^{20} = 1.16 g/cm^3) is higher than that of the starting compounds (0.869 and 1.11 g/cm^3 respectively), thus, the formation of spiro-compounds from two cyclic molecules results in a change of volume corresponding to a 13% shrinkage. Bulk polymerization of 1,4,6-trioxaspiro|4.4|nonane carried out with $BF_3 \cdot Et_2O$ initiator (0.6 vol%) at room temperature for 24 hrs gave a yellow, viscous liquid polymer with 95% yield. The intrinsic viscosity in chloroform was 0.087 dl/g (at 25 °C). Dilatometry during polymerization indicated no shrinkage [51] although the density of the polymer was 1.17 g/cm^3 which would indicate about 0.9% shrinkage; one has to remember that the density depends on the conditions of measurements and is different for bulk and in various solvents.

8.4.2.2.2 1,5,7,11-Tetraoxyspiro|5.5|undecane

1,5,7,11-Tetraoxaspiro|5.5|undecane is a spiroorthocarbonate containing two 1,3-dioxane rings. It may be prepared in almost quantitative yield according to the following scheme [52].

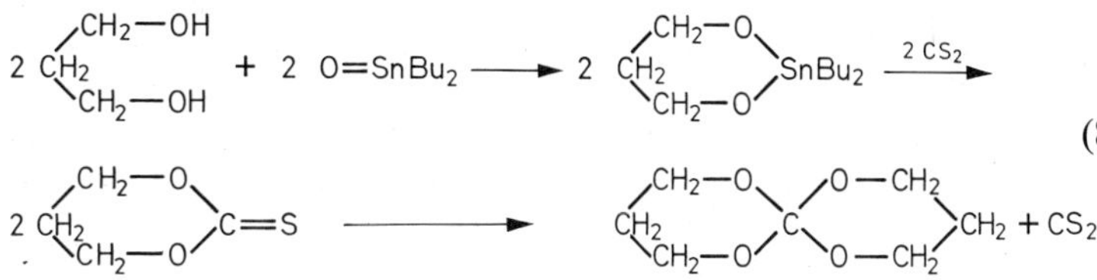

(8-68)

Vacuum sublimation (0.01 mm Hg) gives a product with an m.p. 137–141 °C (purity was not determined).

1,5,7,11-Tetraoxaspiro|5.5|undecane (0.73 g) was polymerized with $BF_3 \cdot Et_2O$ initiator (20 mmol) at 142 °C (above the melting point) for 7.5 hrs. The volume increase was about 2% (70% conversion in the dilatometer) [53]. After purification the polymer was a viscous liquid whose intrinsic viscosity in chloroform at 25 °C was 0.26 dl per g. The ^{1}H-NMR spectrum showed three triplets, attributed to the $-CH_2-O-C(O)O-$ group (δ 4.23), the $-CH_2OCH_2-$ group (δ 3.52) and the $-\overline{CH_2}\underline{CH_2}CH_2-$ group (δ 1.92), indicating mostly a linear structure $\left(\!-O\underset{\underset{O}{\|}}{C}O-CH_2CH_2CH_2OCH_2CH_2CH_2-\!\right)_n$

8.4.2.2.3 3,9-Dimethylene-1,5,7,11-tetraoxyspiro|5.5|undecane

3,9-Dimethylene-1,5,7,11-tetraoxaspiro|5.5|undecane is a spiroorthocarbonate containing two double bonds. It may be prepared starting from cyclopentadiene and acrolein with ~40% yield [54].

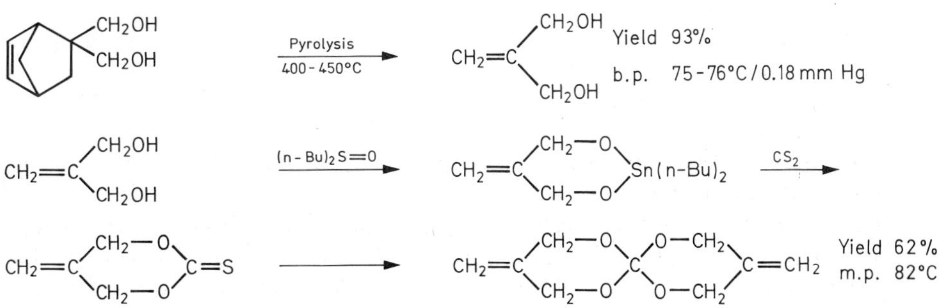

(8-69)

3,9-Dimethylene-1,5,7,11-tetraoxaspiro|5.5|undecane was polymerized in bulk or in solution (methylene dichloride, benzene, toluene) with 1 mol% $BF_3 \cdot OEt_2$ or $SnCl_4$ initiators at temperatures ranging from 25 °C to 110 °C for 4–25 hrs. In all cases high conversions (90–94%) were observed; intrinsic viscosities measured in $CHCl_3$ at 25 °C were in the range 0.09–0.24 dl/g ($\overline{M}_n = 10^3$–10^4). Bulk polymerization gave 10–20% insoluble product, while the polymer prepared in solution was soluble [54]. Soluble polymer turned into cross-linked (insoluble) material after one week at room temperature.

Table 8.12. Densities of monomer and polymer of 3,9-dimethylene-1,5,7,11-tetraoxaspiro|5.5|undecane and the calculated volume change during polymerization [54]

Temp., °C	Density g/cm³		Expansion, %
	Monomer	Polymer	
25	1.220	1.171	4.25
50	1.220	1.145	6.55
70	1.220	1.140	7.00
82	1.155	1.130	2.21
100	1.125	1.110	1.35
115	1.100	1.100	0
135	1.072	1.080	−0.74

The polymerization proceeded with expansion in volume, the extent of which was dependent on temperature. Densities of monomer and polymer and the volume changes during polymerization are given in Table 8.12. The ^{1}H-NMR spectrum of the monomer consists of two bands at δ 4.95 ($\underline{CH_2}$=C< groups) and δ 4.60 ($\underline{CH_2}$O); the spectrum of the polymer is very similar and it also contains two bands at δ 4.80 (CH_2=C< groups) and at δ 4.45 (attributed to $-OCH_2-$ and $-CH_2OCO-$ groups).

8.4.2.2.4 *3-Methylene-1,5,7,11-tetraoxaspiro|5.5|undecane*

3-Methylene-1,5,7,11-tetraoxaspiro|5,5|undecane is an unsymmetric spiroorthocarbonate containing one double bond. The monomer was prepared according to the following scheme [4]:

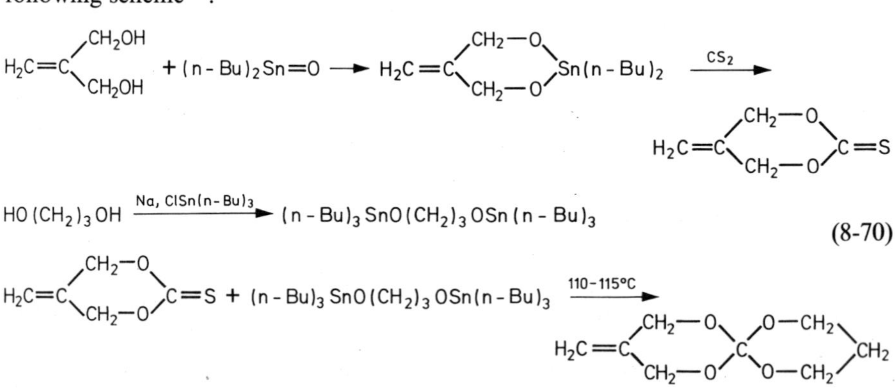

(8-70)

The monomer obtained in 86% yield after purification by recrystallization from hexane and sublimation, melted at 61–62 °C.

Bulk and solution (methylene chloride, benzene) polymerization with $BF_3 \cdot OEt_2$ or $SnCl_4$ initiators at 25 °C gave after 48 hrs 70–90% conversion to polymer with an intrisic viscosity of 0.09–0.12 dl/g ($CHCl_3$, 25 °C) [9].

Using methods similar to those described earlier, several substituted spiroortho-esters were synthetized and polymerized [44]:

(8-71)

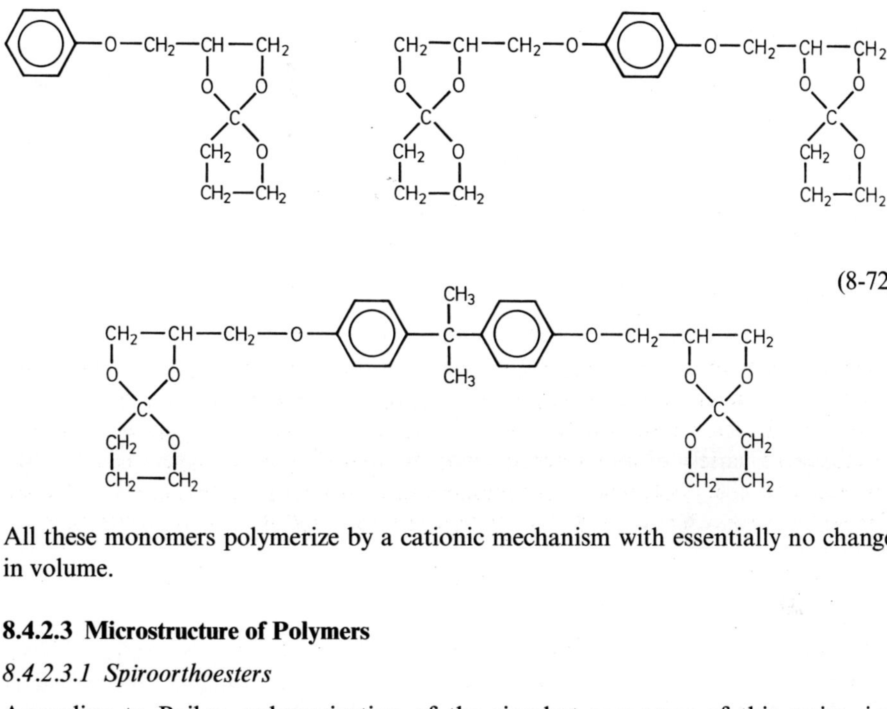

(8-72)

All these monomers polymerize by a cationic mechanism with essentially no change in volume.

8.4.2.3 Microstructure of Polymers

8.4.2.3.1 Spiroorthoesters

According to Bailey, polymerization of the simplest monomer of this series, i.e. 1,4,6-trioxaspiro|4,4|nonane (TOSN) gives a polymer with the following structure [44]:

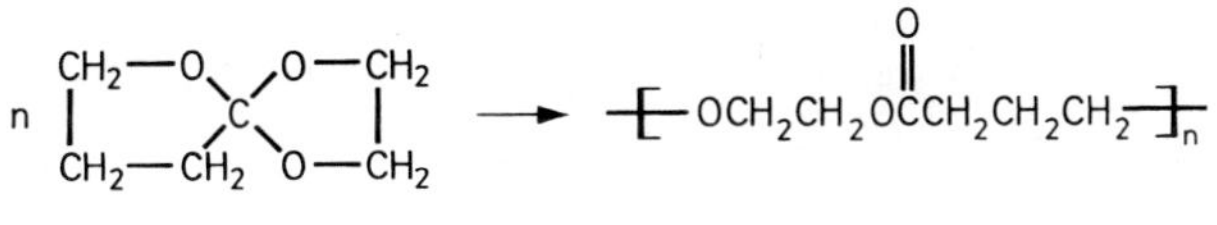

(8-73)

with 10–20% head-to-head units:

$$-\!\!\!\![\,OCH_2CH_2O\overset{\displaystyle O}{\overset{\|}{C}}CH_2CH_2CH_2O\,]\!\![\,OCH_2CH_2CH_2\overset{\displaystyle O}{\overset{\|}{C}}OCH_2CH_2O\,]\!\!\!\!- \qquad (8\text{-}74)$$

This was based on ¹H-NMR spectroscopy although details were not given.

The alternating polyester/ether structure was proposed also for polymers prepared from other spiroorthoesters and the following mechanism was suggested:

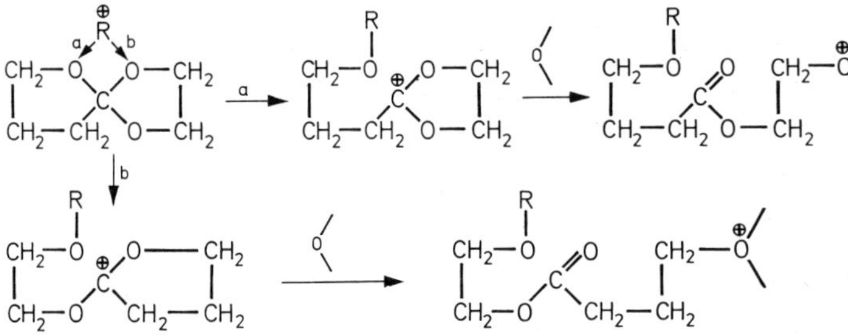

(8-75)

According to Scheme (8-75) the cationic initiator R^+ may react by routes *a* and/or *b* with respective ring opening. Attack and opening in subsequent steps may also proceed in this alternative manner.

Head-to-head sequences are formed when consecutive openings involve route *a* followed by route *b* (or vice versa).

Recently, the microstructure of the polymer and the polymerization mechanism were reinvestigated by one of us and the systems were found to be much more complex than thought originally [55]. The new data came mostly from ^{13}C-NMR studies of the polymerization of 1,4,6-trioxaspiro|4,4|nonane and observation of the chemical shifts of the orthoester carbon atoms, converted during polymerization into carbonyl carbon atoms. The ^{13}C-NMR spectrum of the polymerizing system is shown in Fig. 8.6. The complicated pattern shows up to twelve lines in the region 126–136 ppm (orthoester) which indicates that besides the large signal of the unreacted monomer (127.6 ppm) a variety of orthoester linkages are present in the polymer chains. With

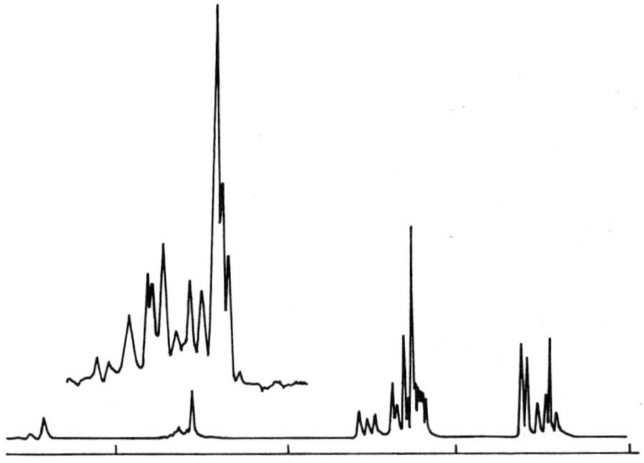

Fig. 8.6. ^{13}C-NMR spectrum of poly(1,4,6-trioxaspiro|4.4|nonane in $CDCl_3$ solvent terminated with $(C_6H_5)_3P$ after 4 days at 0 °C. [Monomer] = 4.5 mol · l^{-1}, $[(CH_3)_3O^{\oplus}SbF_6^{\ominus})$ = 4.6 · 10^{-2} mol · l^{-1}, $[Ph_3P]$ = 1.5 · 10^{-1} mol · l^{-1} [55]

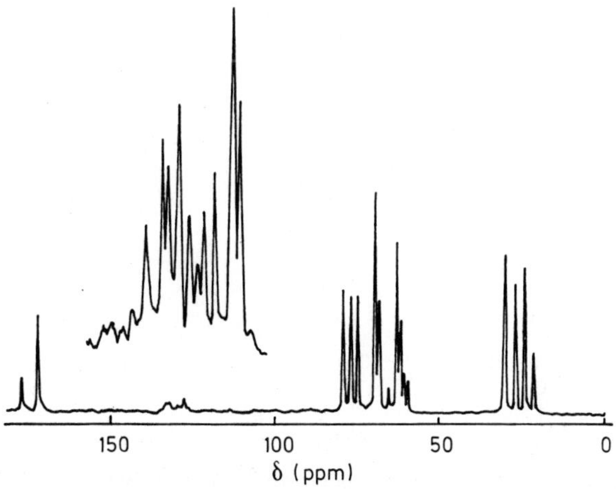

Fig. 8.7. As in Fig. 8.6. Terminated with Ph_3P after 24 hrs at 50 °C. [55]

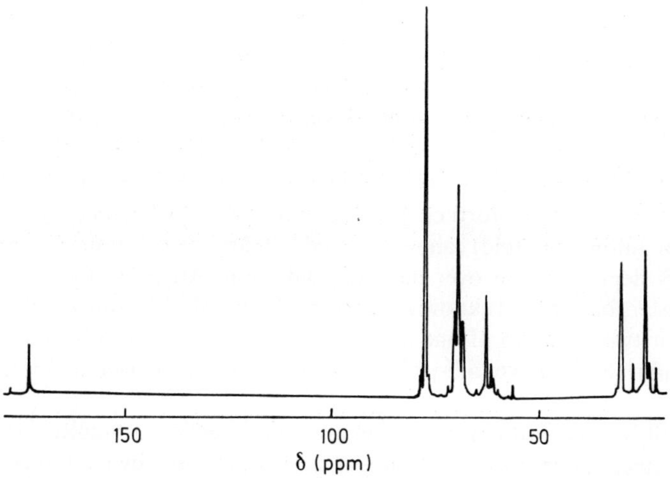

Fig. 8.8. As in Fig. 8.6. Terminated with CH_3OH after 5 days at room temperature [55]

proceeding reaction, monomer disappears but the orthoester bonds (i.e. cycles in the macromolecules) are still present (Fig. 8.7). Only during a later stage of polymerization are these bonds quantitatively converted to ether and ester linkages of the linear poly(ester ether), a structure similar to that proposed by Bailey [3]. The signals of the orthoester carbon atoms vanish completely, whereas the intensity of the carbonyl C-atoms (172–178 ppm) increases proportionally (Fig. 8.8). Thus, polymerization does not proceed exclusively by sequential opening of both rings [Eq. (8-75)] but the opening of one of the two rings in the monomer is followed by the opening of the second ring, accompanied by the formation of stable ester bonds:

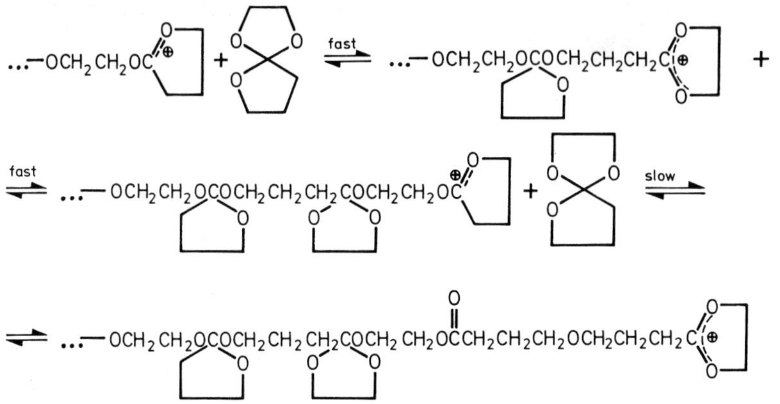

(8-76)

According to model experiments, the nucleophilic reactivity of the oxygen atoms and the ring-opening ability of the two rings in the unsubstituted monomer are similar [55]. Similar proportions of both types of active centers, i.e., isomeric cyclic dialkoxycarbenium cations, were observed by ^{13}C-NMR.

The polymerization of TOSN proceeds very slowly ($k_p \simeq 10^{-4}$ mol · l · s^{-1} at 35 °C) although polymerization of bicyclic orthoesters has been reported by Hall to proceed nearly instantaneously under comparable conditions. The rapid propagation of orthoesters (their monomers being the least nucleophilic among heterocyclics) is in good agreement with our recently proposed structure — reactivity relationship for cationic ring-opening polymerizations [56]. Thus, the slow polymerization of TOSN is due to low strain and because the polymerization by opening only one of the rings that would lead to the poly(orthoester) structure is thermodynamically unfavorable. This reaction (although rapid) cannot proceed in many steps, because it creates strain that favours depropagation over monomer addition. Attack at C-4, C-5, and C-7 leading to ester formation, is known to be very slow, and is the rate-determining step. Thus, one or two openings of one of the rings must be followed by a slow opening of the second ring, as shown in Eq. (8-76), which results in an overall low rate of polymerization.

The initially formed orthoester linkages are isomerized by intramolecular rearrangement to the linear poly(ether-ester), similar to that proposed by Hall [Eqs. (8-60) and (8-61)], e.g.:

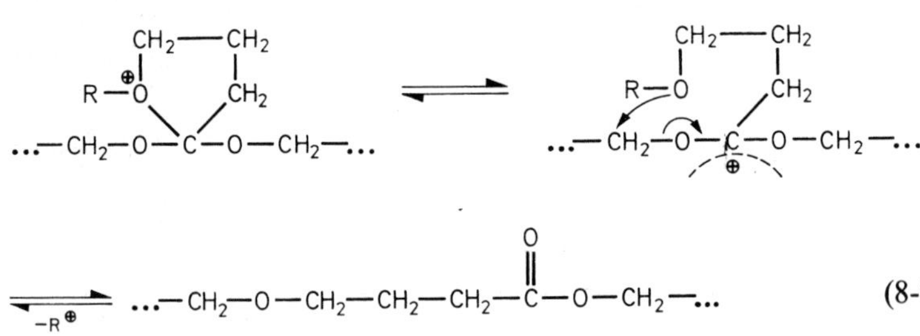

(8-77)

This process leads a polymer with not only head-to-head or head-to-tail units [cf. (8-74)], but also with two consecutive ester and ether linkages. In such units carbonyl C-atoms are separated not by seven atoms, as in Structures (8-72) but by only four atoms.

In the region of the carbonyl C-atoms absorption in the ^{13}C-NMR spectra there are three signals at 177.6, 173.1 and 172.6 ppm. It is very unlikely to expect a difference of 5 ppm for the H-H and H-T units, because the difference in the chain structure is only at the seventh atom to the carbonyl group:

$$\cdots-O\underset{\underset{O}{\|}}{C}CH_2CH_2CH_2OCH_2CH_2O\underset{\underset{O}{\|}}{C}-\cdots \quad HT$$

$$\cdots-O\underset{\underset{O}{\|}}{C}CH_2CH_2CH_2OCH_2CH_2CH_2\underset{\underset{O}{\|}}{C}-\cdots \quad HH \qquad (8\text{-}78)$$

Thus, the two high field signals are probably due to Structures (8-78), whereas the low-field signal is ascribed to structures in which the carbonyl C-atoms are separated by four atoms, e.g. as shown in Eq. (8-77).

Signals due to the irregular structures, like those shown in Eq. (8-77), also have been observed by ^{1}H-NMR spectroscopy.

8.5 References

1. Sumitomo, H., Okada, M.: Adv. Polym. Sci. *28*, 47 (1978)
2. Yokoyama, Y., Hall, H. K.: Adv. Polym. Sci. *42*, 107 (1981)
3. Bailey, W. J. et al.: J. Polym. Sci. Symp. *56*, 117 (1976)
4. Bailey, W. J. et al.: ACS Symposium Series, *59*, 38 (1977)
5. Schuerch, C.: Adv. Polym. Sci. *10*, 173 (1972)
6. Schuerch, C.: in Encyclopedia of Polymer Science and Technology, Suppl. Vol. 1, *S10* (1976)
7. The Carbohydrates, Chemistry and Biochemistry, Pigman, W., Horton, D., Eds., Academic Press, New York and London, 1972, Vol. Ia, Chapter I
8. Zachoval, J., Schuerch, C.: J. Polym. Sci. *C*, *28*, 187 (1969)
9. Uryu, T. et al.: Makromol. Chem. *179*, 1773 (1978)
10. Uryu, T., Koyama, Y., Matsuzaki, K.: J. Polym. Sci. Polym. Lett. *17*, 673 (1973)
11. Kops, J., Schuerch, C.: J. Polym. Sci. *C*, *11*, 119 (1965)
12. Micheel, F., Broddle, O. E., Reinkins, K.: J. Liebigs Ann. *1974*, 124
13. Micheel, F., Broddle, O. E.: J. Liebigs Ann. *1974*, 702
14. Liu, J. W. P., Schuerch, C.: Macromolecules *5*, 656 (1972)
15. Uryu, T. et al.: Makromol. Chem. *178*, 1929 (1977)
16. Korshak, V. V. et al.: Vysokomol. Soed. *3*, 477 (1961), *5*, 1593 (1963)
17. Tu, C. C., Schuerch, C.: J. Polym. Sci., Polym. Lett. *1*, 163 (1963)
18. Ruckel, E. R., Schuerch, C.: J. Org. Chem. *31*, 2233 (1966)
19. Klemer, A.: Apostolides, Ch. Carbonyl Res. *22*, 432 (1972)
20. Zachoval, J., Schuerch, C.: J. Amer. Chem. Soc. *91*, 1165 (1969)
21. Uryu, T., Schuerch, C.: Macromolecules *4*, 342 (1971)
22. Ruckel, E. R., Schuerch, C.: J. Amer. Chem. Soc. *88*, 2605 (1966)
23. Ruckel, E. R., Schuerch, C.: Biopolymers *5*, 515 (1967)
24. Frechet, J. M., Schuerch, C.: J. Amer. Chem. Soc. *91*, 1161 (1969)
25. Liu, J. W. P., Schuerch, C.: J. Polym. Sci. *10*, 2045 (1972)
26. Uryu, T. et al.: Macromolecules *3*, 345 (1970)
27. Vernovic, B., Schuerch, C.: Carbohyd. Res. *14*, 199 (1970)
28. Masura, V., Schuerch, C.: Carbohyd. Res. *15*, 65 (1970)
29. Uryu, T. et al.: Makromol. Chem. *178*, 1929 (1977)

30. Uryu, T. et al.: Makromol. Chem. *180*, 1509 (1979)
31. Ito, H., Schuerch, C.: J. Polym. Sci. *16*, 2217 (1978)
32. Ito, H., Marousek, V., Schuerch, C.: J. Polym. Sci. Pol. Chem. Ed. *17*, 1299 (1979)
33. Murray, T. P., Wiliams, C. S., Brown, R. K.: J. Org. Chem. *36*, 1311 (1971)
34. Kops, J.: J. Polym. Sci. *A10*, 1275 (1972)
35. Sumitomo, H., Okada, M., Hibino, Y.: J. Polym. Sci. *B10*, 1871 (1972)
36. Hall, H. K., Steuck, M. J.: J. Polym. Sci. *A11*, 1035 (1973)
37. Jackman, L. M., Sternhell, S.: "Application of Nuclear Magnetic Resonance Spectroscopy in Organic Chemistry" 2nd ed., Pergamon Press, New York, 1969, p. 238
38. Okada, M., Sumitomo, H., Hibino, Y.: Polymer J. *6*, 256 (1974)
39. Penczek, S., Szymański, R.: Polymer J. *12*, 617 (1980)
40. Okada, M., Sumitomo, H., Sumi, A.: Macromolecules *15*, 1238 (1982)
41. Komada, H., Okada, M., Sumitomo, H.: Macromolecules *12*, 5 (1979)
42. Okada, M., Sumitomo, H., Komada, H.: Macromolecules *12*, 395 (1979)
43. Okada, M., Sumitomo, H., Komada, H.: Makromol. Chem. *178*, 343 (1977)
44. Lee, H., Neville, K.: Encyclopedia of Polymer Science and Technology, Vol. 6, p. 243
45. Bailey, W. J., Sun, R. L.: ACS Div. Polym. Chem. Prepr. *13*, 400 (1972)
46. Bailey, W. J.: J. Macromol. Sci.-Chem. *9*, 849 (1975)
47. Hall, H. K. Jr., Yokoyama, Y.: Polymer Bull. *2*, 281 (1980)
48. Yokoyama, Y. et al.: Macromolecules, *15*, 11 (1982)
49. Endo, T., Saigo, K., Bailey, W. J.: J. Polym. Sci. Pol. Lett. Ed., *18*, 457 (1980)
50. Endo, T. et al.: J. Polym. Sci., Pol. Lett. Ed. *18*, 771 (1980)
51. Bailey, W. J.: J. Elastoplast. *5*, 142 (1973)
52. Sakai, S., Kobayashi, Y.: Ischii, J. Org. Chem. *36*, 1176 (1971)*y*
53. Bailey, W. J., Katsuki, H.: ACS Div. Pol. Chem. Prepr. *14*, 1679 (1973)
54. Bailey, W. J., Endo, T.: J. Polym. Sci. **1***4*, 1735 (1976)
55. Matyjaszewski, K.: J. Polym. Sci., *22*, 29 (1984)
56. Matyjaszewski, K., Penczek, S.: Macromolecules, submitted

9 Polymers from Cyclic Esters

Lactones, i.e. esters of hydroxyacids and their dimers, like glycolide and lactide, are two major groups of cyclic esters used in polymerization. These compounds are used on the large scale and polymerized mostly by anionic or coordinative mechanisms. Four, six-, and seven-membered lactones polymerize by both cationic and anionic mechanisms. Polyesters prepared in this way are however only a small fraction of the polyesters prepared by polycondensation.

Anionic polymerization, especially if carried out in the presence of bulky counter-ions, crown ethers, or cryptates, proceeds under the same conditions much faster than the cationic process for ε-caprolactone and at a similar rate for β-propiolactone. Both cationic and anionic polymerizations of β-propiolactone may proceed under proper conditions as living processes, i.e. in the absence of transfer and termination [1,2,3]. The synthesis of poly(β-propiolactone) $\overline{M}_n = 100000 - 300000$ using $CH_3COCl \cdot SbCl_5$ initiator at 20 °C in CH_2Cl_2 solvent has been reported; the molecular weights were determined by the ratio $\overline{DP}_n = [M]_0/[I]_0$, i.e. indicating living polymerization with quantitative initiation [3].

Both cationic and anionic polymerizations of ε-caprolactone are accompanied by the formation of cycles [4,5] as discussed later in this section. Anionic polymerization leads to living systems whereas side reactions were observed in all cationic processes.

The cationic and coordinative polymerizations are the only routes for polymerization of the six-membered dimers of the glycolic and lactic acids, i.e., glycolide and lactide, leading to polymers used in medicine.

9.1 Initiators

Cationic polymerization of lactones can be initiated by a wide variety of protonic and Lewis acids, and by oxonium and carbenium salts.

Application of H_2SO_4, $FeCl_3$ and $SnCl_4$ leads to rather low-molecular-weight poly(β-propiolactone) with $\overline{M}_n = 800–1000$ ($\eta_{sp}/C = 0.05$ dl/g) [7]. High-molecular-weight polymers ($\overline{M}_n = 10^5$) are formed with $AlCl_3$, oxycarbenium salts ($CH_3CO^+ClO_4^-$) and trifluoroacetic acid [8]. Phosphates of Ti, Zr and V are also efficient initiators of β-propiolactone polymerization [9].

γ-Rays induce the cationic polymerization of β-propiolactone in the solid state [10,11] and in solution [12]. At −78 °C (10^3–10^6 R/hr) crystalline polymer (mp = 110 to 120 °C) was formed with 15–20% yield ([η] = 0.15–0.20 dl/g). UV [13] and visible light with uranium nitrate activator [14] were also used but yields and molecular weights were low.

The cationic polymerization of glycolide and lactide was accomplished with Lewis acids such as SbF_3 and $ZnCl_2$ at rather high (>200 °C) temperatures. The other group of initiators are tin derivatives, like $SnCl_4$, $SnCl_2$, $Sn(Ph)_4$ or stanneous octoate, known also to give good results in the polymerization of lactones [15]. The mechanism of initiation and propagation as well as the structure of the active species are not clear. Besides the usual cationic mechanism the coordinative route is also possible (cf. Sect. 15.1.3.3).

9.2 Peculiarities of Chain Growth

The cationic polymerization of lactones is not yet sufficiently well understood. The structure of the active centers is not known with certainty and the direction of the opening of the lactone ring (O-alkyl (*a*) vs. O-acyl (*b*)) has been established only recently for a few representative monomers [16,17]:

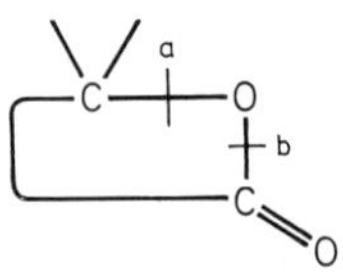

(9-1)

In the cationic polymerization of β-propiolactone (βPL), initiated with dimethyliodonium hexafluoroantimonate (a source of $CH_3^{\oplus}$ cations), the initiator induces rapid and quantitative formation of the active centers (by transfer of the methyl cation). The structure of the end-groups has been determined, the terminal one by cation trapping with triphenyl phosphine. These end-groups indicate O-alkyl ring-opening [16] (route a in Equ. 9.1.):

$$\underline{\underline{CH_3}}O\overset{O}{\overset{\|}{C}}CH_2CH_2O\overset{O}{\overset{\|}{C}}\sim\sim CH_2CH_2O\overset{O}{\overset{\|}{C}}CH_2CH_2\underline{\underline{P}}Ph_3 \qquad \text{O-alkyl} \quad (a) \qquad (9\text{-}2)$$

NMR: 3.67 δ(^{1}H) 24.0 δ(^{31}P)

O-acyl cleavage would lead to the formation of the following structure, not detected in NMR spectroscopy:

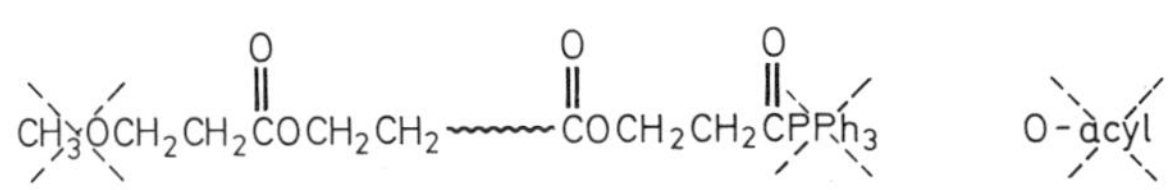

(9-3)

NMR: ~3.20 δ(^{1}H) ~10 δ(^{31}P)

In the polymerization of ε-caprolactone (εCL) initiated with "$CH_3^{\oplus}$" coming from $(CH_3)_2I^{\oplus}SbF_6^{\ominus}$, NMR spectroscopy of the end-groups (3.60δ by ^{1}H-NMR and 23.2δ by ^{31}P-NMR) also indicates O-alkyl opening of the seven membered lactone [17].

The following mechanism is proposed to account for the findings of the polymerization of βPL and εCL initiated with "$CH_3^{\oplus}$":

In Scheme (9-4) the exocyclic O-atom of the carbonyl group (which is more basic than the endocyclic one [18]) is preferentially attacked, giving the 2-methoxyoxetane-2-ylium cation (I). Then the endocyclic C-4 atom is attacked by monomer leading

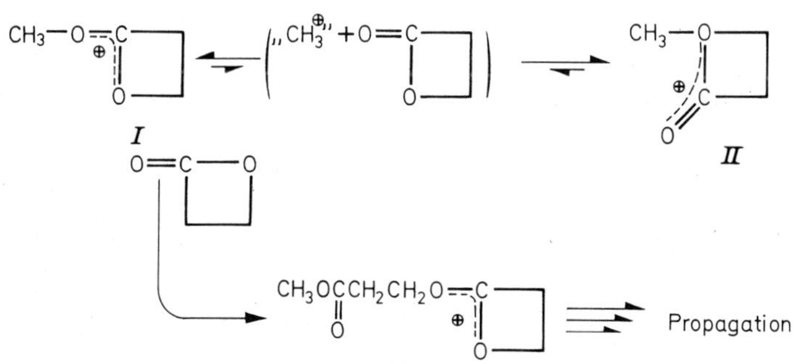

(9-4)

to O-alkyl cleavage Methyl cation attack on the less basic endocyclic O-atom and formation of the 1-methyl-2-oxo-oxetanium cation (II) is less probable. In this type of oxonium cation the O-acyl bond should be broken intra- or bimolecularly leading to CH_3OCH_2— ... end-groups, which have not been observed experimentally. This confirms the exocyclic attack and O-alkyl cleavage.

A slightly different direction of ring-opening was observed in the polymerization initiated with acylium salts. The end-groups: $\underline{CH_3}C(=O)OC(=O)CH_2CH_2$—... (2.20δ:^{1}H) and $\underline{CH_3}C(=O)CH_2CH_2C(=O)O$—... (2.08δ:^{1}H) were observed in nearly equal proportions although ~~~$CH_2CH_2\overset{+}{P}Ph_3$ (24.0δ:^{31}P) end-groups were present. Thus, during the first step both O-alkyl and O-acyl bonds are broken with equal rates. The probability of the latter cleavage is decreased by a factor of two with each subsequent monomer addition, eventually leading to only one type of end-group, in agreement with preferential O-alkyl cleavage. This can be explained by the rapid exchange of the exocyclic acetyl group in the 2-acetoxyoxetane-2-ylium cation (structure III — scheme 9-5). The substitution at the C-4 atoms is less probable (although participating in 50% of all ring-openings at this stage) than in cation I (scheme 9-4):

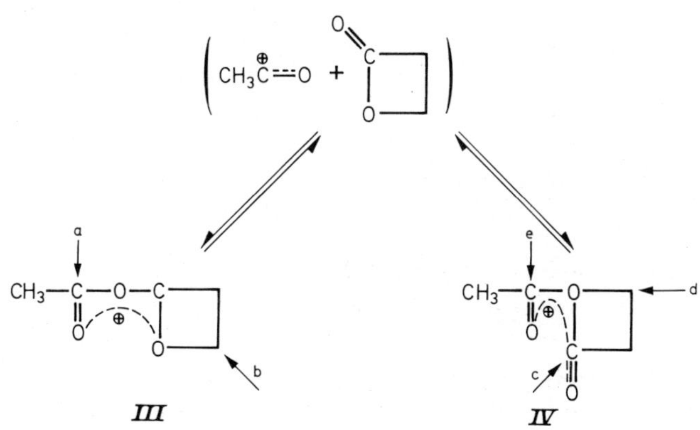

(9-5)

Once structure IV is formed rapid O-acyl cleavage ensues. However, once the O-alkyl linkage is broken no more exchange is possible and ring-opening proceeds as in I. Thus, the proportion of structures like III doubles with each propagation step under the conditions described in Ref. [17] (CH_2Cl_2, 25 °C), leading to $\sim CH_2CH_2\overset{+}{P}Ph_3$ tail-groups (97% after 5 monomer additions). This conversion from the acylium to oxonium cation is accompanied by the formation of the anhydride linkage in the polymer chain; these bonds were observed directly by IR spectroscopy at 1820 cm^{-1}.

9.2.1 Stereochemical and Mechanistic Considerations Pertinent to the Cationic Ring-Opening in Cyclic Esters

The β-substitution stabilizes cation II and increases the proportion of the O-acyl cleavage. Lando et al. [19,20] have found that the optical activity of poly (D-+-β-methyl-β-propiolactone) is lower than that of the monomer, indicating a simultaneous inversion (route I) and retention (II) at the β-C-4 atom.

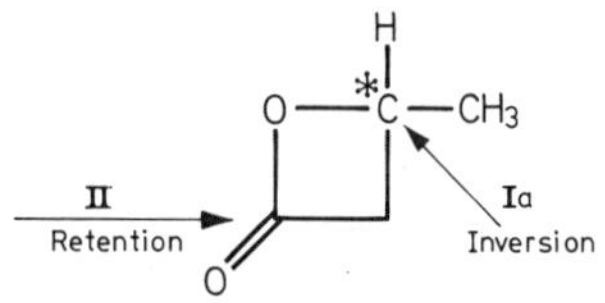

(9-6)

In the cationic polymerization of β-trichloromethyl-β-propiolactone retention (IIc) dominates [21].

Hydrolysis of L-lactide gives lactic acid with the same optical purity as hydrolysis of the polymer of L-lactide. This indicates that cationic polymerization of L-lactide proceeds with retention of configuration on both C* atoms, i.e. by O-acyl cleavage [21,22]:

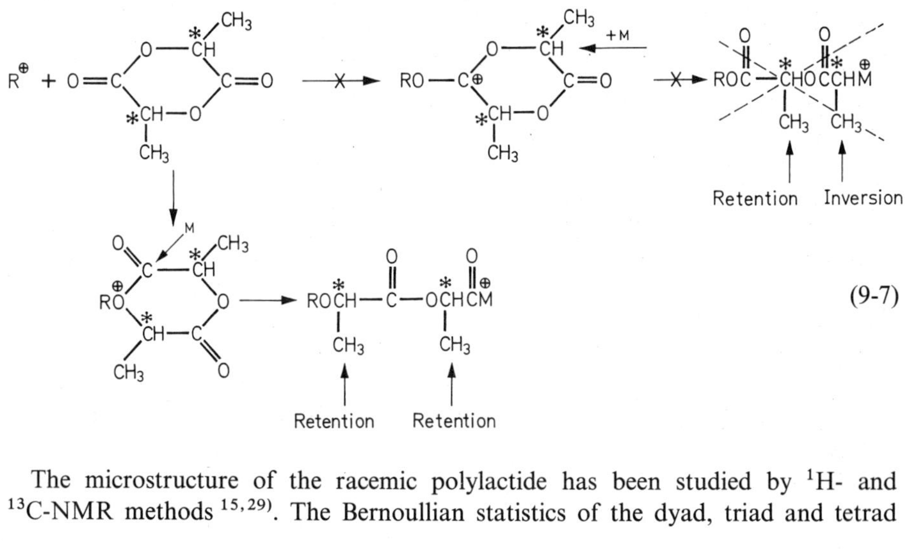

(9-7)

The microstructure of the racemic polylactide has been studied by ^1H- and ^{13}C-NMR methods [15,29]. The Bernoullian statistics of the dyad, triad and tetrad intensities indicate that transesterification is slower than chain growth at least in bulk at 130 °C with 50 ppm stanneous octoate initiator after 260 hrs [15].

Among the reasons for increased proportion of O-acyl cleavage in substituted lactones are: higher negative charge density on the endocyclic O-atom (consequently better stabilization of the cation type II by inductive effect of the alkyl substituent) and steric hindrance preventing bimolecular substitution and O-alkyl cleavage. The O-acyl cleavage can proceed intramolecularly and steric effects are unimportant.

Thus, the direction of the ring-opening may change by the nature of the monomer: For strained 4- and 7-membered unsubstituted lactones it proceeds via O-alkyl cleavage whereas for substituted lactones O-acyl cleavage becomes important.

A number of papers concern the kinetics of cationic lactone polymerization [24,25], and ionic triplets and branched polymeric ions were postulated to explain some unusual kinetic patterns. We think, however, that the fundamentals of these systems are not yet sufficiently understood (e.g. the structure of active species), for sophisticated analyses.

9.3 Side Reactions

Polymerization of βPL with a variety of initiators is a terminationless process. Lyudvig a.o. have mostly contributed to our present knowledge of the influence of reaction variables on the rates of polymerization [24–27]. They obtained quantitative yields in the polymerization of lactones using bulky complex counterions, like $SbCl_6^{\ominus}$. Thus, the decomposition of these anions, leading to transfer and termination in the polymerization of cyclic ethers and acetals, does not lead to side reactions in the polymerization of lactones. This can be ascribed to reversible reactions between acyl halides plus Lewis acids and the acylium cation [28]:

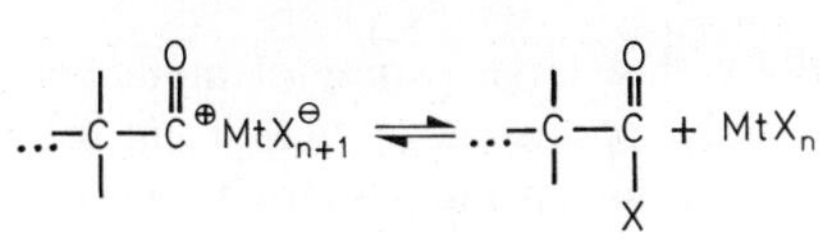

(9-8)

The addition of water or alcohols does not influence the reaction rate [26]; it is remarkable that water present in a concentration exceeding 30 times that of the initiator does not change the rate of polymerization although induction periods appear [26]. The addition of ketones leads to termination [27] and thus to limited yields.

Chain transfer to polymer and macrocyclization, frequently observed in cationic ring-opening polymerization, has also been studied for lactones [4]. In the polymerization of βPL cyclic oligomers were not observed, whereas in the polymerization of ε-CL initially high-molecular-weight polymer is formed which then is slowly degraded to cyclic oligomers, mostly from dimer to pentamer.

Degradation occurs because the monomer is much more basic and nucleophilic than the linear ester unit. Therefore, monomer will be consumed before chain transfer to polymer will occur. Typical GPC traces of the oligomers formed are shown in Fig. 9.1.

To decrease the proportion of cycles the reaction should be carried out at high monomer concentration and the polymerization should be terminated before cycles

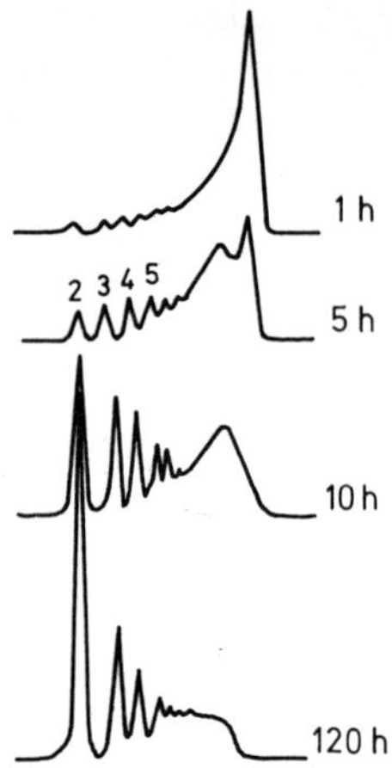

Fig. 9.1. GPC traces in polymerization of ε-caprolactone. $[\varepsilon CL]_0 = 0.214\ mol \cdot l^{-1}$ $[HSO_3CF_3]_0 = 0.038\ mol \cdot l^{-1}$, 100° in xylene [4)]

start to appear. If cycles are desired, long reaction times and low $[M]_0$ should be used. Below $[\varepsilon\text{-}CL]_0 = 0.2\ mol \cdot l^{-1}$ high-molecular-weight polymer cannot be formed and linear polymer formed initially is converted into cycles.

9.4 Copolymerization

Lactones copolymerize cationically with a large variety of monomers [29)]. This can be ascribed to the multiplicity of active species, e.g. equilibrium between oxonium (structure I) and carbenium (intramolecular opening of structure II) species (Scheme (9-4)). The copolymerization of βPL and εCL with cyclic ethers is discussed in Section 15.1.3.3. With 3,3-bis(chloromethyl)oxetane (BCMO) as the standard comonomer, the following order of reactivities of β-propiolactones was reported: β,β-dimethyl ≫ β-methyl > unsubstituted > α-methyl ≫ α,α-dimethyl [30)]. This order reflects the increasing nucleophilicity of the endocyclic O-atom and the higher probability of structure I in (9-4). Our understanding of the copolymerization with BCMO is rather meager; for example in the copolymerization of the BCMO-βPL pair the reactivity ratios change considerably during polymerization [31)]. Also the proportion of βPL homodyads is larger in the copolymers than that calculated by the simple four-parameter equation (cf. Sect. 15.1, Eq. 15-1) [32)]. Two types of active species were proposed to explain these results, one being the usual oxonium ion and the second the coordinative type [33)].

Copolymerizations of lactones with acetals [34)], cyclic amines [35)], and styrene [36)] have also been reported.

9.5 Applications

Polyglycolide, polylactide and their copolymers are produced commercially by cationic polymerization, and used as e.g. synthetic, biodegradable surgical sutures (Dexon® by American Cyanamide or Vitcar® by Ethicon). The mechanical and biological properties of polyester sutures are described in detail in reviews [37,38]. It is stressed that the properties of polyglycolide sutures are much more uniform than those of natural catgut. Initial knot strengths are higher and they are superior to gut *in vivo* during the first 20 days, i.e. the period critical in most surgical operations. Absorption of polyglycolide sutures proceeds via hydrolysis without intervention of enzymes. Studies with labeled polymer indicate that less than 1% remains after 4 months at the implant site. The main degradation products are glycolic acid excreted in urine and CO_2 in expired air. Homopolymers of lactic acid and β-hydroxybutyric acid have unacceptable absorption characteristics; however, a 5–10% lactide content in the copolymer allows more precise regulation of biodegradability of the suture [39].

Glycolide (mp = 84–85 °C, t. subl. = 80–84 °C/1 mm Hg) is polymerized in the melt using Lewis acids (SbF_3, $ZnCl_2$, $SnCl_2$, stanneous octoate) or other metal salt catalysts [40,41,42]:

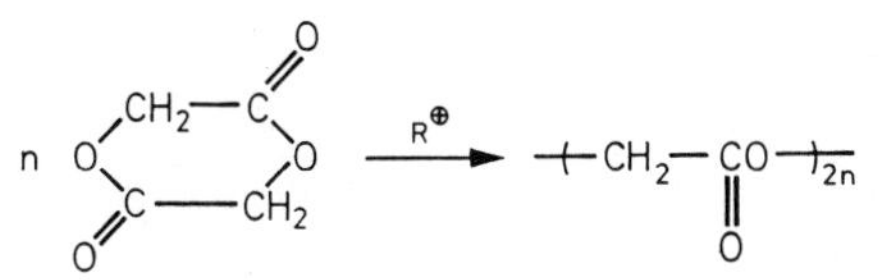
(9-9)

Melt copolymerization requires high temperatures (T > 200 °C). In spite of the six-membered ring it has the extrapolated ceiling temperature T = 1800 °K, well above the decomposition temperature [43]. Polyglycolide (mp = 224–226 °C) is converted into multifilament yarn by usual melt-spinning and -drawing procedures giving products with tenacities (5–10 g/den) close to polyethyleneterephtalate fibres.

Lactide (mp = 96 °C, bp = 150 °C/12 mm Hg) is prepared from lactic acid in two stages. First, the acid is heated to 130 °C; at this stage the low-molecular-weight polymer is formed and water is expelled. Further heating in the presence of zinc dust gives the lactide. The polymer is formed easily with Lewis acids or organometallic compounds ($SnCl_4$, $ZnCl_2$, $AlEt_3$) [21]. Other cationic initiators (tritylium, tropylium, triethyloxonium salts) failed to initiate the polymerization [21]. Poly(L-lactide) is levorotary (ORD, CD); it forms clear films (mp = 170 °C, Tg = 110–115 °C). Light scattering measurements indicate $\overline{M}_w = 30000 - 130000$. Lactide copolymerizes with other lactones, e.g. β-propiolactone (5–25% lactide content), ε-caprolactone (10–90%) and glycolide (30–90%), but it fails to copolymerize with cyclic ethers [21].

Poly(β-hydroxybutyrate) occurs naturally as the optically active high-molecular-weight polyester in some bacteria (e.g. Bacterium megaterium):

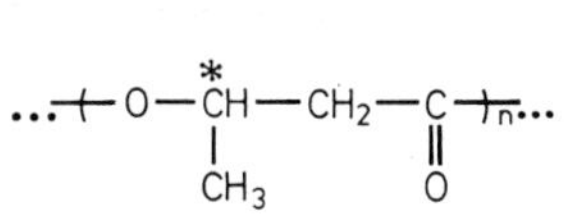
(9-10)

This polymer has also been obtained synthetically via cationic polymerization of D(+)β-methyl-β-propiolactone [19]. The racemic monomer polymerizes predominantly to the isotactic polymer.

Poly(ε-caprolactone) is another practically important polyester formed by ionic polymerization of the cyclic ester. Cationic polymerization requires relatively high temperatures; this enhances proton transfer and decreases the molecular weight, whereas anionic polymerization provides living systems.

PolyεCL has an ability to blend with several other polymers and can be used as an adhesive for polymer films of the same or different composition, as an excellent low-profile additive for fiber-glass-reinforced polyesters and as a pigment dispersant. It improves the impact strength and crack resistance of some thermoplastic polymers.

9.6 References

1. Słomkowski, S., Penczek, S.: Macromolecules *13*, 229 (1980)
2. Deffieux, A., Boileau, S.: Macromolecules *16*, 369 (1980)
3. Khomyakov, A. K., Lyudvig, E. B.: Dokl. Ak. Nauk SSSR *201*, 877 (1971)
4. Yamashita, Y.: Polymer Preprints *20*, 126 (1979)
5. Ito, K., Hashizuka, Y., Yamashita, Y.: Macromolecules *10*, 821 (1977)
6. Gresham, T. L., Jansen, J. E., Shaver, F. W., Gregory, I. T.: J. Amer. Chem. Soc. *70*, 1003 (1948)
7. Okada, M., et al.: Kogyo Kagaku Zasshi *66*, 104 (1968)
8. Cherdron, H., Ohse, H., Korte, F.: Makromol. Chem. *56*, 179 (1962)
9. Jap. Patent 15245 (1961)
10. Okamura, S., Hayashi, K., J. Chem. Phys. *59*, 429 (1962)
11. Hayashi, K. et al.: Makromol. Chem. *68*, 194 (1963)
12. Jap. Patent 13895 (1962)
13. Jap. Patent 6392 (1961)
14. Sakamoto, M., Hayashi, K., Okamura, S.: J. Polymer Sci. *B3*, 205 (1965)
15. Schindler, A., Harper, D.: J. Polym. Sci., Polym. Letters Ed. *14*, 729 (1976)
16. Hofman, A. et al.: IUPAC Symposium "Ring-Opening Polymerization of Heterocycles", Karlovy Vary 1980, p. M28
17. Hofman, A. et al.: in preparation
18. Kirby, A. J., In: Comprehensive Chemical Kinetics, Bamford, C. H., Tipper, C. F. H.: Eds., Vol. *10*, p. 57, Elsevier, Amsterdam 1972
19. Agostini, D. E., Lando, J. B., Shelton, J. R.: J. Polym. Sci. A1, *9*, 2775 (1971)
20. Shelton, J. R., Agostini, D. E., Lando, J. B.: J. Polym. Sci. A1, *9*, 2789 (1971)
21. Schulz, R. C.: Proc. IUPAC Int. Symp. Macromol. Chem., Budapest 1969, p. 185
22. Kleine, J., Kleine, H. H.: Makromol. Chem. *30*, 23 (1959)
23. Lille, E., Schulz, R. C.: Makromol. Chem. *178*, 1901 (1975)
24. Belenkaya, B. G., Lyudvig, E. B.: Vysokomol. Soed. *A20*, 565 (1978)
25. Belenkaya, B. G., Levenko, A. I., Lyudvig, E. B.: Vysokomol. Soedin. *A20*, 559 (1978)
26. Khomyakov, A. K. et al.: Vysokomol. Soed. *A18*, 1699 (1976)
27. Lyudvig, E. B., Sanina, G. S., Khomyakov, A. K.: Vysokomol. Soed. *A16*, 801 (1974)
28. Cook, D.: Can. J. Chem. *37*, 48 (1959)
29. Sazanov, I. N.: Usp. Khimii *37*, 1084 (1968)
30. Yamashita, Y.: Kogyo Kagaku Zasshi *66*, 1493 (1963)
31. Tada, K., Saegusa, T., Furukawa, J.: Makromol. Chem. *102*, 47 (1967)
32. Tada, K., Saegusa, T., Furukawa, J.: Makromol. Chem. *71*, 711 (1964)
33. Yamashita, Y. et al.: Macromolecules *2*, 613 (1969)
34. Yamashita, Y., Kondo, S., Ito, K.: Polymer J. *1*, 327 (1970)
35. Kagiya, T. et al.: Kogyo Kagaku Zasshi *68*, 1741 (1965)

36. Tsuda, T., Shimizu, T., Yamashita, Y.: Kogyo Kagaku Zasshi *67*, 2145 (1964)
37. Frazza, E. J.: In: Encyclopedia of Polymer Sci. and Technology Vol. *Suppl. 1*, p. 587, Wiley, New York, 1976
38. Frazza, E. J., Schmitt, E. E.: J. Biomed. Mat. Res. Symp. *1*, 43 (1971)
39. Cann, J. Jr. et al.: Amer. J. Surg. *128*, 19 (1974)
40. Lowe, C. E.: US Pat, 2668162 (1954)
41. Chup, K. et al.: Makromol. Chem. *100*, 262 (1967)
42. Sania, G. S., Lyudvig, E. B., Petrov, E. S.: Vysokomol. Soed. *B19*, 121 (1977)
43. Lebedev, B. V. et al.: Vysokomol. Soed. *A20*, 29 (1978)

10 Polyamines and Polysulfides

Cationic polymerization of both cyclic amines and sulfides is combined in this section because of the similarities in their polymerization. This similarity is due to the high nucleophilicities of monomers that belong to these groups, the high stability of the active centers and the relatively higher basicity of heteroatoms in the polymer units than in the corresponding monomers. This combination of features leads to very similar synthetic methods used for polymer preparations.

There are two different ways of preparing the simplest polyamine, polyethylenimine. The first method leads to a commercial polymer. The parent monomer (aziridine) is used and the resulting polymer is highly branched due to pronounced chain transfer. The second method, leading to linear polyethylenimine, requires the preparation of polyoxazoline intermediate and is discussed in the next section.

Synthetic aspects as well as the mechanisms of polymerization of these monomers were reviewed by Goethals [1,2,3,4], who largely contributed to these fields.

The synthesis of polyamines can occur only by cationic polymerization. Anionic polymerization is not possible because the positive charge on the α-C atoms in the monomer is too low to be susceptible to attack by anions. On the contrary, the $C—N^{\oplus}\equiv$ bond in the ammonium cation is sufficiently reactive to break when the carbon atom is attacked by a highly nucleophilic monomer.

Cationic polymerization of cyclic amines proceeds usually with a high activation energy and often requires relatively high temperatures, even for the strained four-membered rings, e.g. for 1,3,3-trimethylazetidine $\Delta H^{\neq} = 73\ kJ \cdot mol^{-1}$, $k_p = 1.4 \times 10^{-4}\ mol^{-1} \cdot l \cdot s^{-1}$ at 78 °C [5].

Cyclic sulfides polymerize via anionic and cationic mechanisms.
Although it is relatively easy to prepare high-molecular-weight polysulfides, these products have not yet found applications. This is mostly due to a lack of industrially applicable monomer preparation methods. Neither ethylene sulfide nor propylene sulfide can be prepared by direct reaction of the corresponding olefins with elemental sulfur. Moreover, it is difficult to obtain poly(ethylene sulfide) sufficiently stable at higher temperatures and its odour may become offensive.

Sigwalt et al. prepared high-molecular-weight poly(ethylene sulfide) [6] by anionic polymerization and tested and described its major features. It is a high melting (mp = 205 °C), crystalline material. Poly(propylene sulfide) can easily be obtained by living anionic process, giving an elastomer with a good combination of mechanical properties [7,8].

Polymers with sufficiently high molecular weight, however, could not be prepared by cationic polymerization. The cationic process is accompanied by the formation of large quantities of oligomeric products and the yields are limited due to degradative chain transfer to polymer. In the anionic polymerization chain transfer to polymer does not take place, cycles are absent, termination with polymer chains is excluded, and therefore the yields and $\overline{M}_n$ are higher. Mechanistic features of the cationic synthesis of cyclic sulfide and amine polymers will be discussed briefly.

10.1 Initiating Systems

Polymerization of cyclic sulfides and amines can be initiated not only by the usual cationic and cationogenic compounds like Lewis and protonic acids, carbenium and oxonium salts and esters of strong acids, but also by alkyl halides, which are active enough to induce polymerization of some azetidines [9,10].

Protonic acids are efficient initiators for the polymerization of both sulfides and amines. The polymerization of thiiranes initiated with perchloric acid proceeds without induction periods. Induction periods are present, however, with methyl fluorosulfonate initiator [11]. Secondary sulfonium salts are more reactive than tertiary ones (the opposite is true with oxonium ions) [12] and induce rapid polymerization:

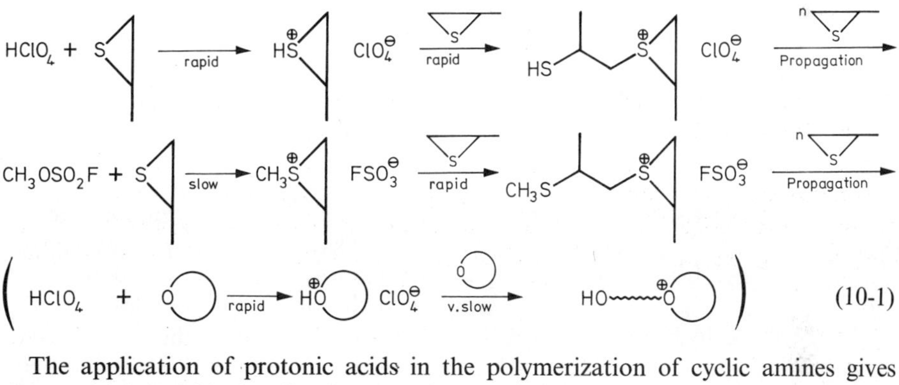

(10-1)

The application of protonic acids in the polymerization of cyclic amines gives rise to a higher yield of cyclic oligomers due to the formation of more basic secondary amine end-groups and rapid end-to-end cyclization. For instance, in the polymerization of 1-t-butyl aziridine, 25% of cyclic pentamer was formed with HSO_3CF_3 initiator [13].

Tritylium salts react reversibly with sulfides via alkylation and irreversibly via hydride transfer, [14], as it has been described earlier for cyclic ethers (cf. Adv. Polymer Sci. *37*):

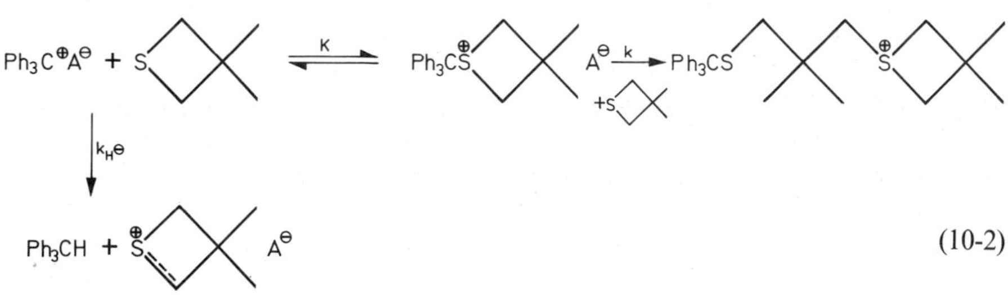

(10-2)

The equilibrium constants K = 2.27 and K = 0.096 $mol^{-1} \cdot l$ at 25 °C in CH_2Cl_2 solvent were determined for the model compounds tetrahydrothiophene and diethyl sulfide, respectively. It has been anticipated [14] that under "usual" polymerization conditions of thietanes 40 to 75% of trityl salt gives 1-tritylsulfonium cations.

Oxonium salts have been used as initiators in the polymerization of sulfides and amines as low-molecular-weight compounds (e.g. $Et_3O^{\oplus}BF_4^{\ominus}$) [15,16] and as dicationically living polyTHF [17,18] in the synthesis of diblock and triblock copolymers. Among alkyl esters and halides, tosylates, iodides, bromides, fluorosulfonates, and triflates were used in the polymerization of azetidines [4,9,10] and aziridines [19,20]. Methyl triflate forms the first alkylation product in the polymerization of 1-t-butylaziridine, which precipitates out of solution [19]:

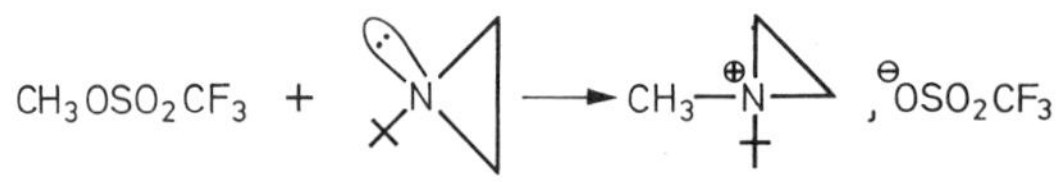

(10-3)

Cyclic sulfides are less nucleophilic than amines, thus alkyl halides and esters of weak acids are inefficient for their polymerization.

Only one growing species is formed per molecule with triflic anhydride.

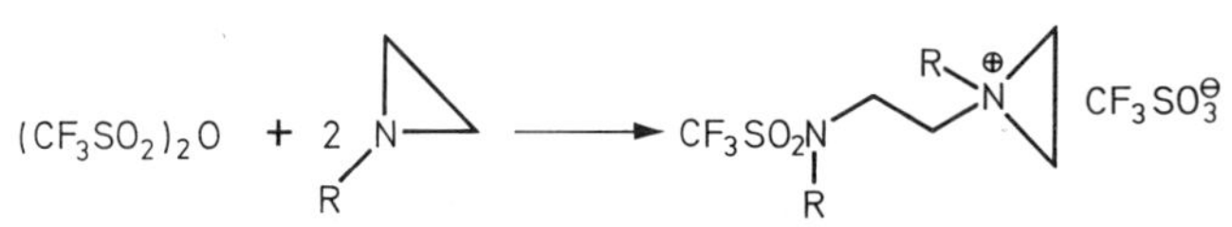

(10-4)

This is in contrast to the already discussed polymerization of cyclic ethers and acetals, where two growing species are formed, yielding dicationically growing macromolecules. The amide group, which could be a potential site for the second growing species, is unreactive. As shown in Table 10.1 the majority of polymers were prepared with Lewis acid initiators, most frequently with the $BF_3 \cdot OEt_2$ complex.

Mixed cyclic anhydrides (e.g. 3-hydroxy-1-propanesulfonic acid sulfone [21]) provide the zwitterionic polymerization of 1-(2-cyanoethyl)azetidine [22] and 1-(2-cyanoethyl)-aziridine [23]. The sulfonate anions are sufficiently stable for the formation of high-molecular-weight polyamines.

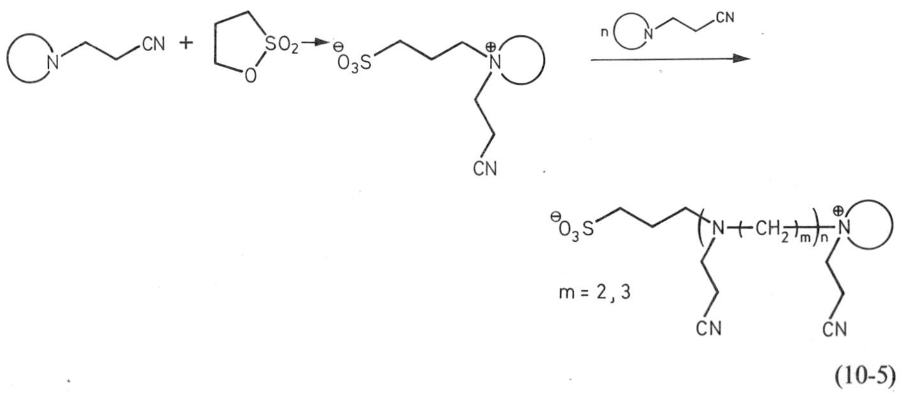

(10-5)

Table 10.1. Examples of polysulfides and polyamines prepared by cationic polymerization

No.	Monomer	Conditions	Properties	Ref.
1	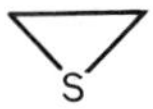	BF_3, H_2SO_4	mp $\simeq$ 170 °C, $\overline{DP}_n \simeq 16$	[24]
2	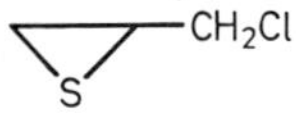	H_2SO_4	dark brown, viscous liquid $\overline{M}_n \simeq 1000$	[25]
3		$BF_3 \cdot OEt_2$, CH_2Cl_2 25 °C, 48 h	soft. p $\simeq$ 100°, Tg = −17 °C η_{inh} = 0.2 dl/g, 90–100% yield	[26]
4	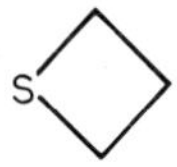	$BF_3 \cdot OEt_2$, bulk or CH_2Cl_2 solv., 10 °C, 24 h	soft. p $\simeq$ 50–60°, Tg = −40 to −60 °C η_{inh} = 0.2–0.6 dl/g, yield 60%	[27]
5	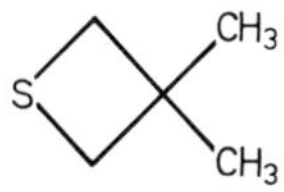	$BF_3 \cdot OEt_2$, 25 °C, 48 h		[28]
6	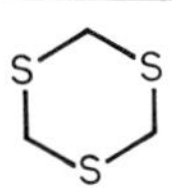	CH_3I, $(CH_3)_2SO_4$, BF_3, SbF_3, 220–230 °C γ-Rays	mp = 256 °C, soluble in sulfolane at 200 °C	[29] [30] [31]
7	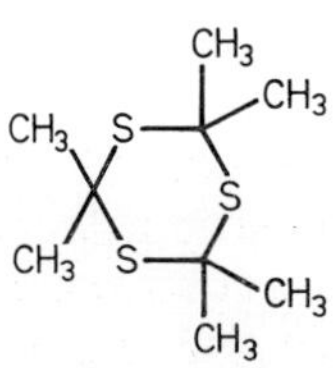	−20 °C, spontaneously	$\overline{M}_n$ = 2500–27000 mp = 120–125 °C	[32]
8	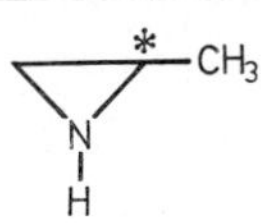	BF_3, $BF_3 \cdot OEt_2$, 80 °C bulk, 2 min	Yield = 100%, [η] $\simeq$ 0.15 dl/g soft. p $\simeq$ 15° (racemic), $\approx$70° (D) $\simeq$80° (L)	[33] [34]
9	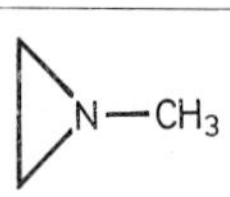	—	η_{sp}/c = 0.15–0.19 dl/g, (1%, H_2O); soluble H_2O, ROH, benzene, acetone	[34]
10	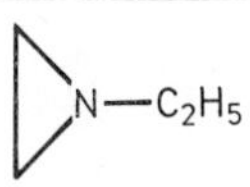	—	η_{sp}/c = 0.22 dl/g (1%, H_2O); soluble H_2O, ROH, benzene, acetone	[34]
11	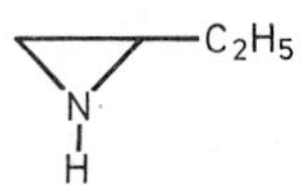	—	η_{sp}/c = 0.05–0.07 dl/g (1%, H_2O); soluble H_2O, ROH	[34]
12	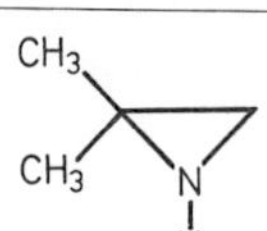	—	η_{sp}/c = 1.07 dl/g (1%, $CHCl_3$); mp = 55 °C; soluble ROH, Et_2O, insol. H_2O	[34] [35]

Table 10.1. (continued)

No.	Monomer	Conditions	Properties	Ref.
13	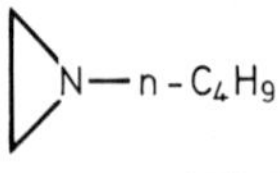	—	soluble ROH, benzene, THF; insoluble H_2O viscous oil, $\overline{M}_n = 2900$,	[34]
14	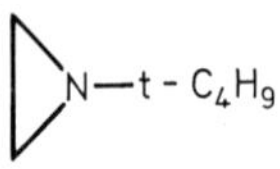	—	mp = 116 °C soluble benzene, insol. H_2O	[34]
15	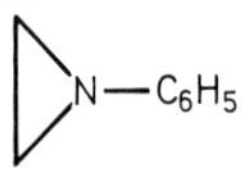	—	mp = 270 °C insoluble H_2O	[34]
16	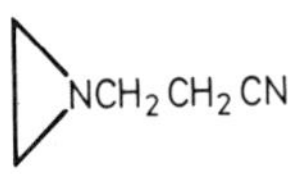	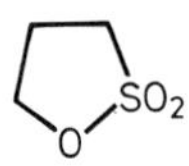 THF, ~60°C, 2 h	$\eta_{sp}/c = 0.137$ dl/g 52% yield, mp = 95–101 °C	[23]
17	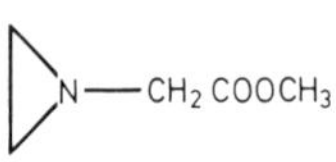			[37]
18	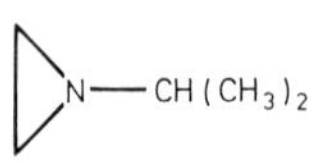		$\overline{M}_n = 24000–40000$	[37]
19	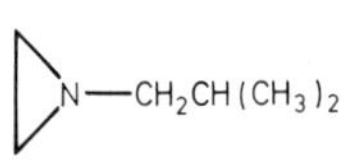			[37]
20	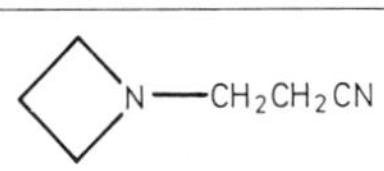	a) CH_3OTos, 3 h, 60 °C b 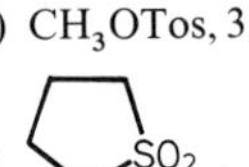, 60°C, 24 h	white powder, yield >90%, mp = 57 °C, $\eta_{sp}/c = 0.20$ dl/g, (a); 0.10 dl/g, (b)	[22]
21	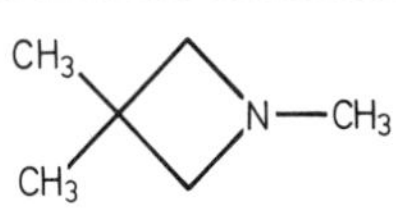	a) $BF_3 \cdot OEt_2$, 100 °C, 24 h b) $HClO_4$, 100 °C, 24 h	a) viscous oil, $\overline{M}_n = 2900$, yield = 93% b) tacky substance $\overline{M}_n = 6300$, yield = 50%	[38] [38]
22	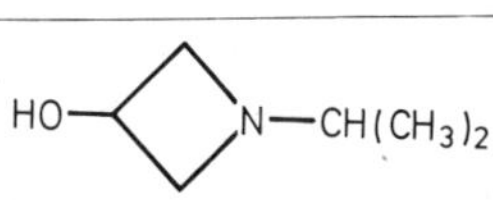	$Et_3O^{\oplus}BF_4^{\ominus}$, bulk, 100 °C, 70 h $Ph_3C^{\oplus}PF^{\ominus}$ $PhCH_2Br$	$\overline{DP}_n = 10$ to 20	[38]
23	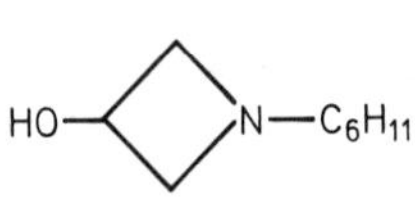	$Ph_3C^{\oplus}PF^{\ominus}$		
24	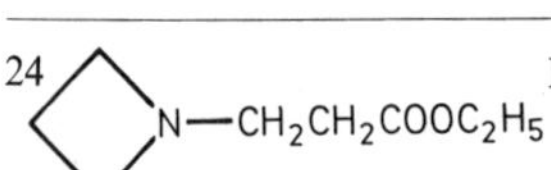	$Et_3O^{\oplus}BF_4^{\ominus}$, T > 80 °C. bulk	$\overline{M}_n = 8000$	[39]

Table 10.1. (continued)

No.	Monomer	Conditions	Properties	Ref.
25	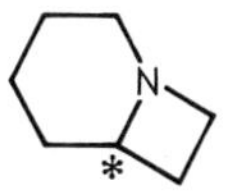	$BF_3 \cdot OEt_2$	mp = 94 °C (D, L), 91–93 °C (racemic) [η] = 0.4 dl/g,	40)
26		$BF_3 \cdot OEt_2$, 20 °C 30 min, bulk	80% yield, viscous liquid, $\overline{M}_n$ = 4000, $\|\alpha\|_D$ = −162°	36)

10.2 Peculiarities of Chain Growth

Cationic polymerization of cyclic amines and sulfides proceeds by onium ions as active species:

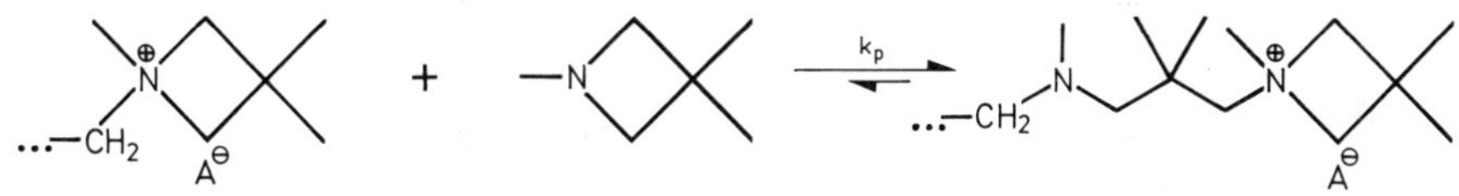

(10-6)

These species were directly observed by ^{1}H-NMR spectroscopy in the polymerization of thietanes [41], azetidines [10,42] and aziridines [19]. For a few monomers polymerization is nearly living (many are "slowly dying" as described by Goethals [18]). These processes were applied in the preparation of block, graft, star-shaped, etc., polymers.

In some other systems propagation is disturbed by chain transfer to polymer, accompanied by formation of cyclics. These processes result in limited yields of linear polymer, as discussed in the next section.

Polymerization of cyclic amines proceeds usually with high activation energies, due to the high strength of the C—N bond. To attain significant rates ($k_p \simeq 10^{-3}$ $mol^{-1} \cdot l \cdot s^{-1}$) requires temperatures from 30 to 100 °C for azetidines [16] and about 200 °C for the less strained bicyclic six-membered amines quinuclidine and triethylenediamine [9]. Cyclic sulfides polymerize more rapidly than amines but much slower than ethers of a comparable structure (e.g. for 3,3-dimethylthietane $k_p = 2 \cdot 10^{-2}$ $mol^{-1} \cdot l \cdot s^{-1}$ at 35 °C, which is at least 10^4 times lower than for 3,3-dimethyloxetane (cf. Adv. Polymer Sci. *37* (1980)).

10.3 Reactions Limiting High Polymer Formation

10.3.1 Macrocyclization

Macrocyclization is one of the processes that limits the preparation of high-molecular-weight polymers from both cyclic sulfides and some cyclic amines. The mechanism of cationic macrocyclization is discussed in detail in Chap. 3.

In the polymerization of cyclic sulfides, cyclization becomes particularly important because of the enhanced basicity (and nucleophilicity) of the linear sulfides in comparison with their parent monomers [43]. The simplest cyclic oligomer, formed in the polymerization of thiiranes, is a dimer-1,4-dithiane or its derivatives. 1,4-Dithiane was first observed by Bell in the polymerization of ethylene sulfide [44]; later Price isolated the styrene sulfide dimer [45] and recently Goethals obtained dimers of isobutylene and cyclohexene sulfides [46]. Structures of these dimers are:

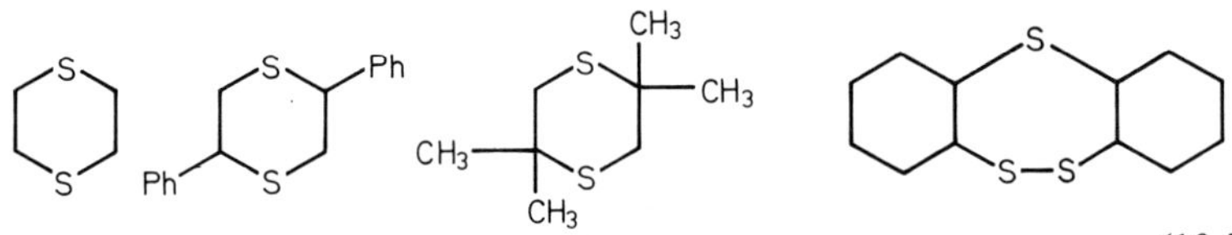

(10-7)

The yields and conditions of macrocyclization are given in Table 3.1 in Chap. 3.

In the polymerization of propylene sulfide and 1,2-butylene sulfide mainly tetramers were observed. Cycles were formed mostly during the slow degradation process that followed rapid polymerizations. Degradation can also be induced by adding cationic initiators to polymer prepared by other mechanisms, e.g. by anionic processes. Thus, poly(trans-2,3-butene sulfide) is rapidly degraded to equimolar amounts of 3,5,6,7-tetramethyl-1,2,5-trithiacycloheptane and trans-butene [47]. Poly(cis-2,3-butene sulfide) forms, however, a mixture of tetramer, trithiacycloheptane derivative and cis-butene [47]. If one is forced to use cationic processes for the synthesis of polysulfides, the reaction conditions should be controlled to avoid macrocyclization. If cyclic products are desired, the kinetics of their formation should be studied to determine optimum yields.

Polymerization of 1-alkyl aziridines is often accompanied by the formation of a dimer, 1,4-dialkyl piperazine:

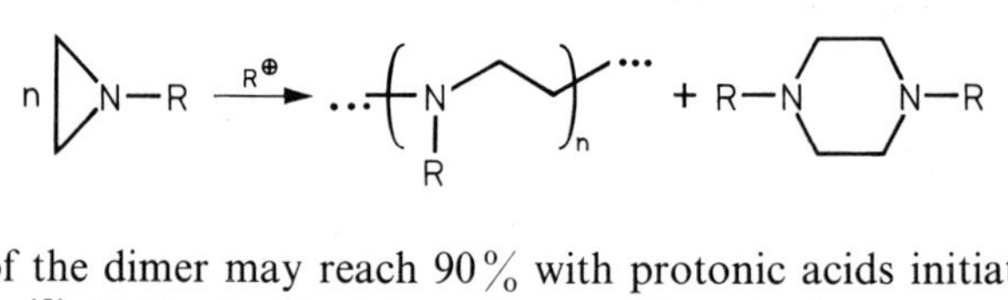

(10-8)

The yield of the dimer may reach 90% with protonic acids initiators in solvents of low polarity [48]. With alkyl halides and esters dimerization stops at the non-strained N-alkyl piperazinium salt stage [49]:

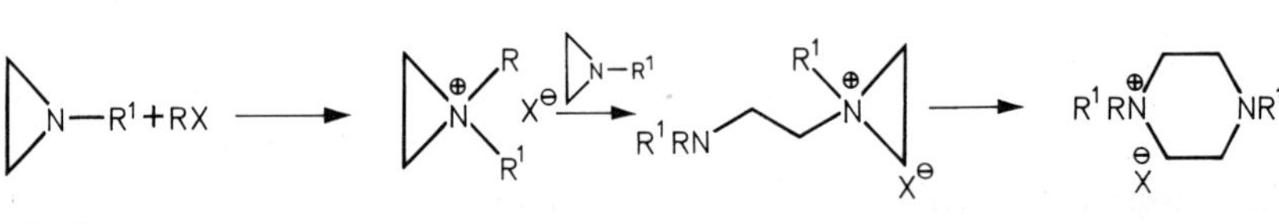

(10-9)

The last izomerization cannot proceed in one step, but should involve either the initial formation of the covalent species or the next monomer addition followed by the attack on the exocyclic α-CH_2 group (cf. also Sect. 15.1.3.2).

The relative proportions of linear polymer and cyclic oligomers depend on the monomer structure, initiator and solvent. High polarity and solvating power favour the formation of the linear polymer, whereas acid initiators increase the proportion of cyclics. 1-Benzyl aziridine polymerization initiated with p-toluenesulfonic acid in ethanol [50] quantitatively forms cyclic tetramer. With triethyloxonium salt cyclic dimer results by degradation of the initially formed linear polymer [51]. 1-Benzyl-2-methyl aziridine does not form cycles [51] when initiated with oxonium salt, whereas 1-(2-cyanoethyl)aziridine [52] and 1-benzyl-2(R)ethyl aziridine [51] yield mixtures of cyclic tetramers and linear polymer:

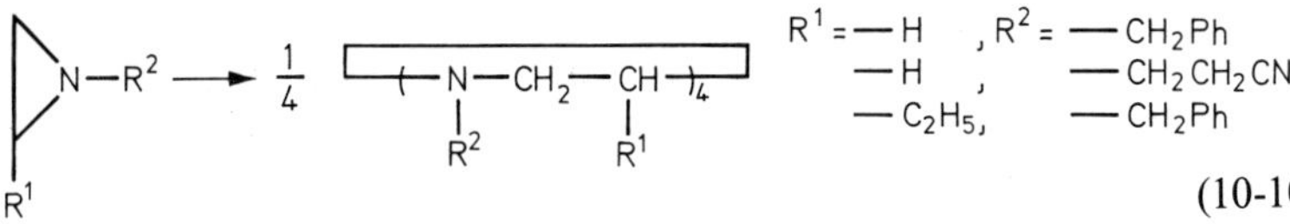

(10-10)

1-Ethyl aziridine, 1-butyl aziridine, 1-allyl aziridine and 1-(2-phenylethyl) aziridine give dimers [50] (R = $-C_2H_5$, $-C_4H_9$, $CH_2 = CH-CH_2-$, $PhCH_2CH_2-$).

1-Propylaziridine, 1-isobutylaziridine, 1-phenylaziridine and 1-(2-hydroxyethyl)-aziridine do not form cyclic oligomers in CH_2Cl_2 solvent with triethyloxonium salt initiator [4].

Recently the formation of cyclic pentamer in the polymerization of 1-t-butyl aziridine was reported. With $HOSO_2CF_3$ 5% yield at —10 °C and 25% yield at +40 °C was found [13].

In contrast to the polymerization of four-membered cyclic ethers that yield a large proportion of cycles (predominantly tetramers), macrocycles were not detected in the polymerization of azetidines or thietanes.

10.3.2 Transfer and Termination Reactions

The preparation of high-molecular-weight polyamines and polysulfides requires the elimination of transfer and termination reactions. Amines are so strong nucleophiles that their polymerization is possible even with such strongly nucleophilic anions as $Br^\ominus$, $Cl^\ominus$ or $F^\ominus$, as shown in the polymerization of 1-azabicyclo|4.2.0|octane [9,53], and in strongly nucleophilic solvents, i.e., water or alcohols [9]. Therefore the danger of termination by recombination of the growing cation with the counteranion or other adventitiously present nucleophiles is rather low for cyclic amines. In the cationic polymerization of less nucleophilic sulfides, some termination was observed, however, with $SbCl_6^\ominus$ anion [15]. $BF_4^\ominus$ and $PF_6^\ominus$ anions were stable during the entire polymerization process for all monomers.

As already mentioned and analysed in some detail in Vol. I (Adv. Polymer Sci. *37* (1980)), the cationic polymerization of cyclic sulfides and amines is accompanied by chain transfer to polymer. This type of transfer could become a termination, if the heteroatoms in the chain are more nucleophilic than those in the monomer:

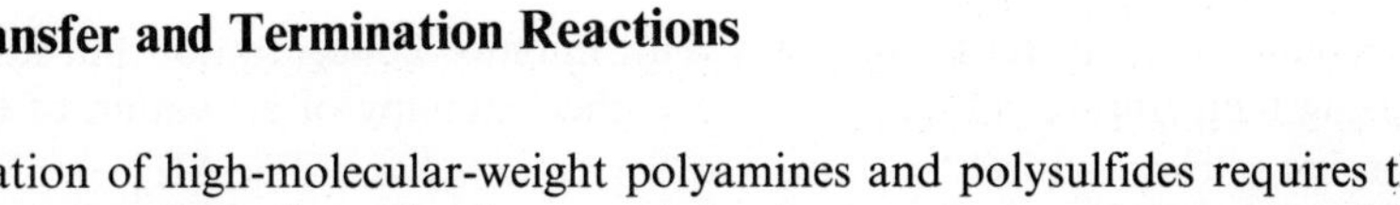

(10-11)

Depending whether transfer to polymer proceeds intramolecularly or bimolecularly, two equations relate the final polymer yield with the concentration of initiator and the ratio of the propagation (k_p) and transfer rate constant (k_t):

$$\text{intra-}^{15)}\text{: } \ln \frac{[M]_0}{[M]_\infty} = \frac{k_p}{k_t} [I]_0 \tag{10-12}$$

$$\text{inter-}^{54)}\text{: } \ln \frac{[M]_0}{[M]_\infty} - \frac{[M]_0 - [M]_\infty}{[M]_0} = \frac{k_p}{k_t} [I]_0/[M]_0 \tag{10-13}$$

Thus, the yield has to be referred to the given reaction conditions to find the influence of the monomer structure on the yield. At $[M]_0 = 1$ mol · l^{-1}, $[Et_3O^{\oplus}BF_4^{\ominus}] = 10^{-2}$ mol · l^{-1}, 0 °C, and in CH_2Cl_2 solvent the following yields were observed in the polymerization of aziridines: 2% (N-ethyl), 12% (N-Phenylethyl), 15% (N-isopropyl), 55% (N-cyanoethyl, and N-benzyl), 100% (N-tert-butyl) [1]. These yields have been related to the effective size of substitutents [54]. With an increase in the size of the substituent the rate constants of propagation and termination decrease, but the latter to a larger extent. This is well illustrated in the polymerization of four-membred rings. The rate constants of propagation decrease from k_p (in $mol^{-1} \cdot l \cdot s^{-1}$) = 0.6 (thietane) to 0.06 (3,3-dimethyl thietane) and 0.002 (3,3-diethyl thietane), but the ratios k_p/k_t increase from $k_p/k_t = 4$ to 20 and 360, respectively, in CH_2Cl_2 at 20 °C [1]. For 1-methyl azetidine at 20 °C $k_p = 5 \cdot 10^{-3}$ $mol^{-1} \cdot l \cdot s^{-1}$ ($k_p/k_t = 250$), whereas for 1,3,3-trimethyl azetidine extrapolation to 20 °C gives $k_p = 10^{-7}$ $mol^{-1} \cdot l \cdot s^{-1}$ and $k_p/k_t \simeq \infty^4$. This high ratio also means a complete conversion, not hampered by termination.

The final yield increases not only with initiator concentration but also with decreasing temperature. This is due to a higher enthalpy of activation of termination than that of propagation. For example, for 1-methyl azetidine $\Delta H_p^{\ddagger} = 57$ kJ · mol^{-1} and $\Delta H_t^{\ddagger} = 63$ kJ · mol^{-1} [55].

10.4 Some Properties and Applications of Polysulfides and Polyamines

Cationic polymerization of sulfides, as described above, proceeds with extensive chain transfer to polymer (inter- or intramolecularly). Thus, it can not be considered as a competitive synthetic method in comparison with anionic polymerizations, addition of dithiols to α,ω-diolefins or condensation of dithiols with α,ω-dihalides.

In contrast to polysulfides, cationic polymerization is the only method to prepare high-molecular-weight polyamines. Polyethyleneimine is the only polymer commercialized among the group of cationically prepared polysulfides and polyamines. Branched polyethyleneimines, are used on the technical scale mostly as flocculants.

Another practically applied polymer containing aziridine groups, is polyTHF with aziridine end-groups, applied in dentistry [56] (cf. Chap. 6). It has also been observed by

Bauman et al. that treatment of wool with polyacrylates containing azetidine side groups, and subsequent crosslinking, considerably decreases shrinkage [57].

The living character of polymerization of some amines is suitable for an efficient preparation of macromolecules with well defined structures (cf. Sect. 15.3).

10.4.1 Polyethyleneimine

Polyethyleneimine (polyEI) is obtained by cationic polymerization of the parent monomer in water solution. A monograph is available in which polyEI and its preparation are described in great detail [59]. Thus, we quote here only the most important facts and the interested reader should consult the source cited above.

These polymers are mostly applied in the paper industry for wet-strength papers usable in neutral or alkaline systems instead of the urea-formaldehyde resins requiring acid conditions for crosslinking. PolyEI increases retention of dyes, pigments and fillers. PolyEI is known as a powerful flocculant used in the clarification of fruit juices and in water treatment. It removes effectively clays, colloidal acids, pectines and tannines from water. Commercially available polyEI have molecular weights in the range $\overline{M}_n = 600$ to 100000 [60].

Polyethylenimine prepared in the usual way, i.e. with protonic acids in water solution at elevated temperatures is highly branched and has $\overline{M}_n < 100000$. Under anhydrous conditions $\overline{M}_n < 3000$. The kinetics of polymerization resembles a step-growth process. At 90% monomer conversion the main reaction product (80%) is the linear dimer [1-(2-aminoethyl) aziridine] [61,62]:

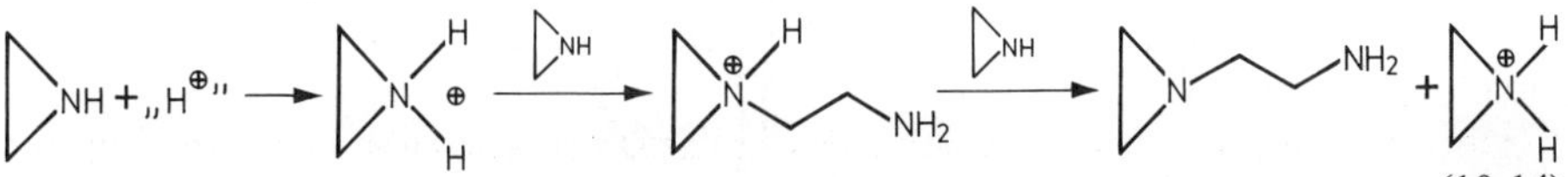

(10-14)

Potentiometric titration revealed that primary amines are more basic than tertiary ones. This is ascribed to inductive effects. Thus, the next step is as follows:

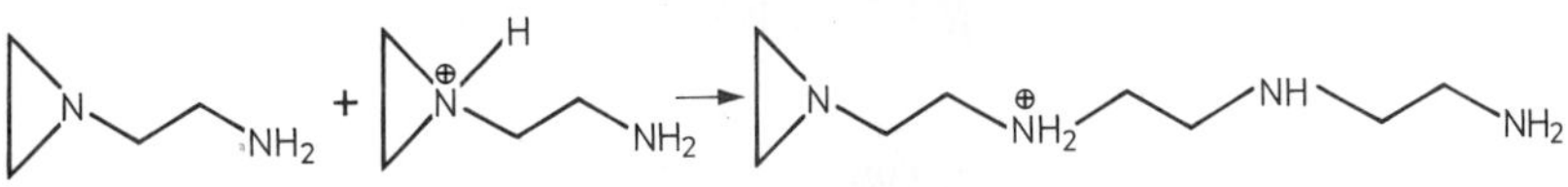

(10-15)

With the progress of the reaction the proportion of the secondary amines increases and their contribution to the overall polymerization becomes more important. At this stage branches and tertiary amine groups are formed:

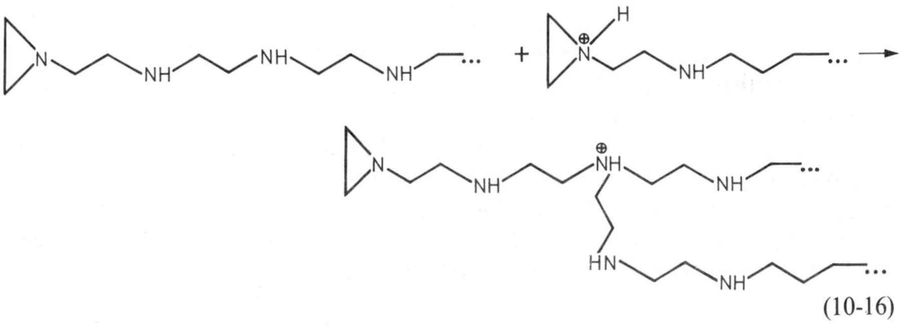

(10-16)

The proportion of primary, secondary and tertiary amino groups in the final polymer was found 1:2:1 indicating a high branching frequency.

Zuk et al. [63,64] have recently reported the preparation of linear polyEI using $HClO_4$ in H_2O solution at low temperature ($T < 40$ °C) and using a two-step procedure. In the first step a linear crystalline oligomers ($n = 1 - 4$) precipitate in form of hydrates, and in the next stage this pure oligomer is converted to high-molecular-weight ($\overline{M}_n > 10^4$) linear polymer. A yield of 40% was reported using the linear trimer as the starting material. Water was believed to favour the reaction by H-bonding between aziridinium cations and the primary amines. The linear structure of this polymer was confirmed by ^{13}C-NMR, spectroscopy where instead of eight signals of the "normal" branched polymer only one signal was observed (cf. Fig. 10.1).

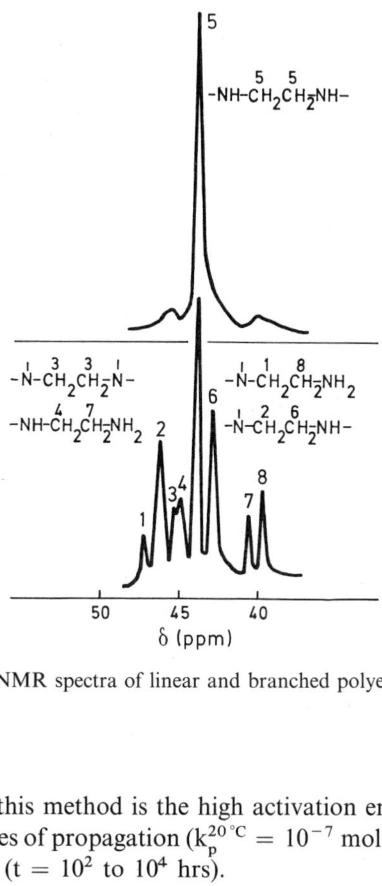

Fig. 10.1. ^{13}C-NMR spectra of linear and branched polyethyleneimines [63]

The disadvantage of this method is the high activation energy $E \simeq 80$ kJ · mol^{-1} which causes the low rates of propagation ($k_p^{20\,°C} = 10^{-7}$ mol^{-1} · l · s^{-1}), and requires extented reaction times ($t = 10^2$ to 10^4 hrs).

Another method, leading to an enhanced formation of linear polyamines, is by the introduction of RCO—, ROCO—, RSO_2—, or $R_2P(O)$— substitutents at N-atoms to replace the easily migrating H-atoms.

Recently, with perfluoroalkyl derivatives [$CF_3(CF_2)_nCO$—, n = 2 and 6] a linear polymer with $\overline{M}_n = 3650$ was synthesized using 1% of a protonic acid [65]. These substitutents can later be removed by hydrolysis, leaving the linear polyethyleneimine.

Polymerization of substituted monomers may also involve migration of the substituent giving polymers differing in their microstructure from polyEI. This has been shown by Araki [66]:

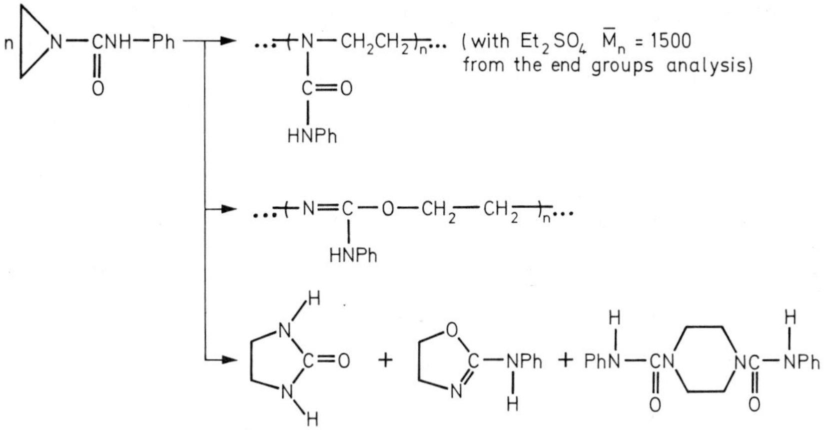

(10-17)

Saegusa et al. invented a new method of preparing linear polyEI [67]. This method involves the izomerization polymerization of cyclic iminoethers (cf. Chap. 12), followed by alkaline hydrolysis, leading to perfectly linear crystalline polymers. Using oxazoline (x = 2) and oxazine (x = 3), polyEI and poly(trimethyleneamine) were prepared:

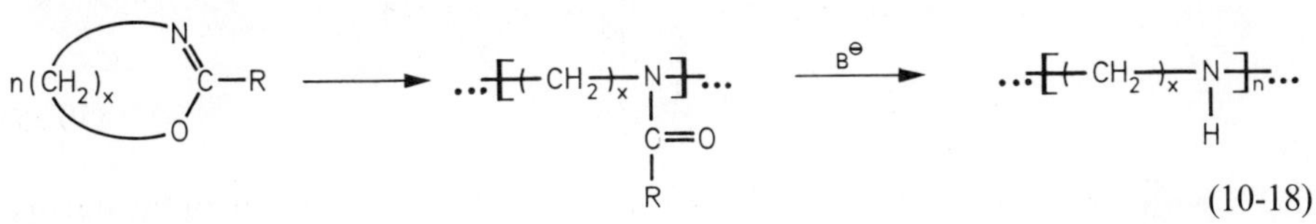

(10-18)

Complete basic hydrolysis was obtained only for R = H, when the formyl group was removed. Molecular weights were somewhat restricted when polymers were prepared according to Saegusa's description.

Recently, very high-molecular-weight. polyEI was prepared by acid hydrolysis of polymer formed from 2-phenyl-2-oxazoline. According to viscosity measurements ($\eta_{sp}/c = 0.15\ l \cdot g^{-1}$) the authors claim $\overline{M}_w = 350000$ [68].

10.5 Microstructure Studies by Degradation of Polysulfides

In Vol. I (Adv. Polymer Sci. *37*) we have discussed the mechanism of polymerization of thiiranes relative to polymer microstructure. The microstructure influences the properties of the polymers; e.g. linear polyethyleneimine is crystalline (mp = 59 °C), whereas the branched one is amorphous.

Recently, Goethals et al. [69] have analyzed the degradation products of poly(cis-2,3-dimethyl thiirane) (polyDMT) and poly(trans-2,3-dimethyl thiirane) to evaluate the microstructure of these polymers. Six different geometric isomers of 3,4,6,7-tetramethyl-1,2,5-trithiepane, the main products of degradation, were separated by GLC. Poly(cis-DMT) leads exclusively to two isomers with both pairs of vicinal methyl groups in trans position to each other (I and II). Poly(trans-DMT) gives isomers with vicinal methyl groups in cis position (III and IV):

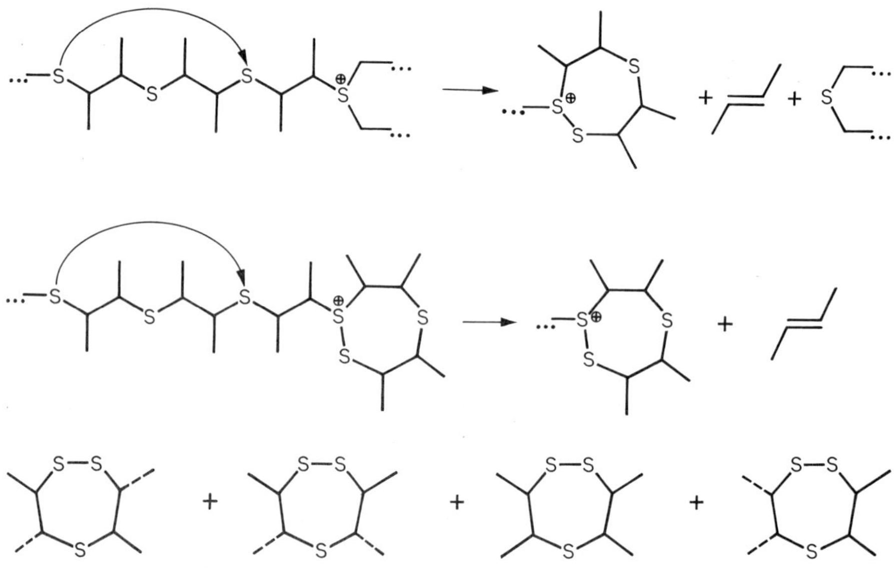

I II III IV

(10-19)

Proportions of these isomers were correlated with the microstructure of the starting polymers. This was proved by degradation studies of the optically active poly(cis-DMT): the higher the optical activity, the higher the corresponding ratios (I/II and III/IV).

These results indicate that the polymers (also the optically active ones) do not exhibit Bernoullian statistics (cf. also Sect. 15.1.2.3.1): the number of the isotactic dyads (or triads) were always higher than expected. The fraction of the diisotactic trithiepane (I or III) varied with the mechanism of polymerization. The polymers obtained by $Et_3O^{\oplus}BF_4^{\ominus}$ initiator led to the smallest proportion of I (or III), equal to 0.58. The highest was found in the polymer obtained with the anionic-coordinative initiator Cd-tartarate (0.68). Transalkylation via a thiiranium intermediate, accompanies degradation and leaves the threo structure unchanged. These results are in good agreement with an analysis based on ^{13}C-NMR spectra.

Recently Aliev et al. [70] have discussed the formation of alkenes in the polymerization of thiiranes and the degradation of the corresponding polymers. A modified mechanism was proposed, assuming that the thiiranium sulfur atom is subject to nucleophilic attack by soft bases [71,72]:

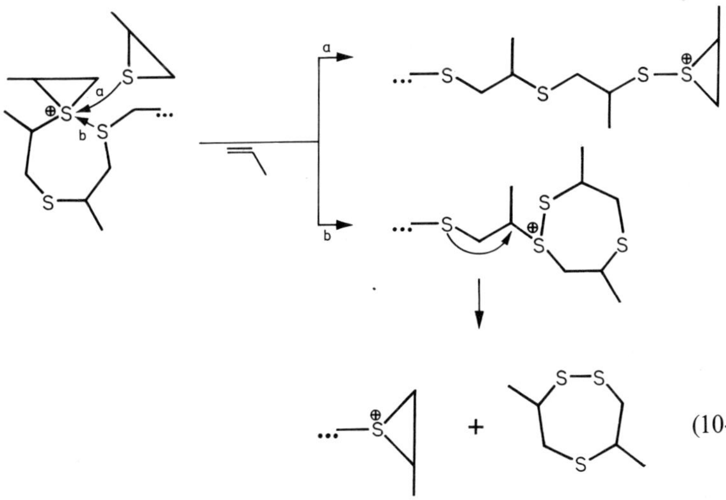

(10-20)

10.6 References

1. Goethals, E. J.: J. Polym. Sci., Polym. Symp. *56*, 271 (1976)
2. Goethals, E. J.: Top. Sulf. Chem. *3*, 1 (1977)
3. Goethals, E. J.: "Polymeric Amines and Ammonium Salts", Pergamon Press, Oxford 1980
4. Goethals, E. J. et al.: ACS Symp. Ser. *59*, 1 (1977)
5. Schacht, E. H., Goethals, E. J.: Makromol. Chem. *167*, 155 (1973)
6. Sigwalt, P.: Chim. Ind., Genie Chim. *104*, 47 (1971)
7. Woodhouse, C. G., Singer, R. D., Smith, J. L.: Adhesive Age *1970*, 21
8. Adamek, S., Wood, B. B. J., Woodhoms, R. T.: Rubber Plast. Age *46*, 56 (1965)
9. Rasvodovskii, E. F. et al.: J. Macromol. Sci.-Chem. *A8*, 241 (1974)
10. Matyjaszewski, K.: Makromol. Chem. in press
11. Ooteghem, D., Goethals, E. J.: Makromol. Chem. *177*, 3389 (1976)
12. Perst, H.: Oxonium Ions in Organic Chemistry, Verlag Chemie, Weinheim 1971
13. Goethals, E. J. et al.: Makromol. Chem. Rapid Commun. *3*, 515 (1982)
14. Florquin, S. M., Goethals, E. J.: Makromol. Chem. *182*, 3371 (1981)
15. Goethals, E. J.: Makromol. Chem. *175*, 1309 (1974)
16. Goethals, E. J. et al.: Polymer J. *12*, 571 (1980)
17. Goethals, E. J., Tezuka, Y.: Europ. Polym. J. *18*, 991 (1982)
18. Goethals, E. J., Munir, A., Bossaer, P.: Pure & Appl. Chem. *53*, 1753 (1981)
19. Munir, A., Goethals, E. J.: J. Polym. Sci., Polym. Chem. Ed. *19*, 1985 (1981)
20. Bossaer, P. K., Goethals, E. J.: Makromol. Chem. *178*, 2983 (1977)
21. Hashimoto, S., Yamashita, T., Kaneda, M.: Polymer J. *6*, 238 (1974)
22. Hashimoto, S., Yamashita, T., Hino, J.: Polymer J. *9*, 19 (1977)
23. Hashimoto, S., Yamashita, T.: J. Macromol. Sci.-Chem. *A17* 559 (1982)
24. Dermer, O. C.: WADC Technical Report 55–447 ASTIA Document No AD 110496 (1960)
25. Furukawa, J., Oda, R.: Bull. Inst. Chem. Res. Kyoto Univ. *30*, 50 (1952); C.A. *47*, 3611 (1953)
26. Empen, J. A., Stille, J. K. in: Macromol. Synth., Coll. Vol. *1*, 269 (1977)
27. Empen, J. A., Stille, J. K. in: Macromol. Synth., Coll. Vol. *1*, 265 (1977)
28. Stille, J. K., Empen, J. A.: J. Polym. Sci. A1, *5*, 273 (1967)
29. Gipstein, E., Wellsch, E., Sweeting, O. J.: J. Appl. Polym. Sci. *10*, 859 (1966)
30. Birkner, H., Steurzenfecker, F.: Ger. Pat. 1202828 (1965)
31. Lando, J. B., Stannett, V.: J. Polym. Sci. B *2*, 375 (1964)
32. Burnop, V. C. E., Latham, K. G.: Polymer *8*, 589 (1967)

33. Minoura, Y., Takebayashi, M., Price, C. C.: J. Amer. Chem. Soc. *81*, 4689 (1959)
34. Goldstein, A., in: Encyclopedia of Polym. Sci and Technology Vol. *1*, p. 734, Wiley-Interscience, New York 1964
35. Jones, G. D. et al.: J. Org. Chem. *9*, 125 (1944)
36. Yamashita, T., Nigo, M., Nakamura, N.: Makromol. Chem. *180*, 1145 (1979)
37. Samyn, G., Smets, G.: IUPAC Symposium "Polymeric Amines", Ghent 1979, preprints p. 3
38. Banthia, K., Schacht, E. H., Goethals, E. J.: Makromol. Chem. *179*, 841 (1978)
39. Łukaszczyk, J., Schacht, E. H., Goethals, E. J.: Makromol. Chem. Rapid Comm. *1*, 79 (1980)
40. Toy, M. S., Price, C. C.: J. Amer. Chem. Soc. *82*, 2613 (1960)
41. Goethals, E. J., Drijvers, W.: Makromol. Chem. *165*, 329 (1973)
42. Goethals, E. J., Schacht, E. M.: J. Polym. Sci., Polym. Letters *11*, 497 (1973)
43. Van Ooteghem, D., Deveux, R., Goethals, E. J.: Int. J. Sulfur, Chem. *8*, 31 (1973)
44. Bell, E. V., Bennett, G. M., Hock, A. L.: J. Chem. Soc. *1927*, 1803
45. Noshay, A., Price, C. C.: J. Polym. Sci. *54*, 533 (1961)
46. Goethals, E. J.: Pure & Appl. Chem. *48*, 335 (1976)
47. Van Craeynest, W., Goethals, E. J.: Europ. Polym. J. *12*, 859 (1976)
48. Dick, C. R.: J. Org. Chem. *35*, 3950 (1970)
49. Dick, C. R.: J. Org. Chem. *32*, 72 (1967)
50. Hansen, G. R., Burg, T. E.: J. Heterocycl. Chem. *5*, 305 (1968)
51. Goethals, E. J., Schacht, E. H., Bruggeman, P.: Int. Symp. Cat. Polymerization, Rouen 1973, Paper C-30
52. Tsuboyama, S. et al.: Tetrahedron Letters *16*, 1367 (1970)
53. Matyjaszewski, K.: Makromol. Chem., in press
54. Penczek, S., Kubisa, P., Matyjaszewski, K.: Adv. Pol. Sci. *37*, 1 (1980)
55. Schacht, E. H., Bossaer, P., Goethals, E. J.: Polymer J. *9*, 3291 (1977)
56. Brauer, G. M.: Polym.-Plast. Techn. Eng. *9*, 87 (1977)
57. Baumann, H. et al.: IUPAC Symposium "Polymeric Amines" Ghent 1979, preprints p. 165
58. Ponomarenko, V. A. et al.: Vysokomol. Soed. *B23*, 235 (1981)
59. Dermer, O. C., Ham, G. E.: Ethyleneimine and other Aziridines, Academic Press, London 1969
60. PEI Polymers . . ., brochure, The Dow Chemical Co., 1974
61. Jones, G. D., MacWilliams, D. C., Bractor, N. A.: J. Org. Chem. *30*, 1994 (1965)
62. Dick, C. R., Ham, G. E.: J. Macromol. Sci.-Chem. *4*, 1301 (1970)
63. Gembitskii, P. A. et al.: Vysokomol. Soed. *20*, 1505 (1978)
64. Gembitskii, P. A. et al.: Vysokomol. Soed. *20*, 2619 (1978)
65. DeMember, J. R., Taylor, L. D.: J. Polym. Sci., Polym. Chem. Ed. *17*, 1089 (1979)
66. Araki, T. et al.: J. Polym. Sci., Polym. Chem. Ed. *16*, 1037 (1978)
67. Saegusa, T., Ikeda, M., Fujii, H.: Macromolecules *5*, 108 (1972)
68. Tanaka, R. et al.: Macromolecules *16*, 849 (1983)
69. Goethals, E. J. et al.: Makromol. Chem. *181*, 2481 (1980)
70. Aliev, A. D., Krentsel, B. A., Alieva, S. L.: Europ. Polym. J. *19*, 71 (1983)
71. Owsley, D. C., Helmkamp, G. K., Spurlock, S. N.: J. Amer. Chem. Soc. *91*, 3606 (1969)
72. Kikuzono, Y. et al.: Tetrahedron *30*, 2197 (1974)

11 Polyamides

Polyamides are produced commercially by two routes. First is the polycondensation of aminoacids or diamines with diacids, and the second route is the ring-opening polymerization of lactams [1]. The ring-opening process can be accomplished with anionic or cationic initiators and also in the presence of water by a hydrolytic process. Nylon-6 (poly-ε-caprolactam) and Nylon-12 (poly-ω-dodecanelactam) are commercially produced mostly by the hydrolytic process, in which the reaction control is easier and the properties of the product better than in polycondensation.

Ring-opening polymerization under anhydrous conditions has been used mostly on the laboratory scale, although the "low temperature" anionic polymerization of ε-caprolactam (εCLM) is a large scale process using bulk polymerization to obtain directly large objects.

Cationic polymerization has not been used commercially due to the poor properties of the final product. Polymers synthesized cationically are much less stable and oxidize faster under the same conditions than those obtained by anionic polymerization [2]. After removal (by water extraction) of traces of strong acid catalysts which would accelerate oxidation, the stability of cationically produced polyεCLM increases but remains lower than that obtained by the anionic process.

Cationic polymerization of lactams shows a number of pecularities, not covered in Vol I (Adv. Pol. Sci. *37*). Therefore, relatively more attention is paid in this section to the polymerization mechanism. These pecularities are discussed separately for different initiating systems.

Cationic process is the only route to the polymerization of N-alkyl lactams; the resulting polymers have much lower melting points than their unsubstituted analogues, because of the absence of the H-bonding. N-alkyl lactams polymerize by mechanisms different from those established for their unsubstituted analogues. Their polymerization is discussed in a separate section.

In the last section of this chapter we describe the hydrolytic process. The mechanism of the hydrolytic polymerization resembles, and in some steps is even identical with the cationic process. The conditions of the commercial hydrolytic polymerization of εCLM are given in more detail.

11.1 Peculiarities of Cationic Polymerization of Lactams

The mechanism of the cationic polymerization of lactams formally resembles the activated monomer mechanism, better known for the anionic process. In this mechanism, in contrast to the more conventional ionic polymerization, the monomer bears the charge and the macromolecules are noncharged:

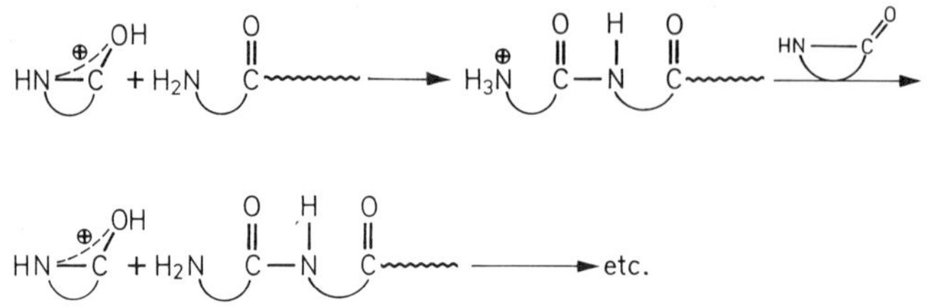

(11-1)

Recently, however, Šebenda [3] has proposed that in polymerization of the highly reactive N-acyl lactams initiated with salts bearing stable complex anions like $AsF_6^{\ominus}$ chain growth proceeds with polymeric acylium cations by attack at the N-atom and not at the O-atom as usually proposed:

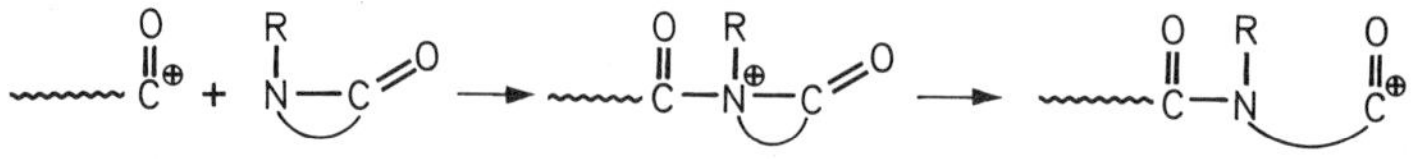

(11-2)

Cationic polymerization of lactams is a complex process consisting of several elementary reactions which influence the kinetics of polymerization and product properties (molecular weights, structure of end-groups).

Unsubstituted lactams polymerize cationically with high activation energies, at elevated temperatures (about 200 °C). With increasing temperature the contribution of termination reactions (i.e. formation of non-reactive amidine end-groups) also increases, which leads to the incomplete conversion.

Cationic polymerization proceeds faster with strained unsubstituted lactams; 50% conversion is reached in bulk polymerization with 1% of HCl at 180 °C after 1 hour with 9-membered lactam and after more than 2 weeks with the less strained 13-membered lactam. With changing the ring size of lactams (e.g. 9 → 13) not only the ring strain is changed ($\Delta\Delta H_p = 7$ kcal · mol^{-1}) but also their basicities and dielectric constant of the medium ($\Delta\varepsilon = 17$) [3].

N-alkyl substituted lactams, although less strained than their unsubstituted analogues, polymerize with higher rates, because formation of amidine (termination) is no more possible.

Cationic polymerization of lactams can be initiated by a large variety of typical cationic initiators like Lewis acids, protonic acids and their salts with primary and secondary amines, as well as by carboxylic acids. The polymerization rates depend on the nature of the initiator. The polymerization of 12-dodecanelactam (13-membered ring) carried out at 270 °C with 3 mol% initiator gave, 95% conversion after 2 hrs, with the complex $HCl \cdot NH_2(CH_2)_{11}CH_3$, but only 20% conversion for lauryl acid and 12-aminododecanoic acid [4]. The structure of the end-groups and polymerization degrees were also dependent on the structure of the initiator. Therefore, we describe the chain growth and termination reactions separately for each initiating system. Then, we discuss the commercial hydrolytic process that proceeds by cationic active species.

11.2 Initiation with Strong Acids

Spectroscopic (NMR, IR) studies show, that protonation (or coordination with Lewis acids) occurs predominantly at the oxygen atom as depicted in Eq. (11-3) [5,6]. In the absence of other strong bases the positively charged C-atom in the protonated monomer is attacked by the N-atom of another lactam and primary amino- and N-acyl lactam end-groups are formed (11-3):

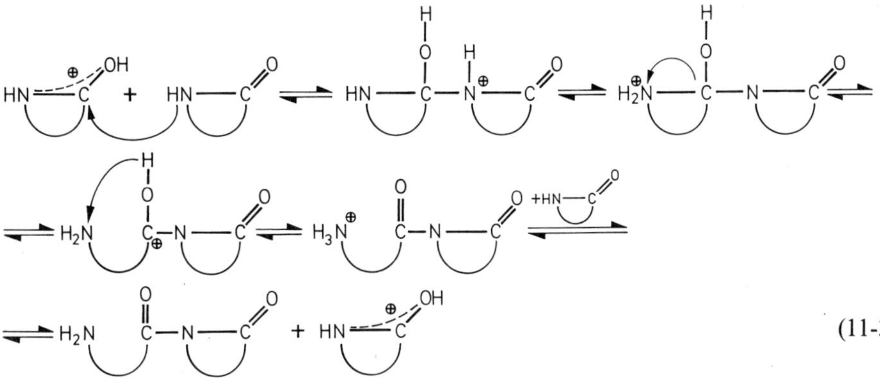

(11-3)

In the next step proton is transferred to the incoming monomer, which becomes ready to participate in chain growth. Propagation involves the most basic primary amino group and protonated (activated) monomer. After protonation of the monomer, the reactive intermediate is formed in which the proton may be located either on the exo-N-atom (*A*), endocyclic N-atoms (*C*) or on hydroxyl O-atom (B):

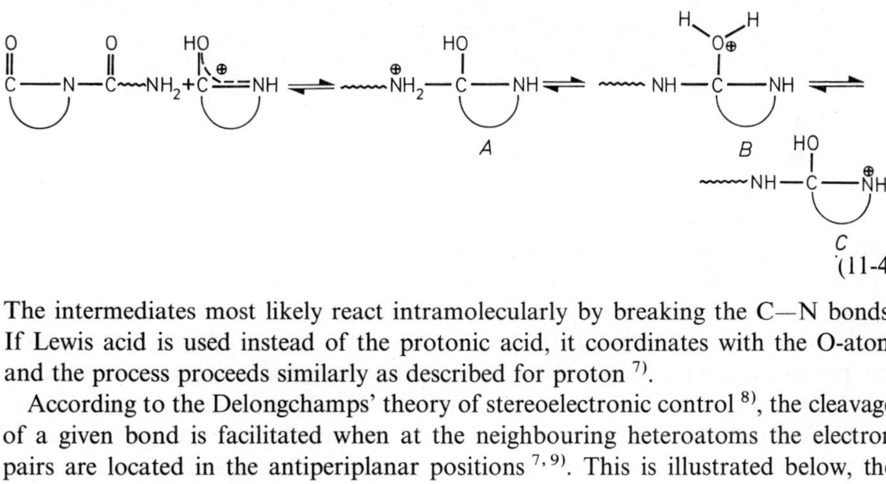

(11-4)

The intermediates most likely react intramolecularly by breaking the C—N bonds. If Lewis acid is used instead of the protonic acid, it coordinates with the O-atom and the process proceeds similarly as described for proton [7].

According to the Delongchamps' theory of stereoelectronic control [8], the cleavage of a given bond is facilitated when at the neighbouring heteroatoms the electron pairs are located in the antiperiplanar positions [7,9]. This is illustrated below, the arrow shows the bond in the antiperiplanar position:

(11-5)

Therefore, depropagation (from *A*) is unlikely, because the electron pair on the endocyclic N-atom cannot be located at the antiperiplanar position (the required position for the electron pair is occupied by a bond that is a part of the ring). The breaking of the C—O bond (*B*) and propagation (*C*) are both allowed stereoelectronically for the same reason. This is visualized by the scheme below, which shows the corresponding conformations and subsequent reaction steps:

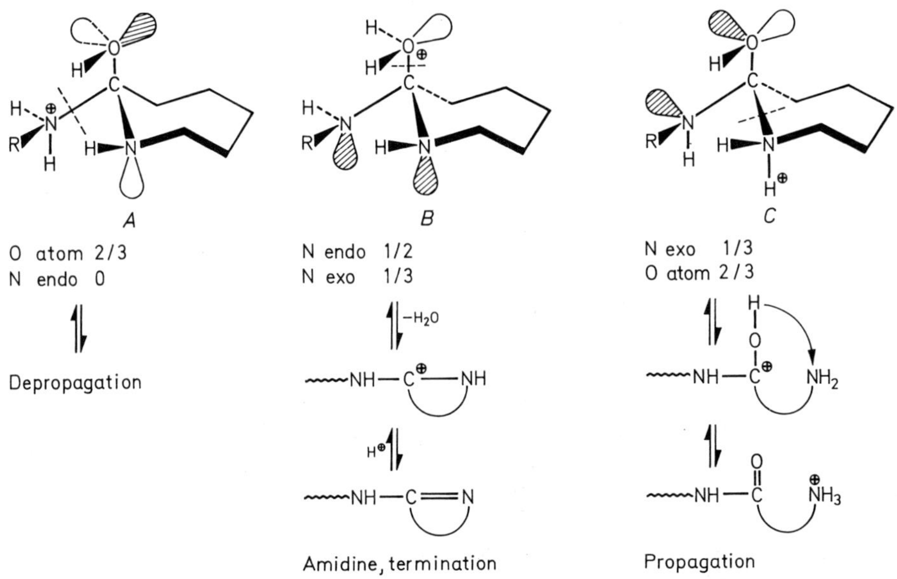

(11-6)

The highest proportion of the antiperiplanar positions of electron pairs is in structure *C*. The propagation step (C) is additionally favoured by the release of ring strain during ring-opening.

Formation of amidines (*B*) is therefore less probable. This route is a termination reaction, because amidines are sufficiently basic to trap protons almost quantitatively [10]. The ratio of propagation (*C*) to termination reaction (*B*) is higher in the hydrolytic process, carried out with excess of H_2O preventing water expulsion from struture *B*. Below 250 °C the proportion of amidine group in polyamides prepared in hydrolytic process is very low [11]. The content of the amidine groups and the rate of polymerization both depend on the nature of initiators. For example, complexes of acids with amines are more effective than acids themselves [9]. The end-groups formed in initiation without amines are N-acyl lactams. The protonation of the N-acyl lactam end group and the subsequent reaction with the basic amino group can lead either to chain coupling or to amidine formation. The latter reaction is favoured by intramolecular H-bonding [9] and it is responsible for much higher proportion of the amidine end groups in the cationic than hydrolytic polymerization [2,12] (11-7).

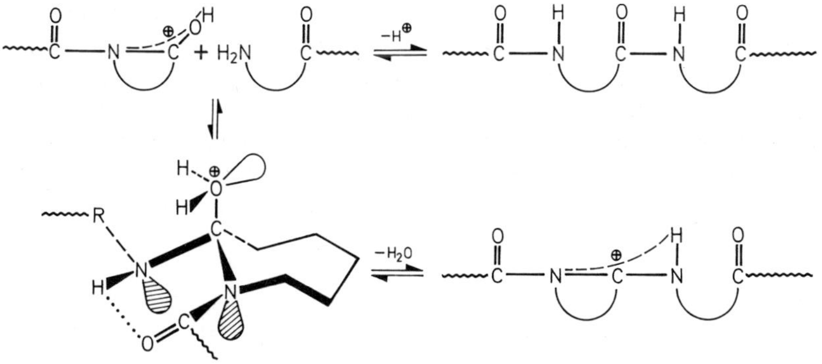

(11-7)

The rates of polymerization depend also on acid strength and the structure of the acid. The apparent rate constants for Lewis acids are higher than for protonic acids which may be related to the ability of coordination with O-atoms in the lactam and to steric effects. The reactivity of Lewis acids increase in the order $AlCl_3 > TiCl_4 < WCl_6$ [11]. The least reactive $AlCl_3$ is 20 times more efficient (at 260°) than lauryl acid [13].

The apparent rate constants increase with the acidity of the carboxylic acids with laurylolactam (13-membered) but they are independent of the acid strenght for capryllolactam (9-membered) [11,14,15]. The polymerization with carboxylic acids is slightly different (c.f. next Section).

11.3 Acidolytic Polymerization

Carboxylic acids used as initiators rapidly incorporate into polymer molecules. This could not be explained by the mechanism described previously for strong acids. Thus, the following acidolytic mechanism has been proposed [15,16]:

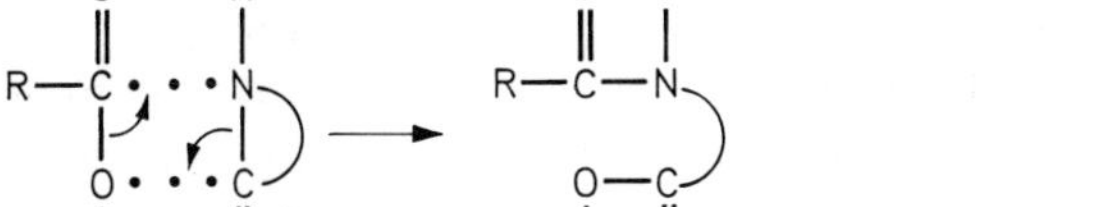

(11-8)

This transacylation of the amide linkage has been assumed to proceed via a four-center mechanism involving two H-bonds [17]:

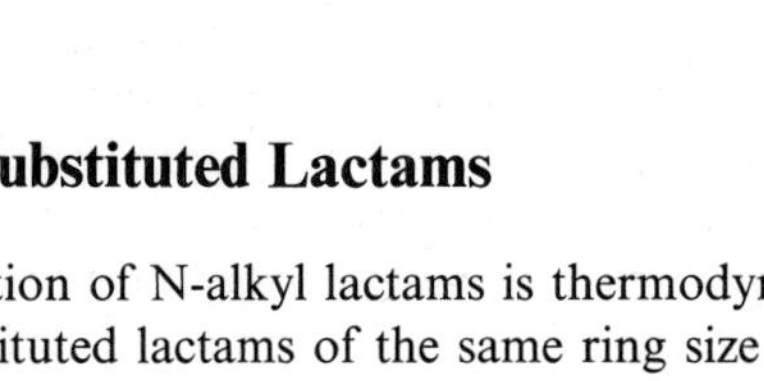

(11-9)

11.4 N-Substituted Lactams

Polymerization of N-alkyl lactams is thermodynamically less favorable than that of the unsubstituted lactams of the same ring size (e.g. N-methyl CLM cannot homopolymerize). However, when polymerization is thermodynamically allowed, it proceeds with higher rates than with the more strained unsubstituted analogues [18, 19]. This can be ascribed to the absence of the amidine formation in the cationic process (no hydrogen atom available for water expulsion).

The cationic polymerization is best initiated by strong acids (e.g. HCl) but also by acyl chlorides and salts of secondary amines [18]. Since the salts are least effective the usual route (cf. Eq. 11-4) of polymerization is of smaller importance:

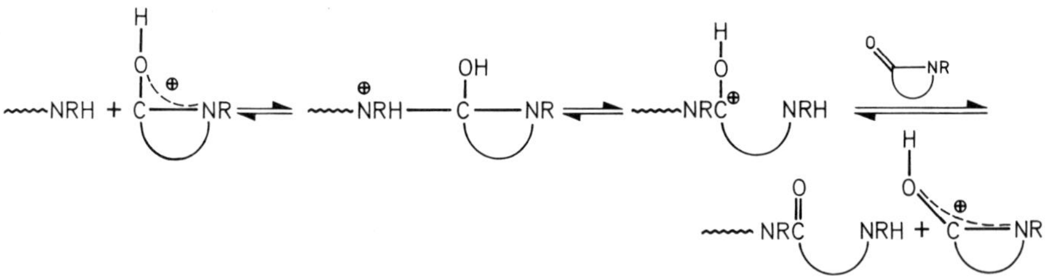

(11-10)

Thus the more important route is condensation of aminoacyl chlorides [19]:

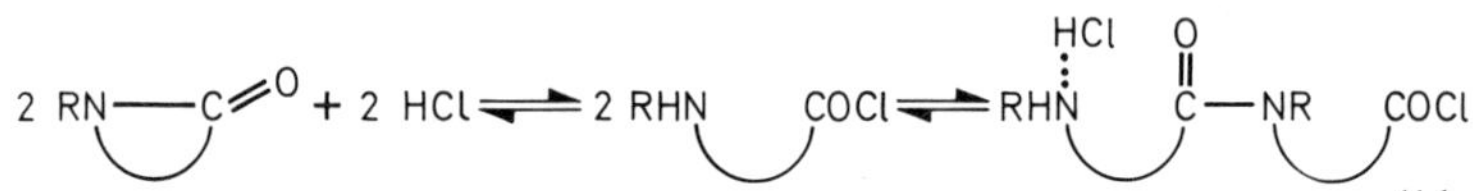

(11-11)

Acyl chloride itself can initiate polymerization indicating that the following route is important:

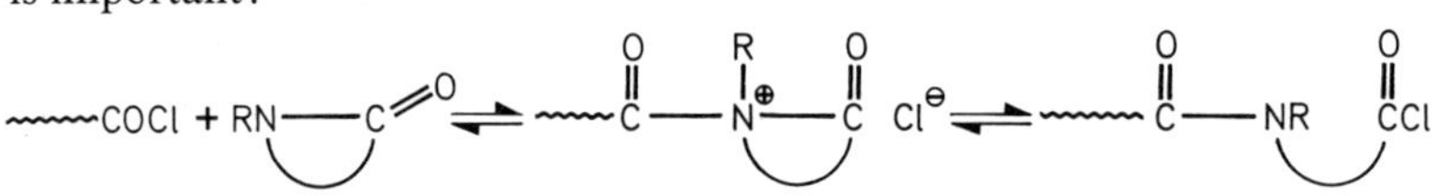 (11-12)

It is also possible that the endocyclic 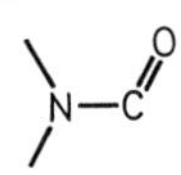 bond breaks intramolecularly and the resulting oxocarbenium cation collapses with the chloride anion to form the acyl chloride. In such a case stronger protonic acids (less nucleophilic anions), e.g. fluorosulfonic or trifluoromethanesulfonic acid, may lead to a considerable increase in the polymerization rates. This pathway is similar to that described in Eq. (11-2).

11.5 Hydrolytic Polymerization

The mechanism of chain growth in hydrolytic polymerization is similar to that of the cationic process under anhydrous conditions [21]. In the first step the amide bond in the lactam is cleaved by H_2O and is converted to primary amine and carboxylic acid functions:

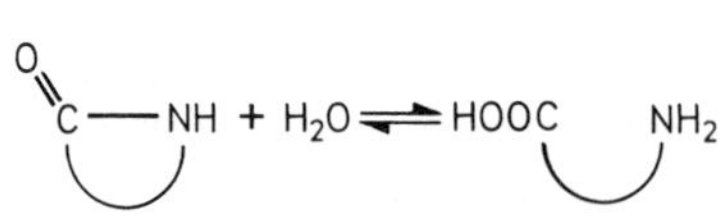 (11-13)

Although water induces the polymerization, the true initiating species is the aminocaproic acid formed by hydrolysis. This acid is usually added to accelerate the polymerization. The preliminary formation of aminocaproic acid by hydrolysis of εCLM explains the increase in polymerization rate with increaseing [H_2O]. Some other carboxylic acid derivatives are also used in practice e.g. the hexamethylene diamine salt of the adipic acid (so called 6,6 salt).

Once the aminoacid is formed polycondensation can contribute to chain growth:

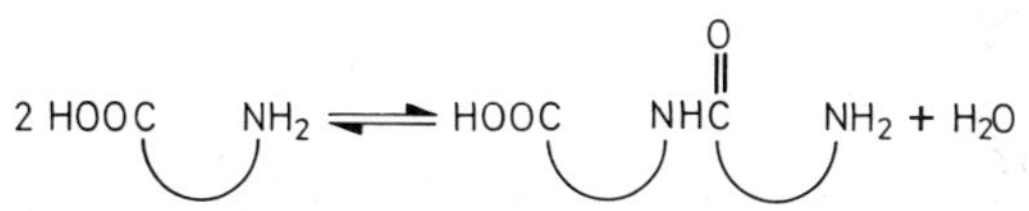

(11-14)

In the polymerization of εCLM only a few percent of lactam is incorporated into the polymer through hydrolysis and bimolecular condensation. Most monomer enters the polymer through stepwise addition, i.e. the cationic chain growth [22]:

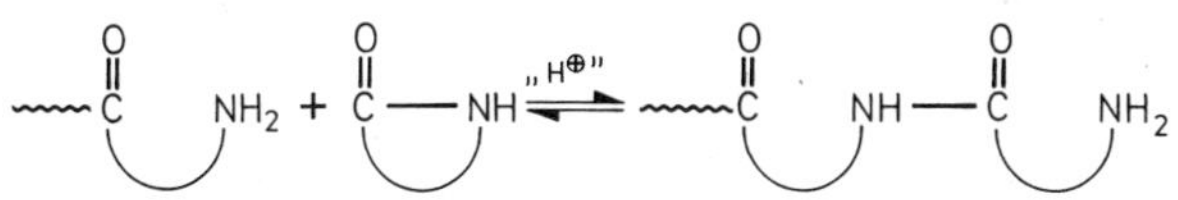

(11-15)

Direct transmidation is very slow without any catalyst added. The specific role of amine and carboxylic groups in this process has been established by Heikens who studied their incorporation during hydrolytic polymerization [23]. These experiments revealed that lactam is added at the amine end group and the reaction is catalyzed by protons, i.e. it follows Scheme (11-4).

In hydrolytic polymerizations carried out below 250 °C the concentrations of amine and carboxylic end groups are equal [24]. At higher temperatures the carboxyl group reacts with the methylene group next to the carbonyl linkage forming some branching [25]. At still higher temperatures the spontaneous formation of amidine groups was observed [26]. Both these reactions are shown in Scheme (11-16):

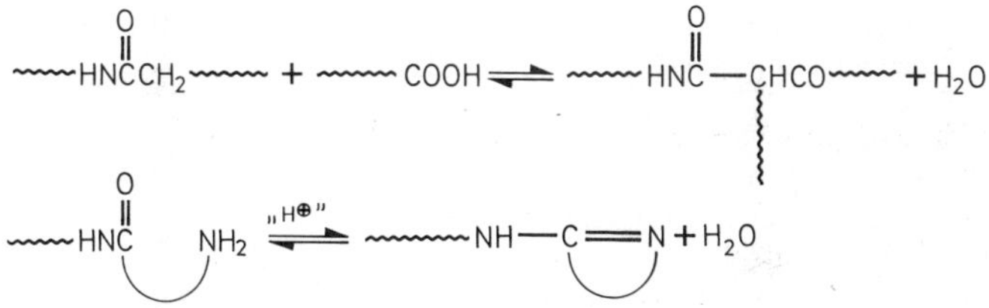

(11-16)

The kinetics and molecular weight distribution in the hydrolytic process were studied and described in detail by Reimschuessel [27,28]. The molecular weights and the equilibrium conversion decrease with an increase of temperature but this influence is not very important due to the narrow temperature range (220–260 °C) applied in practice. Polymerization is usually carried out with 1–5% initiator (H_2O, aminocaproic acid, 6,6 salt) for a few hours. The process involves three steps [1,29]. In the first one 90% polymer with molecular weight $\overline{M}_n$ = 8000 to 14000 is formed. The second step consists of condensation of these chains into polymers with $\overline{M}_n$ = 18000 to

33000. In the last step water is removed and the polymerization is terminated with an amine or an acid (acetic acid is preferentially used). Monomer (8.5%) and oligomers (1.5%) are extracted from the crude product with water at 95 to 100 °C or are removed under vacuum (5–10 mm Hg). The final product with $\overline{M}_n = 20000–36000$ is washed with water and cut into chips.

Nylon-6 (poly-ε-caprolactam) is mainly used for fiber production and applied in textiles, tire cords and carpeting. It is also used as extrusion and molding resin.

Polydodecanelactam (polymer of the 13-membered lactam) is produced by hydrolytic polymerization with 0,01 to 1% phosphoric or boric acid and water as the catalytic system. The process is conducted at 270–300 °C for 4–8 hrs under 5–60 atm. The conversion of the low molecular weight prepolymer into the high molecular weight final product requires the careful removal of residual water remaining in the system.

11.6 References

1. Snider, O. E., Richardson, R. J.: In "Encyclopedia of Polymer Science and Technology", Wiley-Interscience, New York 1969, Vol. 10 p. 347
2. Lanska, B., Šebenda, J.: V Int. Mikrosymp. "Fortschritte in Ionen Polymerisation", Praga 1982, Common. No 24
3. Šebenda, J., Puffr, R.: In "Cationic Polymerization and Related Processes", Goethals, E. J. (Ed.), Academic, Press 1984, p. 261
4. Froyer, G., Sekiguchi, H.: Makromol. Chem. *184*, 2199 (1983)
5. Huisgen, R. et al.: Chem. Ber. *90*, 1437 (1957)
6. Homer, R. B., Johnson, L. D., in "The Chemistry of Amides", Zabicky, J. (Ed.), Wiley, New York 1970
7. Puffr, R., Šebenda, J.: IUPAC Symp. "Ring Opening Polymerization of Heterocyclics" Karlovy Vary 1980, com. No M24
8. Deslongchamps, P.: Tetrahedron *31*, 2463 (1975)
9. Bertalan, G. et al.: 26th International Symposium on Macromolecules Mainz, 1979, Preprints, Vol. I p. 177
10. Csuros, Z. et al.: Makromol. Chem. *137*, 9 (1970)
11. Šebenda. J.: in "Progress in Polymer Science", Jenkins, A. D. (Ed.), Pergamon Press, Oxford 1980, Vol. 6, p. 123
12. Puffr, R., Šebenda, J.: J. Polym. Sci., Polym. Chem. Ed. *12*, 21 (1974)
13. Puffr, R., Šebenda, J.: Polyamidy 1975, Sb. prednasek, 121 (1975)
14. Puffr, R., Šebenda, J.: J. Polym. Sci. Symp. *42*, 21 (1973)
15. Puffr, R., Šebenda, J.: Colln. Czech. chem. Commun. *40*, 3339 (1975)
16. Lanska, B., Šebenda, J.: Colln. Czech. chem. Commun. *40*, 1524 (1975)
17. Kransnoselsky, M., Patai, S.: J. Chem. Soc. (B) *1969*, 24
18. Masar, B., Šebenda, J.: Colln. Czech. chem. Commun. *39*, 110 (1974)
19. Masar, B., Šebenda, J.: Colln. Czech. chem. Commun. *40*, 93 (1975)
20. Lewis, J. R., Reynolds, R. J. W.: Chem. Ind. *1951*, 958
21. Rothe, M., Bertalan, G.: ACS Symp. Ser. *59*, 129 (1977)
22. Hermans, P. H., Heikens, D., van Velden, P. F.: J. Polym. Sci. *30*, 81 (1958)
23. Heikens, D., Hermans, P. H., van der Vant, G. M.: J. Polym. Sci. *44*, 437 (1960)
24. Reinisch, G.: Z. f. Chem. *14*, 165 (1974)
25. Gohlke, U., Reinisch, G.: Faserforsch, Text. Tech. *24*, 282 (1973)
26. Csuros, Z. et al.: Makromol. Chem. *137*, 17 (1970)
27. Reimschuessel, H. K., Dege, G. J.: J. Polym. Sci. A-1, *9*, 2343 (1971)
28. Reimschuessel, H. K., Nagasubramanian, K.: Chem. Eng. Sci. *27*, 1119 (1972)
29. Sweeny, W., Zimmermann, J.: in "Encyclopedia of Polym. Science and Technology", Wiley-Interscience, New York, 1964, Vol. 10, p. 483.

12 Cyclic Imino Ethers

A number of cyclic monomers containing the imino ether linkage —N=C(R)—O— have been successfully polymerized by the ring-opening mechanism. Particularly the derivatives of 2-oxazoline (OXL) [1] and 5,6-dihydro-4H-1,3-oxazine (OXZ) [1] but also 2-iminotetrahydrofuran (ITH) [2] and ethylene iminocarbonate (EIC) [3] have been studied:

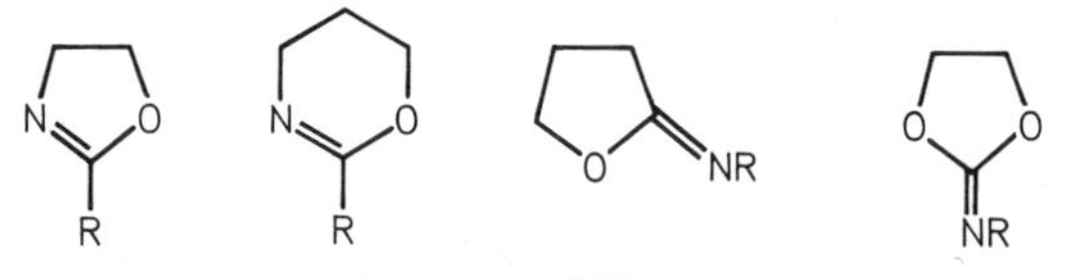

The above listed monomers polymerize exclusively by cationic ring-opening polymerization. Polymerization of cyclic imino ethers has been reviewed by Saegusa [1,4]. In this section we follow mostly the conclusions of the Kyoto group summarizing the applied synthetic methods. Subsequently we describe initiating systems, peculiarities of chain growth, side reactions, copolymerization systems and possible applications, in this order.

12.1 Initiators

Cyclic imino ethers are highly nucleophilic monomers, second only to cyclic amines. Their polymerization can be initiated not only by strong protonic and Lewis acids like: H_2SO_4, $HOSO_2C_6H_5CH_3$, $SnCl_4$, BF_3 or strong alkylating agents like $(C_2H_5)_3O^{\oplus}BF_4^{\ominus}$ and $C_2H_5OSO_2CF_3$, but also by weak cationogenic compounds such as alkyl halides e.g. CH_3I, $C_6H_5CH_2Br$ and $C_6H_5CH_2Cl$. 2-Methyl-2-oxazoline is nucleophilic enough to be polymerized in the presence of $Br^{\ominus}$ or $I^{\ominus}$ counterions i.e., it successfully competes with $Br^{\ominus}$ and $I^{\ominus}$ during propagation. With $Cl^{\ominus}$, however, polymerization proceeds 40 times slower than with $Br^{\ominus}$ or $I^{\ominus}$. Evidently a large proportion of alkyl chlorides are formed during propagation that add monomer much slower than the oxazolinium cation [5]:

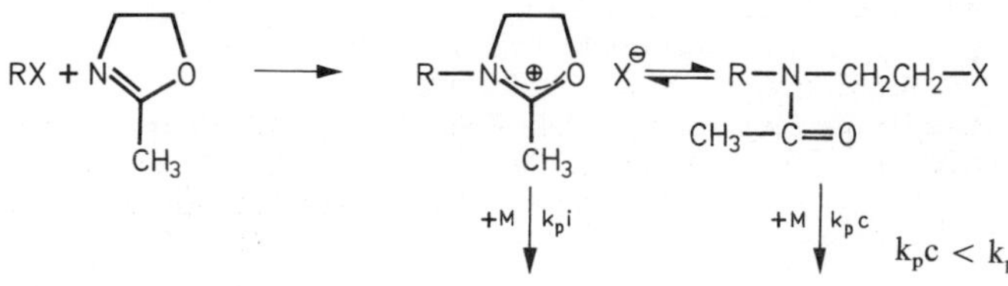

$k_pc < k_pi$ (12-1)

The alkyl iodide initiated polymerization of the less nucleophilic OXL proceeds 10^2 times slower than those of tosylates [6]. Apparently, OXL produces with RI less reactive covalent species in large proportions by recombination of the oxazolinium cation and iodide counteranion.

With strong cyclic electrophiles the polymerization of oxazolines proceeds via zwitter-ions; e.g. 2 methyl OXL with o-sulfobenzoic anhydride [7]:

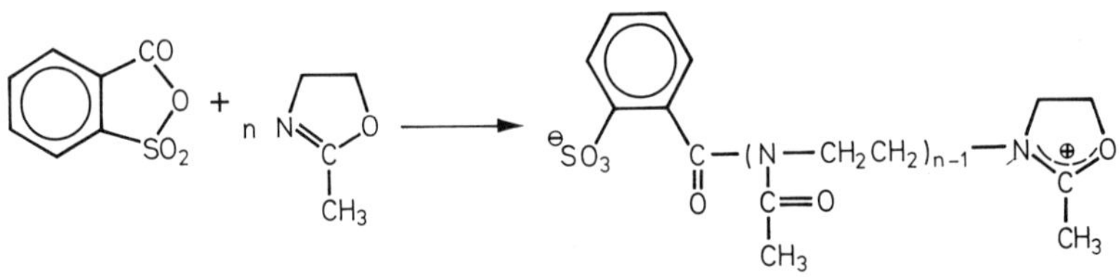

(12-2)

Polymers with $\overline{DP}_n = 20$ were formed after 2 h at 60 °C. Zwitter-ionic copolymerization of oxazolines is discussed in Section 15.2.

12.2 Propagation

The polymerization of five- and six-membered cyclic imino ethers including the substituted ones, goes usually to completion (cf. Table 12-1). In addition to the relief of ring strain, the driving force of these polymerizations is the isomerization of the reactive imino ether linkages (—N=C(—)—O—) to thermodynamically more stable amide bonds >N—C(—)=O. Apparently, the free energy change of isomerization compensates for the thermodynamically unfavorable opening of the six-membered ring:

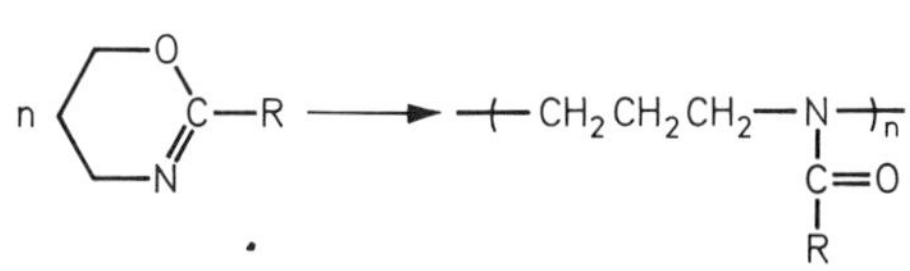

(12-3)

Unfortunately, no information is available on the thermodynamics of the polymerization of these monomers.

Polymerization studies of OXL by Litt et al. [14], Kagiya et al. [9] and Tomalia et al. [15] have later been further developed by Saegusa et al. The latter authors observed directly for the first time by NMR spectroscopy the active species, and determined the rate constants of propagation and initiation in the presence of different initiators [4].

Cyclic imino ethers polymerize cationically by opening the O-alkyl bond in the onium cation, which is attacked by the most nucleophilic N-atom of the monomer in an S_N2 reaction:

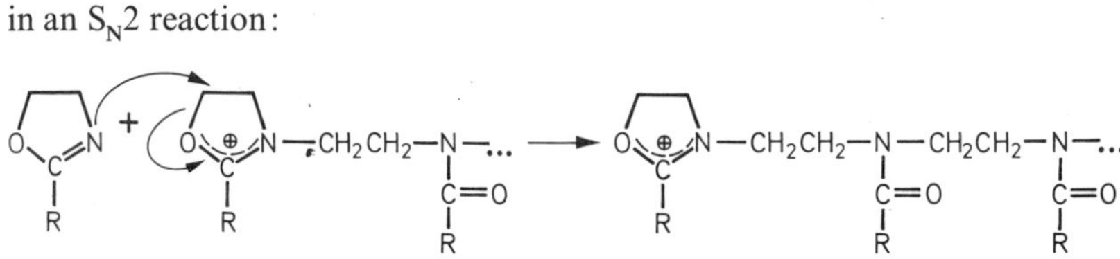

(12-4)

Table 12.1. Properties and conditions of preparations of polyoxazolines and polyoxazines

Monomer	Conditions	Product properties	Ref.
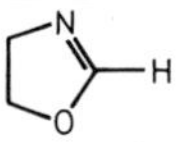	80°C, DMF, 5 h 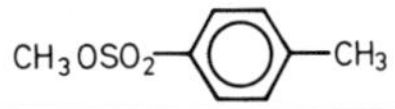	83% yield, mp = 208–210°C; η_{sp}/c = 0.22 dl/g	8)
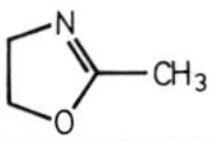	70°C, CH_3CN, 90 h; $SnCl_4$	100% yield, mp = 140–150°C; η_{sp}/c = 0.18 dl/g	9)
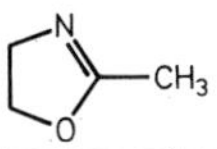	110°C, bulk, 48 h; $(CH_3)_2SO_4$	100% yield, mp = 200°C; η_{sp}/c = 0.89 dl/g	10)
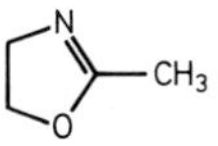	60°C, CH_3CN, 2 h 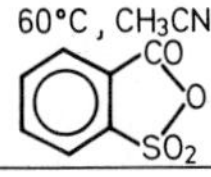	75% yield, $\bar{M}_n$ = 1 900	7)
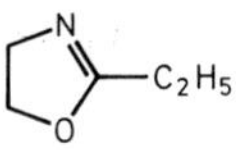	120°C, bulk, 6 h 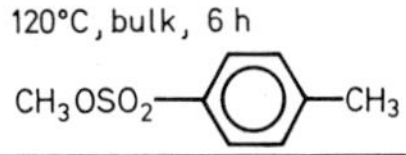	100% yield, η_{sp}/c = 3.05 dl/g	10)
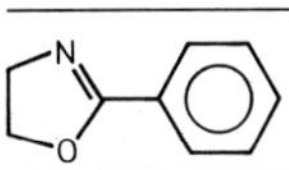	70°C, bulk, 30 h; $BF_3 \cdot OEt_2$	47% yield, mp = 100–110°C; η_{sp}/c = 0.1 dl/g	9)
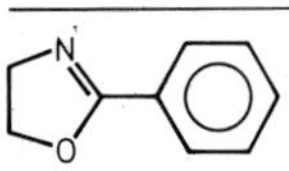	105°C, CH_3CN, 10 h 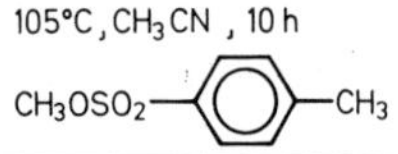	97% yield, mp = 136–138°C; $\bar{M}_n$ = 4 100	11)
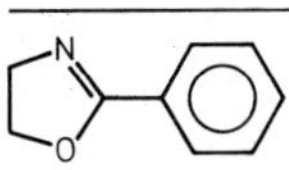	140°C, 48 h; $(CH_3O)_2SO_2$	$\bar{M}_w$ = 350.000	38)
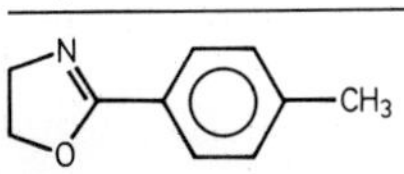	105°C, CH_3CN, 10 h 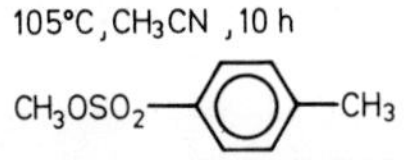	91% yield, mp = 138–140°C; $\bar{M}_n$ = 4 300	11)
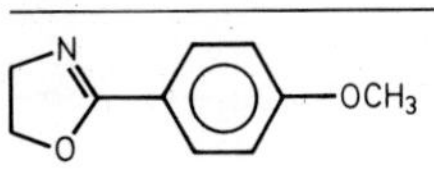	105°C, CH_3CN, 10 h 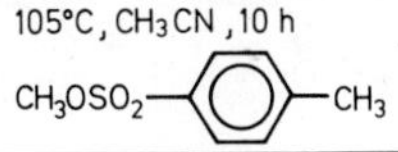	94% yield, mp = 229–232°C	11)
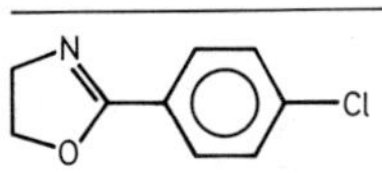	105°C, CH_3CN, 10 h 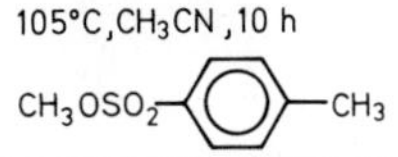	99% yield, mp = 268–270°C	11)
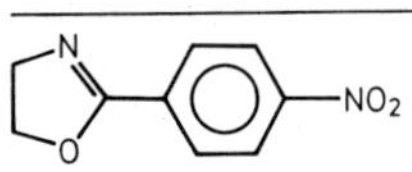	105°C, CH_3CN, 10 h 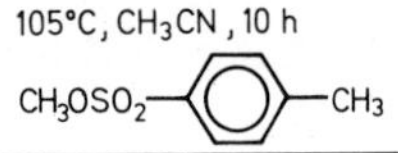	99% yield, mp = 245–250°C; pale yellow powder	11)
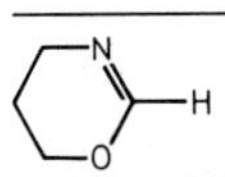	80°C, DMF, 5 h; CH_3I	72% yield, mp = 121–124°C; $\bar{M}_w$ = 4 840	12)
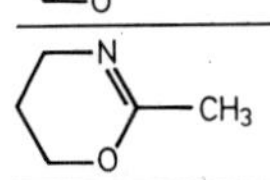	160°C, bulk, 24 h; 20% oleum	100% yield, Tg = 16°C; η_{sp}/c = 0.3 dl/g	13)
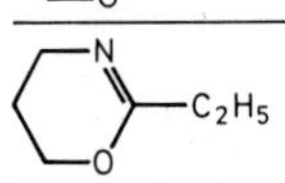	160°C, bulk, 24 h; 20% oleum	100% yield, Tg = 8°C; η_{sp}/c = 0.3 dl/g	13)
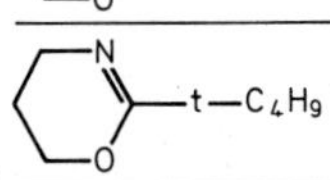	160°C, bulk, 25 h; $(CH_3)_2SO_4$	100% yield, mp = 177°C; η_{sp}/c = 0.43 dl/g	13)
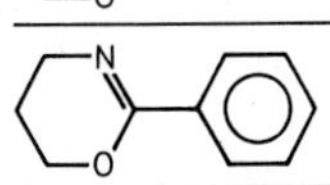	160°C, bulk, 6 h; $(CH_3)_2SO_4$	100% yield, mp = 165°C; Tg = 72°C, η_{sp}/c = 2.37 dl/g	13)

Table 12.2. Rate constants of polymerization of some cyclic imino ethers in CH_3CN extrapolated to 35 °C, counterion OTS^- (ClO_4^- for 2PhOXL)

Monomer	k_p, $mol^{-1} \cdot l \cdot s^{-1}$	Ref.
2-Oxazoline	$2 \cdot 10^{-3}$	6)
2-Methyl-2-oxazoline	$5 \cdot 10^{-5}$	5)
2-Phenyl-2-oxazoline	$3 \cdot 10^{-6}$	9)
2-Phenyl-5,6-dihydro-4H-1,3-oxazine	$4 \cdot 10^{-7}$	4)

Polymerization proceeds rather slowly; the rate constants of propagation (extrapolated to 35 °C) are shown in Table 12.2. The enthalpy of activation of polymerization reaches $\Delta H_p \simeq 25$ kcal · mol^{-1} ($\simeq 100$ kJ · mol^{-1}) so that polymerizations have to be conducted well above room temperature (usually at about 100 °C).

12.3 Side Reactions

As shown in Table 12.1, the molecular weights of polyOXL are rather low. The mechanism of termination reactions were not systematically studied. According to Litt [16], extensive degradative chain transfer takes place leading to enamine ether end-groups (*A*). These groups may react slowly with active species, giving a graft or star-shaped polymer with several branches, and thus increasing the final molecular weight:

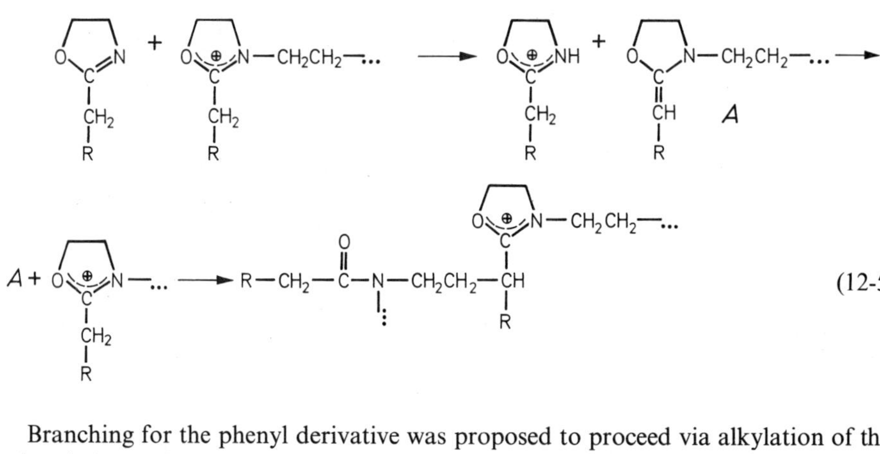

(12-5)

Branching for the phenyl derivative was proposed to proceed via alkylation of the phenyl ring or formation of vinyl amide or propenyl amide linkage. However, high-molecular-weight polymer was synthesized recently from 2-phenyl-OXL [38] at 140 °C.

Chain transfer was not observed. With lower-molecular-weight polymers ($\overline{DP}_n = 5$ to 20), derived from OXL, 4-methyl- and 4-ethylOXL. ^{13}C-NMR spectroscopy revealed the expected end-groups, namely fragments of initiating and terminating species [17].

12.4 Copolymerization

Random copolymerization of different oxazolines is described in Section 15.1.2.2. The reactivity ratios increase as expected with monomer basicity [18]. Statistical copolymers of cyclic imino ethers with other groups of monomers are not known, although oxazolines and oxazines form readily alternating copolymers with a number of electrophilic monomers by spontaneous zwitterionic polymerization. The mechanism and examples of this process are discussed in Section 15.2.1.

Oxazolines have been used in block and graft copolymerization. Addition of 2-methyl-OXL to living polyTHF yielded block copolymer [19]. Mostly on the basis of solubility data the formation of block copolymers of 2-methyl-2-oxazoline and styrene oxide (initiation with I_2) were claimed [20].

Similarly, the reaction of 2-phenylOXL (POXL) with living dicationic poly-1,3-dioxepane (polyDXP) resulted in triblock copolymer [21]:

$$^{\oplus}\text{polyDXP}^{\oplus} + \text{n POXL} \rightarrow$$
$$\text{(polyPOXL)} \text{ (polyDXP)} \text{ (polyPOXL)} \quad (12\text{-}6)$$

Transformation of the hydroxyl-terminated polybutadienes by reaction with tosyl chloride leads to ditosylates, which upon addition of OXL yield triblock copolymers [22]. Alkaline hydrolysis gives a rubbery material ([η] = 0.33 dl/g) containing 39 mol% polybutadiene:

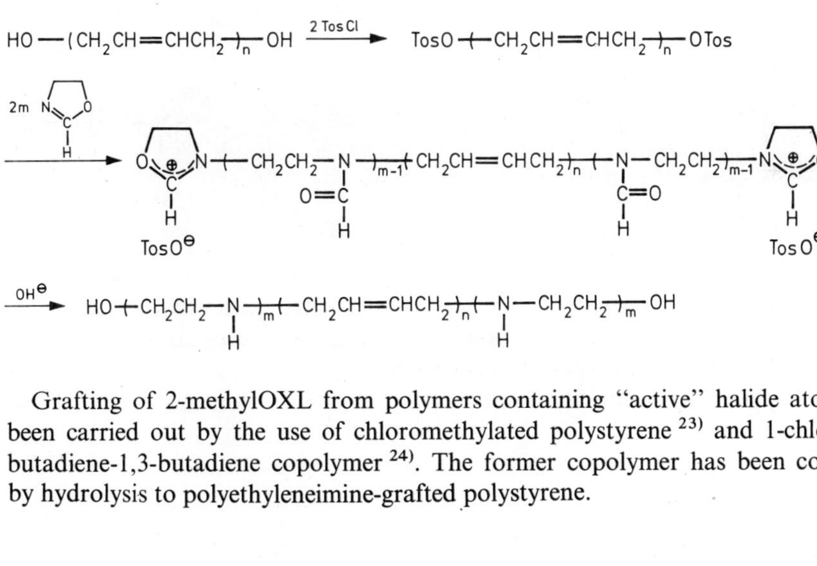

(12-8)

Grafting of 2-methylOXL from polymers containing "active" halide atoms has been carried out by the use of chloromethylated polystyrene [23] and 1-chloro-1,3-butadiene-1,3-butadiene copolymer [24]. The former copolymer has been converted by hydrolysis to polyethyleneimine-grafted polystyrene.

12.5 Some Applications

Polyoxazolines have not yet been commercialized, but several derived polymers found various uses in polymer synthesis. Particularly, the conversion of these polymers into linear crystalline polyethyleneimine, unavailable otherwise, has interesting potentials.

The hydrolysis proceeds quantiatively under mild alkaline conditions only with the N-formylated polymers [25,26]:

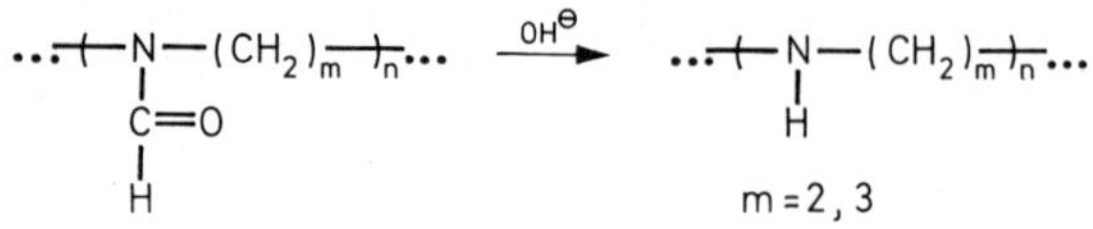

(12-9)

Hydrolysis of polymers derived from 2-substituted monomers requires strong alkaline conditions and the yield does not exceed 90% [27]. However, high-molecular-weight polymer from 2-phenylOXL was converted to linear polyEI using 35% HCl at 100 °C after 5 days [38].

The linear polyalkyleneimines could be used as functional polymers, e.g. polymeric carrier for enzymes or nucleic acid bases [28,29]. It was recently reported, however, that higher-molecular-weight polymers required to for this application could not be prepared [30] (cf. however Ref. [38]).

Graft copolymers of ethyleneimine on polyepichlorohydrine or polystyrene chelated with heavy metal ions like $Hg^{2\oplus}$, $Cu^{2\oplus}$ and Cd^{2+} may have possibility in ion-exchange membranes for the transport of ions [31].

Polyoxazolines can be converted not only to poly(alkyleneimines) but also to poly(N-methyleneimines) and related polymers [32]:

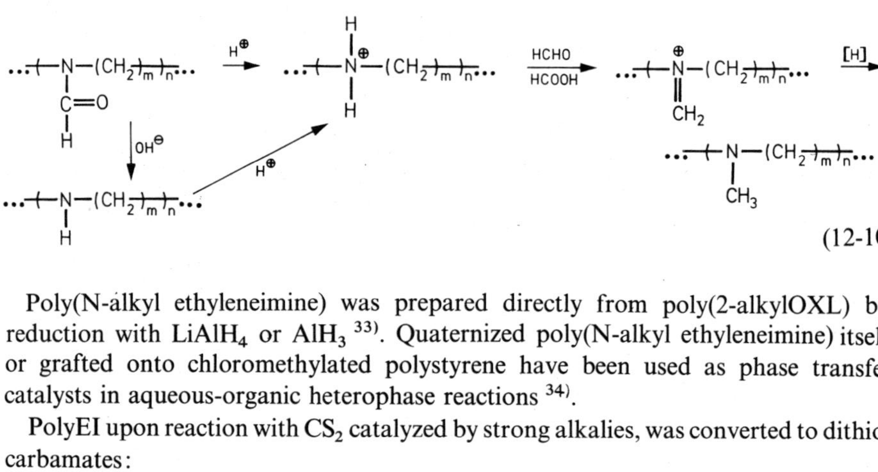

(12-10)

Poly(N-alkyl ethyleneimine) was prepared directly from poly(2-alkylOXL) by reduction with $LiAlH_4$ or AlH_3 [33]. Quaternized poly(N-alkyl ethyleneimine) itself or grafted onto chloromethylated polystyrene have been used as phase transfer catalysts in aqueous-organic heterophase reactions [34].

PolyEI upon reaction with CS_2 catalyzed by strong alkalies, was converted to dithiocarbamates:

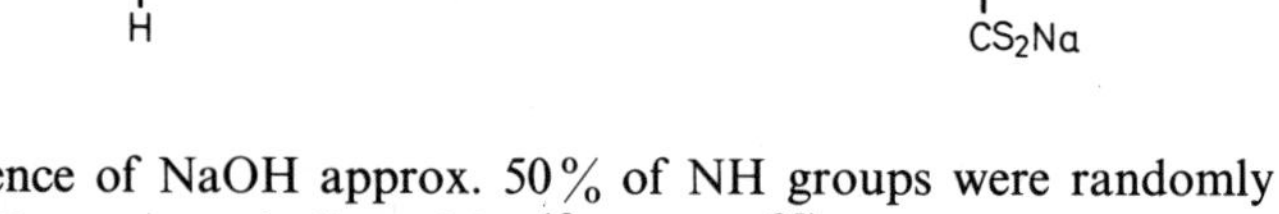

(12-11)

In the absence of NaOH approx. 50% of NH groups were randomly replaced by $-N-CS_2H$ groups as indicated by ^{13}C-NMR [35].

Cationic ring-opening polymerization of optically active 4-methylOXL and subsequent hydrolysis resulted in optically active polypropyleneimine [36]. The optically active polyethyleneimine was effective in the assymetric synthesis of cyanohydrins [37].

12.6 References

1. Saegusa, T., Kobayashi, S.: Encyclopedia of Polymer Science and Technology, Mark, H. F., Bikales, N. F., Eds., Wiley-Interscience, Vol. Suppl. *1*, p. 220, New York 1976
2. Mukaiyama, T., Sato, K.: Bull. Chem. Soc. Jap. *36*, 99 (1963)
3. Mukaiyama, T. et al.: J. Org. Chem. *27*, 3337 (1962)
4. Saegusa, T.: Makromol. Chem. *175*, 1199 (1974)
5. Saegusa, T., Kobayashi, S., Yamada, A.: Makromol. Chem. *177*, 2271 (1976)
6. Saegusa, T., Ikeda, H., Fujii, H.: Macromolecules *6*, 315 (1973)
7. Han, M. J., Chang, J. Y., Lee, Y. Y.: Macromolecules, *15*, 255 (1982)
8. Saegusa, T., Ikeda, H., Fujii, H.: Polym. J. *3*, 35 (1972)
9. Kagiya, T. et al.: J. Polym. Sci., Polym. Letters *4*, 441 (1966)
10. Bassiri, T. G., Levy, A., Litt, M.: J. Polym. Sci., Polym. Letters *5*, 871 (1967)
11. Kobayashi, S., Tokuzawa, T., Saegusa, T.: Macromolecules *15*, 707 (1982)
12. Saegusa, T., Nagura, Y., Kobayashi, S.: Macromolecules *6*, 495 (1973)
13. Levy, A., Litt, M.: J. Polym. Sci., Polym. Letters *5*, 881 (1967)
14. Litt, M., Levy, A.: Belgian Pat. 666, 829 (1965)
15. Tomalia, D. A., Scheetz, D. P.: J. Polym. Sci., A-1, *4*, 2253 (1966)
16. Litt, M., Levy, A., Herz, J.: J. Macromol. Sci. Chem. *A9*, 703 (1975)
17. Guhaniyogi, S., Ivin, K. J., Lille, E. D.: Polymer *18*, 345 (1977)
18. Kagiya, T. et al.: J. Macromol. Sci., Chem. *A6*, 1631 (1972)
19. Saegusa, T., Fujii, H., Ikeda, H.: Japan Pat. 72,000 (1972); C.A. *78*, 1370454 (1974)
20. Seung, S. L. N., Young, R. N.: J. Polym. Sci., Polym. Letters *18*, 89 (1980)
21. Kubisa, P., unpublished results
22. Saegusa, T., Ikeda, H.: Macromolecules *6*, 805 (1973)
23. Saegusa, T., Kabayashi, S., Yamada, A.: Macromolecules *8*, 390 (1975)
24. Saegusa, T., Yamada, A., Kobayashi, S.: Polym. J. *11*, 53 (1979)
25. Saegusa, T., Ikeda, H., Fujii, H.: Macromolecules *5*, 108 (1972)
26. Saegusa, T., Nagura, Y., Kobayashi, S.: Macromolecules *6*, 495 (1973)
27. Saegusa, T., Kobayashi, S.: Kobunshi *22*, 159 (1973)
28. Kiefer, H. C. et al.: Proc. Nat. Acad. Sci. USA *69*, 2155 (1972)
29. Overberger, C. G., Morishima, Y.: J. Polym. Sci., Polym. Chem. Ed. *18*, 1247 (1980)
30. Ludwick, A. G., Overberger, C. G.: J. Polym. Sci. Polym. Chem. Ed. *20*, 2139 (1982)
31. Saegusa, T. et al.: J. Appl. Polym. Sci. *23*, 2343 (1979)
32. Tanaka, R. et al.: J. Polym. Sci., Polym. Letters *16*, 13 (1978)
33. Saegusa, T. et al.: Macromolecules *11*, 435 (1978)
34. Saegusa, T. et al.: Polymer J. *11*, 1 (1979)
35. Ivin, K. J., Lille, E. D.: Makromol. Chem. *179*, 591 (1980)
36. Saegusa, T., Kobayashi, S., Ishiguro, M.: Macromolecules 7, 958 (1974)
37. Tsuboyama, S.: Bull. Chem. Soc. Jap. *39*, 698 (1966)
38. Tanaka, R. et al.: Macromolecules *16*, 849 (1983)

13 Polysiloxanes

Ring-opening ionic polymerization of cyclic siloxanes provides a large group of technically important products. The commercial success of these polymers is due to their chain flexibility and heat stability over broad temperature range. The combination of these properties has been exploited in polymers that are unrivaled by other products as high temperature/high performance oils and rubbers, heat stable coatings, non-toxic solids and liquids used in medicine, etc.

These polymers are mainly prepared by the hydrolysis of dichlorodimethyl (or diphenyl) silanes, followed by further polycondensation of the resulting silanoles which are highly unstable in acidic media. The starting dichlorodimethylsilanes are prepared by a direct reaction of methyl chloride with Si-Fe alloy (Rochow's process). The early development of Si-containing polymers was based on the pioneering works of Russian scientists, mainly Andrianov [1] and of American groups at General Electric Co. [2,3] and Dow Corning [4].

The hydrolysys product of dichlorodimethylsilane are cyclic monomers: 2,2,4,4,6,6-hexamethyltrisiloxane (D_3) and 2,2,4,4,6,6,8,8-octamethylcyclotetrasiloxane (D_4):

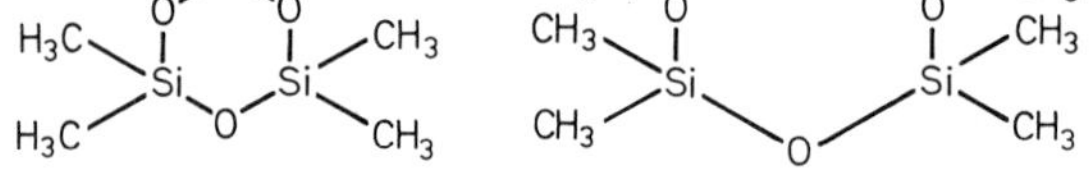

D3 D4 (13-1)

D_3 is the only coplanar molecule and thus the only strained one among cyclic siloxanes.

D_4 can be obtained in high yield and is used on the technical scale mainly for the preparation of high polymers and reactive oligomers with well defined endgroups. Some characteristics of D_4 are as follows: MW = 296, bp. = 175 °C/760, 74 °C 20 mm Hg, mp. = 17.5 °C, D^{20} = 0.9558, n_D^{20} = 1.3968 [5].

Both D_3 and D_4 can be polymerized either anionically or cationically. The main advantages of the anionic polymerization of D_3 are formation of a narrow MWD polymer and formation of linear polymer virtually free of cyclic products (the process can be stopped before the linear-cyclic polymer equilibrium is reached). The cationic polymerization is better suited for the preparation of telechelics, particularly with acetoxy or alkoxy end-groups. Moreover the polymerization of monomers containing functional groups like $\equiv$Si—H, can also be performed only cationically, on account of the lability of the $\equiv$Si—H bond toward bases.

13.1 Synthesis of Polydimethylsiloxanes (PDMS) from D_4

A number of cationic polymerizations of D_3 and D_4 have been described in the literature. More recently CF_3SO_3H (TfOH) has been used in CH_2Cl_2 solvent for both D_3 and D_4 providing satisfactory conditions of the cationic synthesis of polydimethylsiloxanes. It seems, on the basis of analysis of literature data, that TfOH is the most suitable initiator, at least for small-scale laboratory synthesis. In this system the formation of linear polymer is accompanied by cyclic products and their proportion, as it will be discussed later in this section, depends on the initial monomer concentration ($[D_4]_0$). At higher $[D_4]_0$, the proportion of cyclic products is lower. Thus, it is desirable to conduct polymerizations at highest possible $[D_4]_0$.

The proportion of cyclic products is unaffected by other variables (i.e., $[TfOH]_0$, temperature). It may depend on the nature of the solvent used, due to the differences in the interactions between solvents and D_4 or PDMS. However, as mentioned above, for high yields of linear polymer as little solvent as possible should be used.

Thus, practically, a concentrated solution of purified D_4 in CH_2Cl_2 solvent is polymerized by CF_3SO_3H. Traces of water should carefully be removed. The ratio $[D_4]_0/[CF_3SO_3H]_0$ yields the required polymerization degree. The polymer is precipitated in CH_3OH and the solvent is removed from the oily or solid polymer under vacuum. Monofunctional chain-stoppers (like hexamethyldisoloxane) can also be used to control the polymer $\overline{M}_n$.

13.2 Telomerization

Polymerization of D_3 and D_4 has also been used in preparation of various telechelics mostly with —OH, —OR, —OCOR, or —Cl end-groups.

Sommer et al. [9] have prepared linear PDMS with acetoxy end-groups by telomerization of D_4 with diacetoxydisiloxanes in the presence of H_2SO_4:

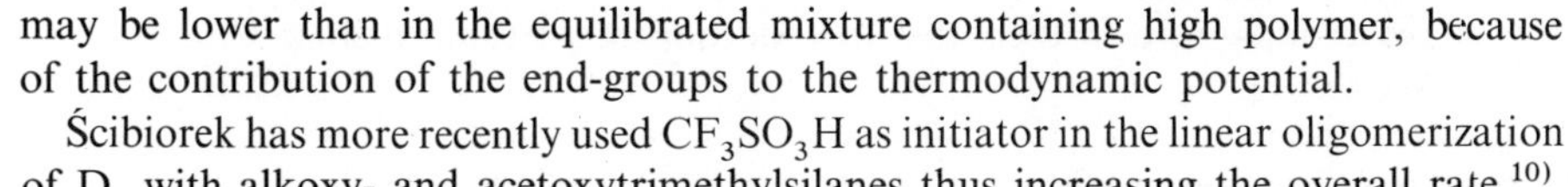

(13-2)

The preparation of oligomers with other end-groups have also been described. The content of cyclic products in these products usually has not been determined but it may be lower than in the equilibrated mixture containing high polymer, because of the contribution of the end-groups to the thermodynamic potential.

Ścibiorek has more recently used CF_3SO_3H as initiator in the linear oligomerization of D_3 with alkoxy- and acetoxytrimethylsilanes thus increasing the overall rate [10]. The process could be conducted at room temperature and was completed within a few hours. Telechelics with two acetoxy end-groups were prepared the same way. The best conditions for these syntheses are bulk oligomerizations at room temperature with 10^{-3} to 10^{-2} mol · l^{-1} CF_3SO_3H initiator, the $[D_4]_0$/telogen$]_0$ ratio depending on the desired chain length.

13.3 Polymerization of Alkylhydrosiloxanes

Polymerization of alkylhydrocyclosiloxanes has been known for over forty years, but information on their polymerization is very limited.

These monomers cannot be polymerized by the anionic mechanism, because of the susceptibility of the Si—H bond to cleavage in the basic medium. This is an important group of reactive polymers because the Si—H bond can be used for further reactions, e.g., with olefins and perfluoroolefins.

Only very recently Lasocki et al. [11] studied in some detail the cationic polymerization and copolymerization of these monomers. Preliminary kinetic measurements indicate that the kinetic external orders change with polymerization conditions and the polymerization medium. Polymerization initiated with methanesulfonic acid in CH_2Cl_2 solvent, is mostly first order in initiator and the internal order in monomer is usually one [11].

Estimation of the reactivity of various cyclosiloxanes of the general formula:

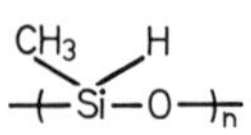

where n = 4, 5, 6 and 7, gave the following order of reactivity: $D_7^H > D_6^H > D_4^H > D_5^H$ (where $D_n^H = (CH_3HSiO)_n$) [11]). This order is based on the measurements of overall rates, including a variety of elementary reactions i.e., initiation, propagation and depropagation. Thus, judgment concerning true reactivities may be premature.

In the range of cyclotetrasiloxanes the overall rate of polymerization increases with the number of CH_3HSiO units in the ring [10]:

$$D_4 > D_3D^H > D_2D_2^H > DD_3^H > D_4^H$$

Copolymerization of hydrosiloxanes was studied without taking into account reversibility (cf. Sect. 15.1), and perhaps this was a reason for the observed dependence of reactivity ratios on copolymerization conditions. Thus, in the copolymerization of D_4^H with D_4 $r_1 = r_2 = 1$ for 22 wt % of comonomers in CH_2Cl_2 solvent, whereas $r_1 = 1.26$ and $r_2 = 0.82$ for 50 wt %, and 2.2 and 0.3 respectively in bulk.

McGrath has recently prepared linear polydimethylsiloxanes with reactive amino, carboxyl or trifluoroacetoxy end-groups and $\overline{M}_n$ in the range from 500 to 3000. D_4 was cooligomerized with tetrahydrotetramethyltetracyclosiloxane (DH_4) using CF_3COOH initiator in the presence of telogens (e.g. linear polysiloxanes with acetamide end-groups). Polymers with ≡SiH groups in the chain were prepared (cf. 13.3), the ≡SiH groups were further used for addition of perfluoroolefins [12].

13.4 Macrocyclization

In the polymerization of the cyclic siloxanes, D_3 and D_4, the formation of linear polymer is accompanied by cyclic oligomers. Thus, polydimethylsiloxane prepared at temperatures not exceeding 200 °C contains up to 15–18 vol% cyclic compounds. This gives [D] $\simeq$ 2.2 mol · l^{-1} [where D = $\text{+}Si(CH_3)_2O$—] and therefore this is the lowest critical concentration of dimethylsiloxane units in the monomer used to produce any linear polymer. The majority (70–80%) of these cyclic oligomers are smaller size rings from D_5 to D_9.

Studies of the cationic oligomerization and polymerization of D_3 and D_4 have shown an important difference in the way they form cyclic oligomers. D_3 gives both linear and cyclic oligomers with every polymerization degrees but those having 3 as the least common multiple (i.e. a series of D_{3n}) are in large excess. Moreover, in the polymerization of D_3 the concentration of cyclic oligomers formed at lower monomer conversions is higher than their equilibrium concentration (kinetic enhancement due to kinetic control); this was not observed in the polymerization of D_4. For D_4, macrocyclics are under thermodynamic control at any monomer conversion. Thus the early stages of cyclooligomerization, at least of D_3, proceeds mostly by end-to-end closure. Chojnowski considered both ring-expansion and back-biting and rightly rejected both mechanisms for D_3 [13,14]. Ring-expansion should lead exclusively to cyclic polymers but this is not observed, back-biting could not produce the early enhancement in rings. Back-biting during the polymerization of D_3 would produce a monotonous distribution of ring-sizes at the later stages of polymerization by reshuffling the original distribution formed by end-to-end closure. D_4 does not show kinetic enhancement or a regular D_{4n} pattern.

This difference between D_3 and D_4 can be illustrated and explained by a common scheme in which we consider the fate of a given linear oligomer of $\overline{DP} = nx$:

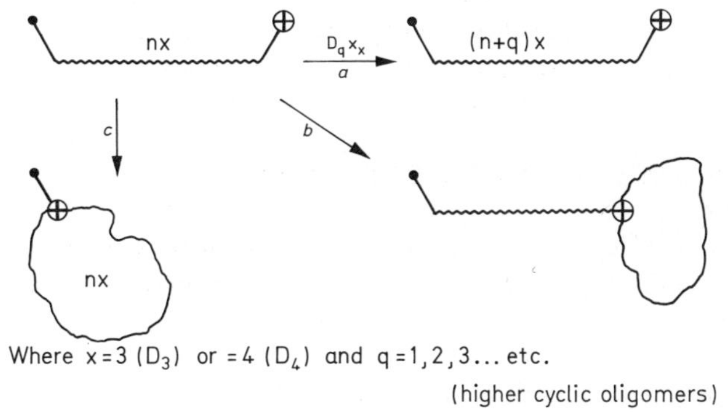

(13-3)

Reaction *a* is the addition of monomer or any cyclic oligomer to the growing species, reaction *b* is back-biting and reaction *c* is end-to-end closure. When x = 3, the monomer (D_3) is more strained and more nucleophilic than any other component in the system. Thus the dominant process is reaction *a* (where $D_{qx} = D_3$) which rapidly gives high polymer. At lower conversions (small n) reaction *c* also takes place. Reactions *a* with D_{qx}, where q > 1, and back-biting (*b*) are slow (lower nucleophilicity of the units) so that the early linear and cyclic polymers obey the 3n pattern.

The situation is different with D_4. This monomer is strainless and its basicity is not higher than that of the linear unit. Thus, the relative rates of reactions are different than the ones described above. Although the rate of formation of cyclics with $\overline{DP}_n = 4n$ is as high as that in the polymerization of $D_3(3n)$, the rate of their disappearance (reaction *a*) is much higher relative to the rate of monomer consumption. The rate of back-biting (reaction *b*) is also relatively much higher, leading to both cyclic and linear oligomers of any polymerization degree. Thus reshuffling occurs earlier in the 4n series and the distribution of cyclics is close to that observed at equilibrium at much lower monomer conversions.

13.5 Elements of the Mechanism of D_3 and D_4 Polymerization Pertinent to High Polymer Synthesis

Cationic polymerization of cyclic siloxanes could not be adequately treated in Vol. I of this monograph (Adv. Polymer Sci. *37* (1980), because the nature of the elementary reactions was obscure and rate constants were unknown. Recently more detailed information has become available, mostly provided by the work of two groups, one in Łódź (Chojnowski a. o.)[6] and the other in Paris (Sigwalt and Sauvet)[7,8]. In our further analysis we will mostly rely on the results of these groups. Their success stems from the use of triflic acid as initiator, following other works in cationic polymerization of heterocyclics.

In earlier works almost any initiator able to initiate cationic polymerization was tested i.e., protonic acids[15–18], anhydrides of inorganic acids[19,20], metal and metaloid halides[21,22].

Opinions differ whether aprotic acids can initiate the polymerization of cyclic siloxanes. According to more recent careful work, the source of protons is indispensable. Earlier it was claimed that $AlCl_3$[23], $FeCl_3$[24], $SnCl_4$[25,26], $TiCl_4$[27] and BF_3[27] can induce polymerization in the absence of added cocatalyst. We think that by the use of proton traps, as it was shown by Kennedy and Chou[28] and Gandini et al. for other processes[29], this question could be decided. If the polymerization of cyclic siloxanes is induced by aprotic initiators it should not be inhibited by a proton trap.

In the first, more carefully studied system H_2SO_4 was used as an initiator. Andrianov's school[23] and then Kojima et al.[30] reported complicated kinetic data that did not provide sufficient information relative to a coherent polymerization mechanism. In these studies H_2SO_4, was used because earlier Patrode and Wilcock observed that this acid polymerizes both D_3 and D_4 to high polymers[15]. More recently, as indicated above, stronger perfluoroalkylsulfonic acids have been used and these works provided much more quantiative information.

The controversy that arose since the beginning of mechanistic studies is focused on the nature of chain growth. The major differences (with some further details to be discussed later in this section) can be reduced to the following two assumptions:

1) acid (HA) reacts with a cyclic siloxane and gives a difunctional linear species $HO(SiO)_xA$. These species undergo further polycondensation and give cyclic and linear macromolecules, regenerating one molecule of HA in every step,
2) acid (HA) reacts with a cyclic siloxane producing (somehow) active cationic species on the macromolecule; propagation proceeds by a chain process and cyclics are formed by end-to-end closing and/or back-biting. The chain process may involve various active species i.e., silicenium and oxonium cations, esters and activated esters, or protonated monomer.

Recent workers, consider a multiplicity of growing centers competing (or cooperating?) in chain growth. Their contribution to monomer consumption is determined by monomer structure and polymerization conditions. We rely in this paragraph on the opinions expressed by Sigwalt and Sauvet (polymerization of D_4 with CF_3SO_3H)[7,8] and Chojnowski et al. (polymerization of D_3 by the same initiator)[6,31-33]. Although Chojnowski's papers came first we take the polymerization of D_4 for the basis of our discussion since information on the polymer formed are more diversified for this system. Summaries of these views have been presented[6,8].

The major facts relative to the polymerization of D_4 can be summarized as follows:

- polymerization of D_4 ($[D_4]_0 \sim 1.0\ \text{mol} \cdot l^{-1}$, $[\text{TfOH}]_0 \sim 5.8 \cdot 10^{-3}\ \text{mol} \cdot l^{-1}$) in CH_2Cl_2 solvent at r.t. leads to a mixture of high polymer (HP) (97000; 28%), low polymer ((LP) 8%), macrocycles (9%), D_5-D_9 fraction (23%), and 32% of unreacted D_4. It seems to the present writers, that this mixture is not at equilibrium, as evidenced by the presence of two linear fractions, which should give only one peak at equilibrium. The rate of D_4 consumption is $6.4 \times 10^{-4}\ \text{mol} \cdot l^{-1} \cdot s^{-1}$; i.e. half of the monomer used reacts in approx. 10^3 s under the above conditions. The polymer composition was determined by GPC. Polymerization of D_3 provided virtually the same GPC profile. Apparently, at equilibrium the linear polymer gives only one peak with no observable shoulder;
- according to the Sigwalt and Sauvet statement, the number of end-groups in the linear fraction corresponds to the number of acid molecules, i.e. one acid molecule produces one linear macromolecule. Some H_2O can, however, be present at equilibrium, contradicting this simple belief. Macrocyclics are simultaneously formed; their concentration is invariable, and in agreement with the Jacobson-Stockmayer theory,
- during the polymerization of D_4 first two cyclic fractions (including D_4, D_5-D_9, and macrocycles) and linear LP appear, then HP begins to form. At later stages it seems as if D_4 were converted directly to HP but in reality all components are in a dynamic equilibrium and it is unclear which way D_4 enters the HP molecules,
- the proportions of the four major polymer fractions (HP, LP, macrocyclics, D_5-D_9) and monomer at equilibrium depend on $[D_4]_0$ in agreement with thermodynamic requirements, as discussed in Chap. 2 of this volume. Thus, the higher $[D_4]_0$, the higher the proportion of HP and the lower the proportion of $[D_4]$ at equilibrium ($[D_4]_e$). When $[D_4]_0$ is close to the bulk concentration over 75% of siloxane units at the end of polymerization belong to the HP fraction, 10% to D_4 and the rest (a total of $\sim$15%) to LP, macrocycles and the D_5-D_9 fraction. Changing the temperature the equilibrium composition will change only slightly, because the rings involved, including D_4, are nearly strainless. Thus, the above described conditions are almost optimum for high-molecular-weight polymer preparation,
- the kinetics show fractional external orders for both D_3 and D_4. Sigwalt and Sauvet succeeded to come close to explain these orders, whereas Chojnowski rightly pointed out that extensive hydrogen bonding may significantly change substrate concentrations (mostly acid)[32,33]. Thus it may be premature to rely strictly on the observed orders in any of these systems.
- ^1H- and ^{19}F-NMR studies of model compounds and systems close to polymerization conditions indicate the presence of hydrated TfOH and triflic esters of

silanols no free silanols were detected, but we cannot be sure on this bases that they are indeed absent.

13.6 Chemistry of Polymerization

13.6.1 Initiation

Although the reaction product between TfOH and D_4 (hydrogen-bonded or protonated D_4) could not be isolated or detected, it seems reasonable to suggest a simplified scheme that includes the protonated monomer:

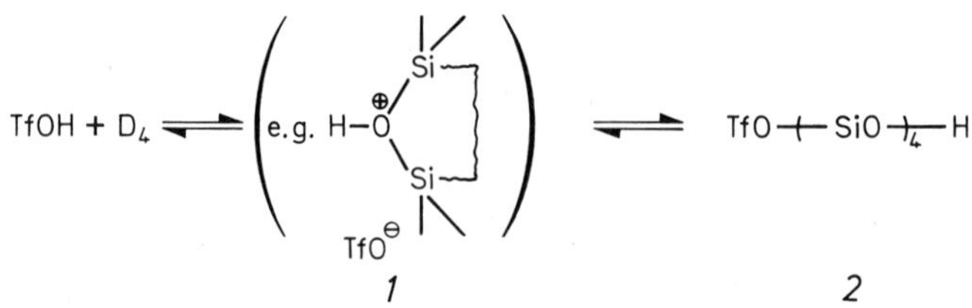

(13-4)

(Sigwalt and Sauvet prefer to write H-bonded D_4 instead of *1*) Subsequently *2* may:

- add activated monomer *1*,
- dimerize to linear dimer by acid-catalyzed hetero- or homocondensation
- form a tertiary oxonium ion (*4*) by direct monomer addition to (*2*) or, according to both Chojnowski and Sauvet, to (*3*) having an ester bond activated by TfOH:

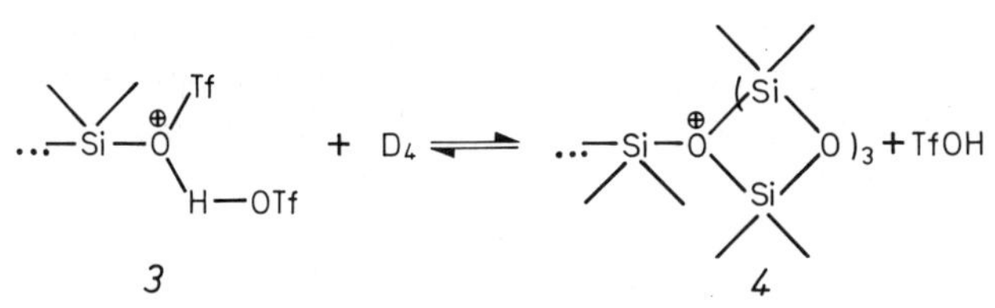

(13-5)

Since the ester is the least basic component in the mixture, we do not favour extensive formation of *3*.

13.6.2 Propagation

Contribution to chain growth of the various active species coexisting in the system depends on the monomer ring strain. Thus, propagation may proceed by addition of activated monomer to the silanol chain end:

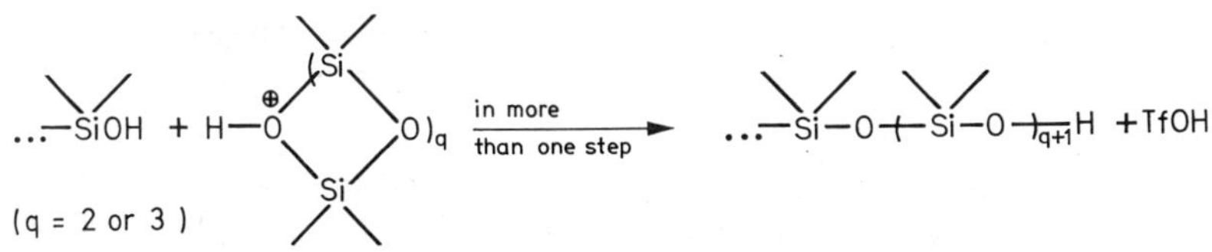

(13-6)

Similar processes have been described for cyclic acetals and cyclic ethers (Chap. 4), and are mentioned as a possibility by Chojnowski for D_3 but are ignored by Sigwalt and Sauvet for D_4.

Tertiary oxonium ions *4* may propagate by the usual onium mechanism or they may react by coupling with a silanol or ester end-group; however these latter reactions do not involve monomer consumption. Similarly, homo- or heteropolycondensation of macromolecules with silanol or silyltriflate end-groups like *2* may occur with expulsion of an acid or anhydride molecule. In this way acid is recycled to the system. Chojnowski and Wilczek [32,33] have studied homocondensation, heterocondensation and acidolytic cleavage reactions via model systems and concluded that these reactions are fast as compared with the overall rate of monomer consumption. Thus, the steady-state concentration of some components (e.g. TfOH) can be established.

The relative contribution of the specific chain growth mechanism (i.e., activated monomer vs. oxonium ion addition) may depend on ring strain of monomer, nucleophilicity of anion and solvating power of solvent (ability to stablize ions). Many of these factors have been quantiatively determined in the polymerization of cyclic ethers and acetals, where the concentrations of the tertiary and secondary oxonium ions were simultaneously determined by the phosphine cation-trapping method (cf. Adv. Polymer Sci. *37*). This method seems to be also applicable in the polymerization of siloxanes, but has not yet been evaluated.

The activated monomer polymerization is particularly sensitive to ring strain. The higher the ring strain, the higher the contribution of the activated monomer mechanism, mostly because the onium ion $\rightleftharpoons$ macroester equilibrium is shifted toward the macroester side without appreciably changing the protonation of the monomer (provided that basicities do not change markedly). Thus, D_3 should be much more susceptible than D_4 to activated monomer growth. Some data on the kinetics of D_3 and D_4 polymerization can be interpreted this way. For instance, the ratio $[H_2O]_0/[TfOH]_0$ influences differently the rate of polymerization of D_3 and D_4. The rate of D_3 polymerization increases up to 10^3 times with increasing $[H_2O]_0$, whereas the rate of D_4 polymerization is practically unaffected according to Sigwalt and Sauvet*. Wilczek estimated that by increasing the $[H_2O]_{added}$ from 0 to $4 \cdot 10^{-2}$ mol $\cdot\, l^{-1}$ in the polymerization of D_4 ($[D_4]_0 = 2.0$ mol $\cdot\, l^{-1}$, $[TfOH]_0 = 7 \cdot 10^{-4}$ mol $\cdot\, l^{-1}$) increases ($\equiv$ Si-on] from negligible to $2 \cdot 10^{-2}$ mol $\cdot\, l^{-1}$ and also increases significantly the [TfOH] (the concentration of the macroester decreases at the same time) [35]. Thus, an increase in $[H_2O]$ may increase simultaneously [$\equiv$Si-on] and

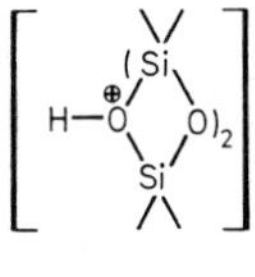

which in turn increase the overall rate of polymerization.

Onium ions will be very short-lived in the polymerization of D_3, even when water is not added, because the strained onium ions rapidly collapse with anions. There is an analogy with the polymerization of ethylene oxide and THF with $TfO^{\ominus}$. In

* This observation is challanged by Chojnowski, who is assuming that adventitious $[H_2O]_0$ was to low to increase the rate with increasing $[H_2O]$ added [34].

the former (strained monomer) tertiary oxonium ions are practically absent, whereas in the latter (less strained monomer) well above half of the initiator can be converted to tertiary oxonium ions (cf. Vol. I Adv. Pol. Cci. *37*).

Thus, the proposed mechanisms of chain growth, i.e. one polycondensation and three polymerization mechanisms (activated monomer, oxonium ion and activated ester) may proceed in various proportions, as dictated by conditions. To determine the relative extent of these contributions conditions have to be devised under which this multiplicity is reduced. The use of an acid like $HMtX_{n+1}$ ($HX \cdot MtX_n$) that has a complex anion and is unable to form covalent bonds will simplify the system by excluding ester formation. Another simplification would be to use TfOH in a nonpolar solvent (e.g. CCl_4); this should highly decrease the concentration of ions.

13.7 References

1. Andrianov, K. A.: Zhur. Obschei. Khim. *16*, 487 (1946)
2. Rochow, E. G.: U.S. Pat. 23 80995
3. Rochow, E. G.: An Introduction of the Silicones, Wiley and Sons Inc., N.Y. 1951
4. Barry, A. J. et al.: Dow Corning, U.S. Pat. 24953 62
5. Dictionary of Organic Compounds, Vol. 4 p. 2559, London, Eyre and Spottiswoode Publ. Ltd. *75*, 2932 (1953)
6. Chojnowski, J.: Wilczek, L., Rubinsztain, S. "Cationic Polymerization and Related Processes". E. J. Goethals, Ed. Academic Press 1984, p. 253
7. Lebrun, J. J., Sauvet, G., Sigwalt, P.: Makromol. Chem., Rapid Commun. *3*, 757 (1982)
8. Sauvet, G., Lebrun, J. J., Sigwalt, P.: IUPAC, "Cationic Polymerization and Related Processes", E. J. Goethals, Ed. Academic Press 1984, p. 237
9. Sommer, L. H., English, W. D., Ansul, G. R., Vivona, D. N.: J. Am. Chem. Soc., *77*, 2485 (1955)
10. Ścibiorek, M., Chojnowski, J.: Europ. Polym. J. *17*, 413 (1981)
11. Graczyk, T., Lasocki, Z.: Bull. Acad. Polon. Sci., Ser. sci. chim. *26*, 917 (1978), *27*, 181 (1979)
12. Chujo, Y., McGrath, J. E.: Polymer Preprints *27* (2), 47 (1983)
13. Chojnowski, J., Mazurek, M., Scibiorek, M., Wilczek, L.: Makromol. Chem., *175*, 3299 (1974)
14. Chojnowski, J., Scibiorek, M., Kowalski, J.: Makromol. Chem. *178*, 1351 (1977)
15. Patnode, W., Wilcock, D. F.: J. Am. Chem. Soc., *68*, 358 (1946)
16. Hurd, D. T.: J. Am. Chem. Soc., *77*, 2998 (1955)
17. Andrianov, K. A., Jakushkina, S. E.: Vysokomol. Soedin., *1*, 613 (1959)
18. Kogan, E. V., Ivanova, A. G., Reichsfeld, V. O., Smirnov, N. I., Gruber, V. N.: Vysokomol. Soedin., *5*, 1183 (1963)
19. Voronkov, M. G.: Gietierociklicheskie reakcji rashcheplenia siloksanowych sviaziej, Izd. AN SSSR, 1962, p. 100
20. Moedritzer, K., Van Wazer, J. R.: J. Am. Chem. Soc., *86*, 802 (1964)
21. Müller, R., Mross, D.: Z. anorg. allgem. Chem., *324*, 78 (1962)
22. Margrave, T., Sharp, K. G., Wilsom, D. W.: J. Am. Chem. Soc., *92*, 1530 (1970)
23. Andrianov, K. A., Jakushkina, S. J.: Izv. AN SSSR, Ser. Khim. *1962*, 1396
24. Gruber, V. N.: Vyskomol. Soedin., *1*, 1194 (1959)
25. Chernyshev, A. J., Jastriebov, V. V.: Vysokomol. Soedin., *11A*, 525 (1969)
26. Prut, E. V., Trofimova, T. M., Enikolopian, J. S.: Vysokomol. Soedin., *6*, 2102 (1964)
27. Nöll, N.: "Chemistry and Technology of Silicones", Academic Press, New York and London, 1968
28. Kennedy, J. P., Chou, R. T.: Polym. Prepr. *20*, 306 (1979)
29. Moulis, J. M., Collomb, J., Gandini, A., Cheradame, H.: Polym. Bull. *3*, 197 (1980)

30. Kojima, K., Tarumi, N., Wakatuki, S.: Nippon Kagaku Zasshi, *76*, 1205 (1955)
31. Chojnowski, J., Wilczek, L.: Makromol. Chem., *180*, 117 (1979)
32. Wilczek, L., Chojnowski, J.: Macromolecules *14*, 9 (1981)
33. Wilczek, L., Chojnowski, J.: Makromol. Chem., *184*, 77 (1983)
34. Chojnowski, J.: private communication 1984
35. Wilczek, L., Ph. D. Thesis, Łódź 1983

14 Phosphorus-containing Polymers

Polymerization of cyclic monomers containing P-atoms in the ring has been recently reviewed by Penczek and Łapienis elsewhere [1]. The intensively studied cyclic phosphazenes (PZ) were recently reviewed several times, mostly by Allcock [2,3,4]. In this chapter we discuss only the main features of the cationic polymerization of these compounds.

The following groups of P-containing compounds were cationically polymerized: 2-substituted-2-oxo-1,3,2-dioxophospholanes (ODPL) and the corresponding phosphorinanes (ODPR), 2-substituted-1,3,2-dioxaphospholanes (DPL) and phosphorinanes (DPR), 2-substituted-1,3,2-oxazaphospholidines (OAP) and 2-substituted-1,2-oxaphospholanes (OPL):

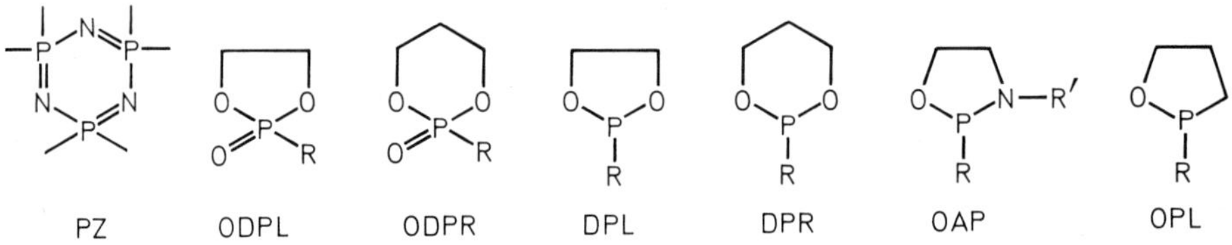

(14-1)

No information is available on the cationic polymerization of cyclic phosphines.

The cationic polymerization of the above listed monomers usually leads to low- and medium-molecular-weight products ($\overline{M}_n < 10^4$). In contrast to cationic processes anionic polymerizations lead to higher-molecular-weight polymers ($\overline{M}_n \simeq 10^5$) [5].

Polyphosphazenes hove been extensively developed. Initially poly(dichlorophosphazenes) were synthesized in the form of insoluble highly crosslinked polymers. Later it was found that benzene or THF soluble high-molecular-weight polymer was formed before 70% of the cyclic trimer was converted into polymer [2]. Soluble poly(dichlorphosphazenes) take part in nucleophilic substitution reactions. Replacement of the Cl atoms proceeds without chain cleavage with various nucleophiles resulting in products with promising properties [3]. Thus, polymers with $—OCH_3$ and $—OC_2H_5$ substituents are elastomers (Tg = —70 °C and —84 °C, respectively); products with $—OCH_2CF_3$ substituents are hydrophobic, microcrystalline fiber-forming polymers (Tg = —66 °C); derivatives with $—OC_6H_5$ substituents are microcrystalline thermoplastics (Tg = —8 °C); CH_3NH-substituents provide water solubility and $C_6H_5NH—$ substituents glassy properties; aryloxy derivatives lead to liquid crystalline behaviour.

Although no other phosphorus-containing polymers synthesized cationically have been commercialized, they exhibit interesting properties due to their composition, i.e., low flamability, flame retardancy and high efficiency to complex heavy metal ions.

14.1 Initiating Systems

Heterocyclics containing P-atoms are usually strong nucleophiles. For example the basicity of 2-methoxy-ODP ($pK_a = 3.1$) is much higher than that of cyclic ethers or sulfides. Therefore, a wide range of initiators, i.e., carbenium or oxonium salts, Lewis and protonic acids, and relatively nonreactive alkyl halides and organoaluminum compounds have been used. Usually, reactions were carried out in a N_2 atmosphere, although no special precautions (e.g. vacuum) were used to avoid contamination with water.

Typical reaction conditions and product properties are shown in Table 14.1. We have also included in this Table systems with "spontaneous" initiation. It is very likely that spontaneous cleavage of the C—O linkage produces cationic growing centers and phosphonate anions.

Initiation, similarly to propagation discussed in the next section, often proceeds by nucleophilic attack of the tricoordinated P-atom of the monomer [11]. For monomers with tetracoordinated P-atoms (ODPL, ODP) the reaction proceeds by attack of the nucleophilic phosphoryl O-atom [24]. The mechanisms of these reactions were studied by NMR spectroscopy, and reaction products were directly observed for the system shown in Eq. (14-3) when R was CH_3:

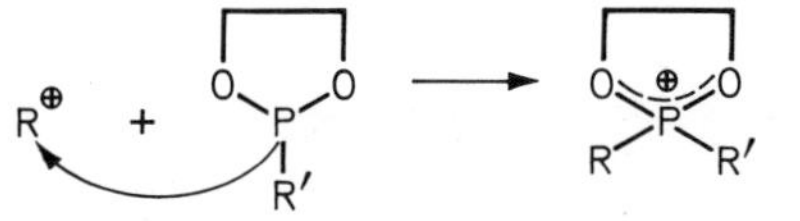

(14-2)

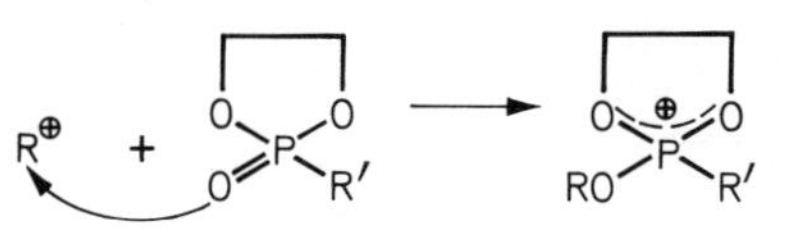

(14-3)

Polymerization of phosphazenes is initiated either thermally (t > 230 °C) or by the use of Lewis acids. In the former system it is belived that the heterocyclic cleavage of the P—Cl bond is followed by a nucleophilic attack of the monomer molecule leading to the phosphonium cation [4]:

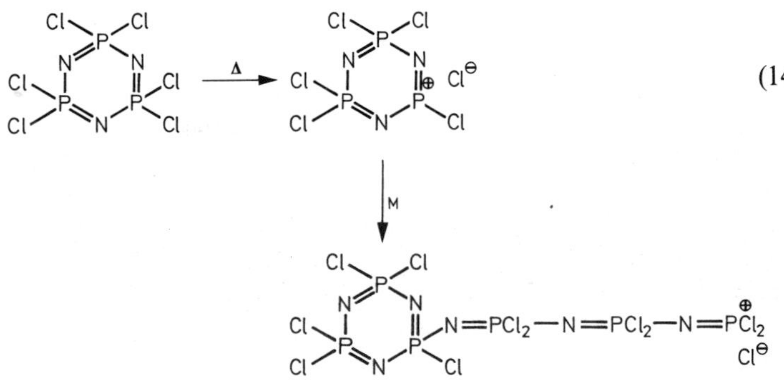

(14-4)

Table 14.1. Cationic Polymerization of Phosphoroorganic Cyclic Monomers

No.	Monomer	Polymerization Conditions	Product properties	Ref.
1	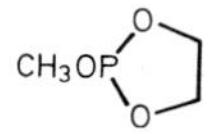	$AlCl_3$, 180 °C, 6 h	yield 95%, $[\eta] = 1.24$ dl/g, colourless hard solid	6,7)
2	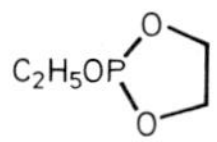	$AlCl_3$, 150 °C, 8 h	$[\eta]_{inh} = 1.01$ dl/g, colourless solid	7)
3	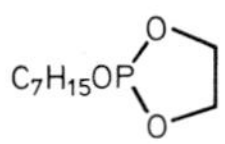	CH_3I, 160–170 °C, 75 h	$\eta_{sp}/c = 0.047$ dl/g, colourless viscous liquid	8)
4	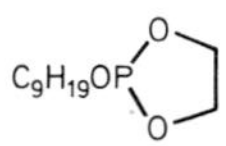	CH_3I, 190–200 °C, 45 h	$\eta_{sp}/c = 0.038$ dl/g, colourless viscous liquid	8)
5	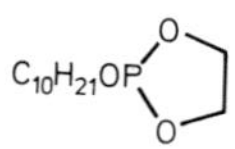	CH_3I, 190–200 °C, 30 h	$\eta_{sp}/c = 0.054$ dl/g, colourless viscous liquid	8)
6	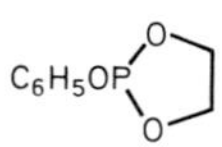	$C_6H_5CH_2I$, 135 °C, 262 h	yield 35%, $\overline{M}_n = 2500$, colourless solid	9)
7	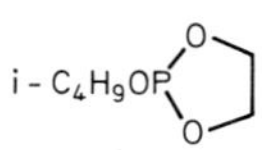	i-C_4H_9Cl, $AlCl_3$, 193 °C, 20 h	yield 50%, transparent solid, soft.p. = 100 °C	10)
8	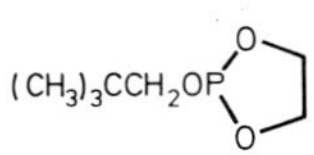	CH_3I, 100 °C, 300 h	jelly-like product	11)
9	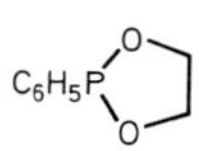	$BF_3 \cdot Et_2O$, 150 °C, 5 h	$\overline{M}_n = 6600$, transparent resin	12,13,14)
10	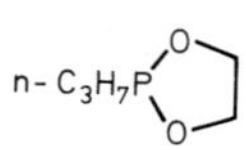	$FeCl_3$, 140–180 °C, a few hrs	yield 100%, $\eta_{sp}/c = 0.07$ dl/g, viscous jelly state	14)

11	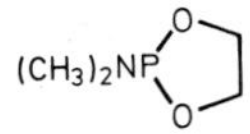	CH_3I, 120 °C, 100 h	yield 100%, white powder insoluble in $CHCl_3$	15)
12	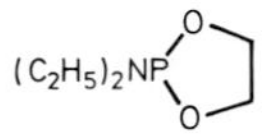	CH_3I, 120 °C, 240 h	yield 58%, $\overline{M}_n$ = 1600, yellow solid	15)
13	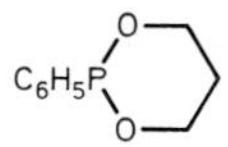	without catalyst, 180 °C, 10 h	yield 95%, η_{sp}/c = 0.14 dl/g, solid, soft.p. = 60 °C	12, 14, 16, 17)
14	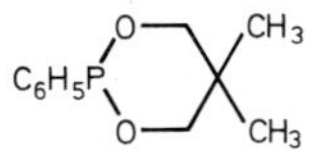	CH_3I, 130 °C, 24 h	$\overline{M}_n$ = 16000, T_g = 4 °C, m.p. = 160 °C	16, 18)
15	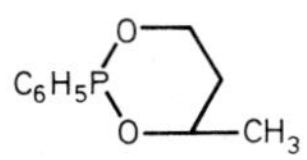	CH_3I, 130–140 °C, 4h	$\overline{M}_n$ = 3200, soft.p. ≃ 60–70 °C	17)
16	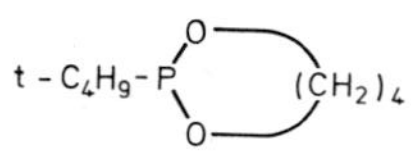	without catalyst, 25 °C	$\overline{M}_n$ = 1700	19)
17	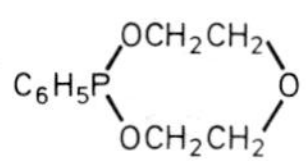	$C_6H_5CH_2Cl$, 130 °C, 195 h	yield 93%, $\overline{M}_n$ = 2800	13, 20)
18	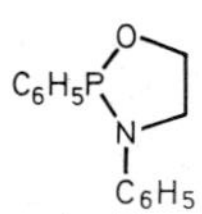	CH_3I, 100 °C, 25 h	yield 64%, η_{sp}/c = 0.04 dl/g, brittle glass-like m.p. ≃ 72–78 °C	21)
19	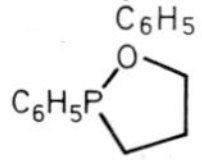	CF_3SO_3,CH_3, 60 °C, 26 h	yield 90%, $\overline{M}_n$ = 3900, white powder	22)
20	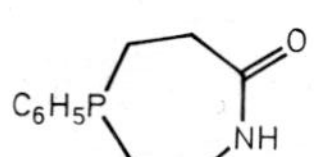	$SnCl_4$, 233 °C, 15 h	yield 85%, η_{sp}/c = 0.08 dl/g, brown glassy solid m.p. = 92–97 °C	23)

Table 14.1. (continued)

No.	Monomer	Polymerization Conditions	Product properties	Ref.
21	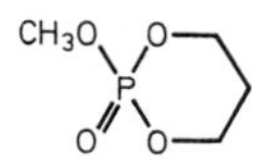	$(C_6H_5)_3C^{\oplus}$, $PF_6^{\ominus}$, 125 °C, 2 h	yield 45%, $\overline{M}_n$ = 1200, viscous oil	[24]
22	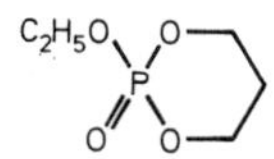	$(C_6H_5)_3C^{\oplus}$, $PF_6^{\ominus}$, 105 °C, 13 h	yield 51%, $\overline{M}_n$ = 800, viscous oil	[25,26]
23	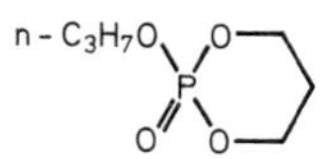	$(C_6H_5)_3C^{\oplus}$, $PF_6^{\ominus}$, 140 °C, 4.5 h	yield 77%, $\overline{M}_n$ = 800, viscous oil	[25]
24	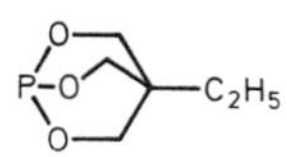	$C_8H_{17}I$, 200 °C, 30 h	yield 16%, $[\eta]_{inh}$ = 0.06 dl/g, solid polymer	[27]
25	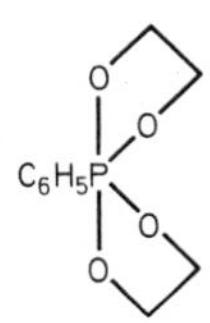	$BF_3 \cdot OEt_2$, 60 °C, 80 h	yield 21%, $\overline{M}_n$ = 4100	[28]
26	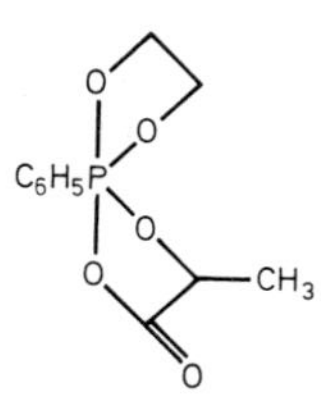	without catalyst, 120 °C, 48 h	yield 55%, $\overline{M}_n$ = 7500	[29,30]

Lewis acids increase the stability of the counteranion by formation of complex anions [31, 32]:

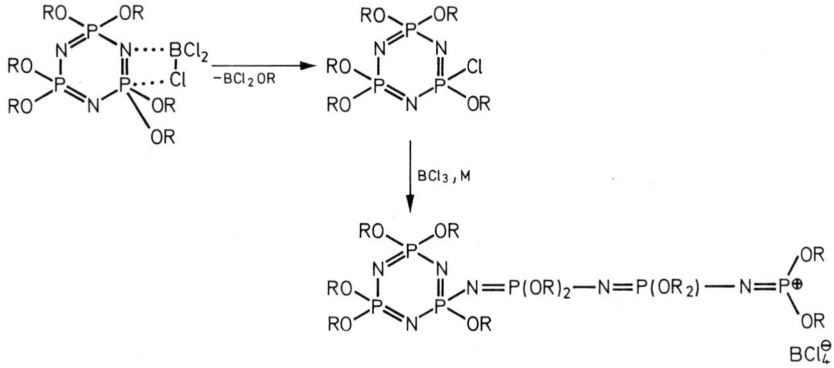

(14-5)

14.2 Peculiarities of Chain Growth

The Arbuzov-Michaelis rearrangement [33] is the usual route of propagation of monomers with tricoordinated P-atoms. First, cationation on P-atoms takes place and then the most labile C—O bond is broken in the endocyclic (*a*, chain growth) or exocyclic position (*b*, transfer), as shown for the 2-alkoxy-1,3,2-dioxaphospholane [11]:

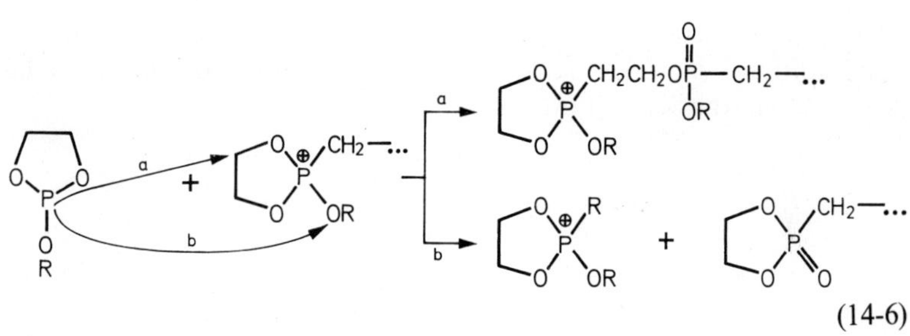

(14-6)

Thus, isomerization proceeds simultaneously with chain growth. Similar isomerized structures were observed for poly(2-phenyl-1,2-oxaphospholane) [22]:

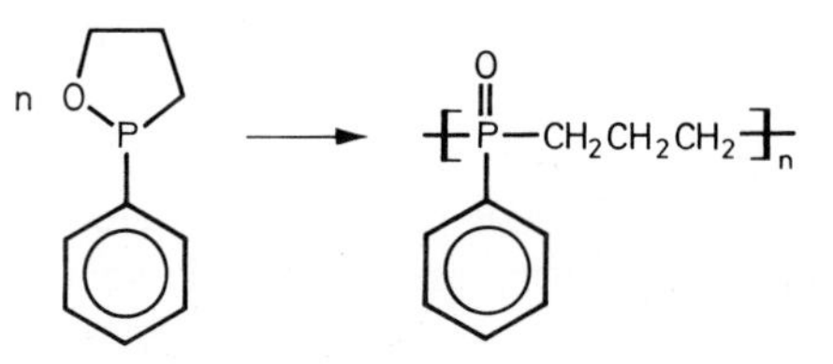

(14-7)

In the polymerization of ODPL and ODPN propagation is accompanied by an exchange of phosphoryl O-atoms leading to the expected polymer structure [24, 25]:

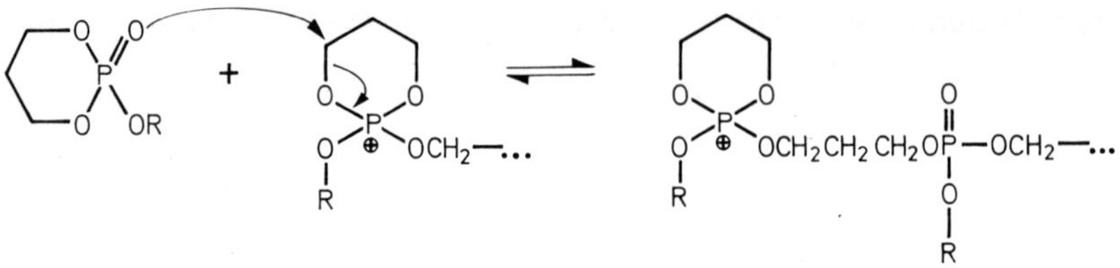

(14-8)

Polymerization rates are usually low and higher temperatures are required. The rate constant of propagation of 2-ethoxy ODP is $k_p = 6 \cdot 10^{-6}\ mol^{-1} \cdot l \cdot s^{-1}$ at 25 °C [25].

In OAP monomers, polymerization occurs by cleavage of the C—O bond, which is weaker than the C—N bond [21]. The same behaviour was reported earlier for cyclic iminoethers (cf. Chap. 12):

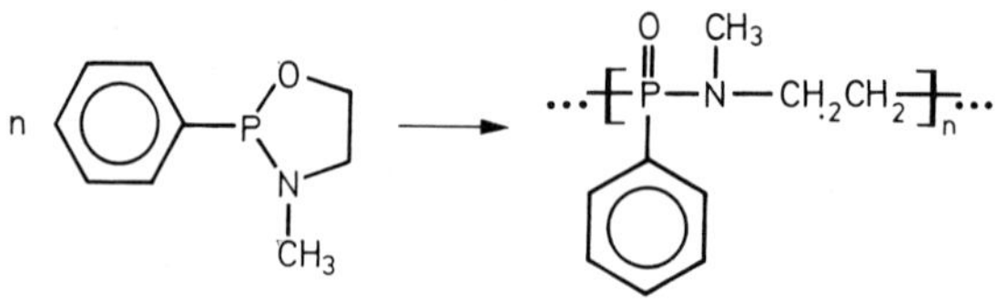

(14-9)

14.3 Side Reactions

Transfer reactions responsible for the decrease of molecular weights were described. One of these reactions already has been shown in Eq. (14-6) (route *b*). "Wrong monomer addition", in which nucleophilic attack proceeds on the exocyclic group, may also result in transfer of β-protons [24, 25], e.g.:

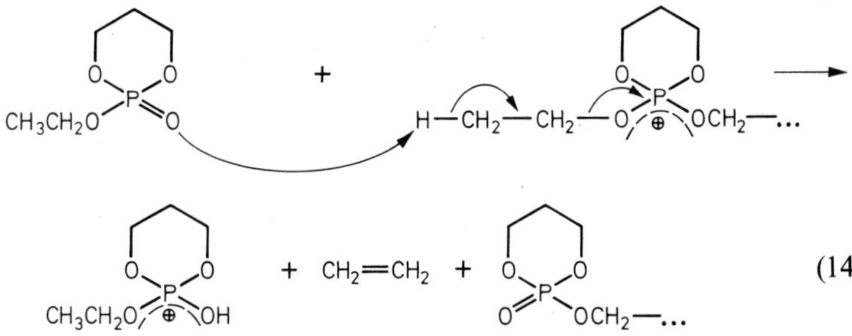

(14-10)

The cyclic end-groups, formed in the reactions by Eqs. (14-5) and (14-9), were observed by ^1H- and ^{31}P-NMR spectroscopy. The NMR data were in agreement with independently measured $\overline{M}_n$, assuming that every macromolecule contains one cyclic end-group.

The polymers are often more basic than their parent monomers, as observed for 2-methoxy-ODPR (pK_a = 3.1 and 3.8 respectively). Thus, the phosphoryl O-atoms in the chain compete successfully with monomer molecules for growing centers.

This type of degradative transfer to polymer limits the yields if the formation of branched strainless ions is irreversible [25]:

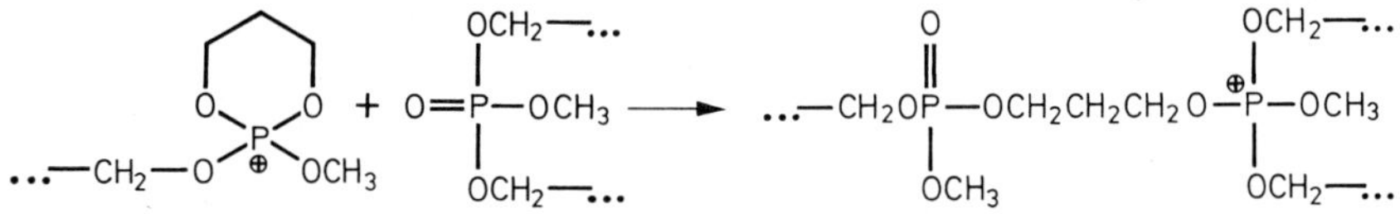

(14-11)

A similar reaction occurs in the polymerization of phosphazenes, where transfer to polymer proceeds at higher conversion (>70%) and leads to the formation of insoluble, crosslinked polymers [3].

Three types of polymer units were observed in the polymerization of 2-methoxy, 2-phenoxy- and 2-phenyl-DPL [6,7,9,12,15,34]:

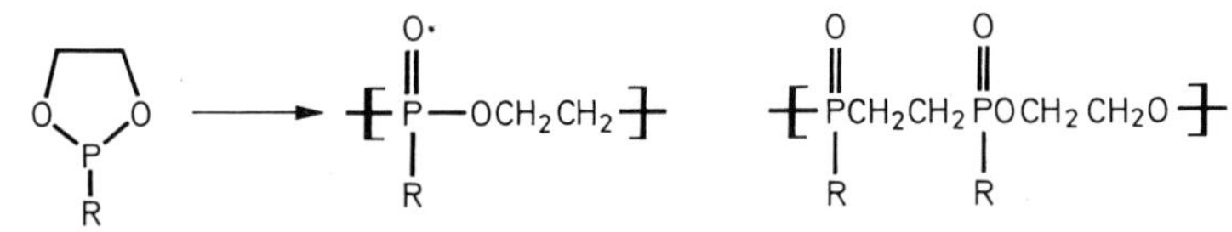

I *II* *III* (14-12)

The formation of the unusual structures II and III was initially explained by monomer isomerization [6,7] and by spontaneous reaction of monomer with polymer chains [12]. Later, however, more detailed kinetic analysis led to the conclusion that the isomerized structures are formed in reactions with reactive end-groups [9]. This is shown below for the polymerization of 2-phenoxy-DPL initiated with benzyl bromide:

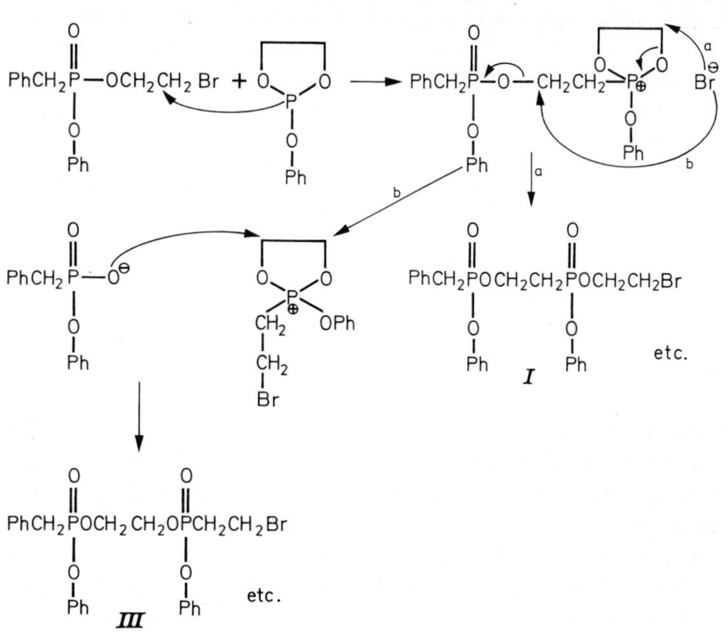

Route *a* leads to the usual structure I, whereas attack at the exocyclic methylene C atom (facilitated by the electron withdrawing phosphonate group) leads, after attack of the phosphonate anion, to structure III. The subsequent reaction with monomer leads to structure II:

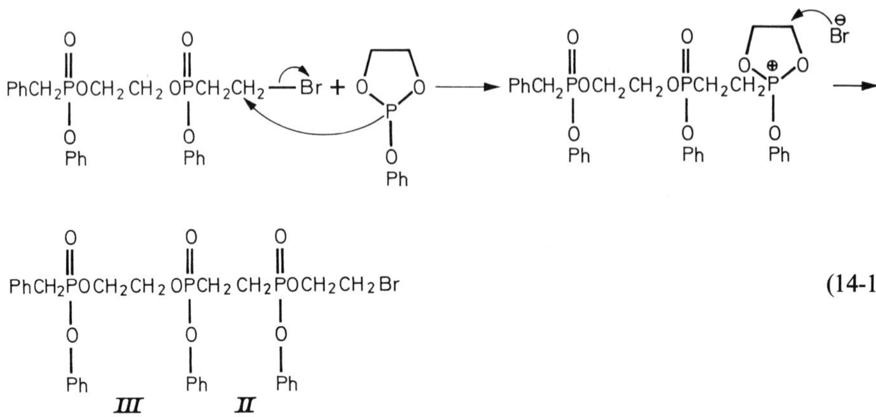

(14-14)

14.4 Copolymerization

P-atom-containing cyclic monomers were copolymerized mostly by the use of anionic initiators [1]. Information concerning cationic copolymerization is rather scarce. It has been reported, however, that 2-ethoxy-2-thio-1,3,2-dioxaphospholane copolymerizes with 1,3,5-trioxane [35]. Using $BF_3 \cdot OEt_2$ at 65 °C copolymer containing up to 20% units with P-atoms were formed (their content in the monomer feed was only 5%). The higher proportions of the more basic phospholane in the product are in agreement with structure/reactivity correlations decribed in Section 15.1. Molecular weights were not reported. Reduced viscosities (0.2% solution in $CHCl_3$ at 25 °C) were in the range from 0.07 to 0.42 [6].

2-Methoxy- and 2-ethoxy-1,3,2-DPL were copolymerized with monomers containing both phosphine and amide linkages:

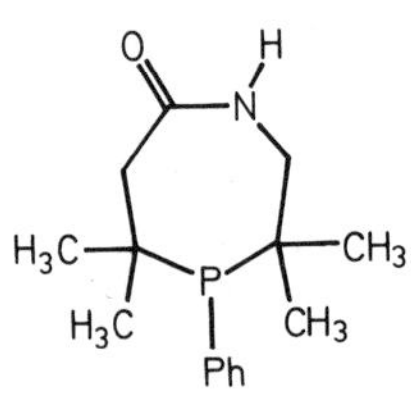

(14-15)

At 202 °C using Et_2AlCl initiator, copolymer containing 83 and 57% of P-units respectively was formed after 15 hrs in 33 and 63% yields. In products exhibited low viscosities ($\eta_{sp}/c = 0.07$ dl/g) [23].

The phosphepane shown in Eq. (14-15) was also copolymerized with a number of different lactones and lactames using H_3PO_4 catalyst. Polymeric solids (m.p. = 80 to 150 °C) were formed in 60–90% yield. Molecular weights were higher than for the preceding systems (η_{sp}/c = 0.13–0.31 dl/g).

A number of cyclic P-atom-containing monomers were copolymerized zwitterionically with electrophilic comonomers. Examples are discussed in Section 15.2.2.2.

14.5 References

1. Penczek, S., Łapienis, G.: In "Ring Opening Polymerization" Ivin, K. J., Saegusa, T., Eds., Elsevier Appl. Sci. Publ., 1984
2. Allcock, H. R.: Angew. Chem., Eng. Ed. *16*, 147 (1977)
3. Allcock, H. R.: Makromol. Chem. Suppl. *4*, 3 (1981)
4. Allcock, H. R.: "Phosphorus Nitrogen Compounds", Academic Press, New York 1972
5. Libiszowski, J., Kałużynski, K., and Penczek, S.: J. Polym. Sci., Polym. Chem. Ed. *16*, 1275 (1978)
6. Shimidzu, T. et al.: Bull. Chem. Soc. Jap. *39*, 562 (1966)
7. Shimidzu, T. et al.: J. Polym. Sci. *B*, *3*, 871 (1965)
8. Petrov, K. A. et al.: Vysokomol. Soedin. *4*, 246 (1962)
9. Vogt, W., Ahmad, N. V.: Makromol. Chem. *178*, 1711 (1977)
10. McManimie, R. J., U.S. Pat. 2,893,961 (1959), CA *53*, 20908
11. Kawakami, Y., Miyata, K., Yamashita, Y.: Polym. J. *11*, 175 (1979)
12. Harwood, H. J., Patel, N. K.: Macromolecules *1*, 233 (1968)
13. Podovik, A. N., Evstaf'ev, T. V.: Vysokomol. Soedin. *6*, 2139
14. Makaiyama, T., et al.: J. Org. Chem. *29*, 2572 (1964)
15. Yamasjita, Y.: J. Polym. Sci. Symp. *56*, 447 (1976)
16. Singh, G.: J. Org. Chem. *44*, 1060 (1979)
17. Petrov, K. A., Nifant'ev, E. E., Sopikova, I. I.: Vysokomol. Soedin. *2*, 685 (1960)
18. White, D. W., Phosphorus *1*, 33 (1971)
19. Dutasta, J. P., Guimaraes, A. C., Robert, J. B.; Tetrahedron Lett. *1977*, 801
20. Kobayashi, S., Huang, M. Y., Saegusa, T.: Polym. Bull. *4*, 185 (1981)
21. Fujisawa, T., Yokota, Y., Mukaiyama, T.: Bull. Chem. Soc. Jap. *40*, 147 (1967)
22. Kobayashi, S., Suzuki, M., Saegusa, T.: Polym. Bull. *4*, 315 (1981)
23. Ogata, N., Saito, S.: Makromol. Chem. *164*, 95 (1973)
24. Łapienis, G., Penczek, S.: Macromolecules *7*, 166 (1974)
25. Łapienis, G., Penczek, S.: Macromolecules *10*, 1301 (1977)
26. Munoz, A., Vives, J.-P.: C.R. Acad. Sci. Ser. *C*, *253*, 1693 (1961)
27. Furukawa, J., Saegusa, T., Nagata, H.: Jap. Pat. 67 06,273 (1967) CA 67:44435
28. Kobayashi, S., Kaku, M., Saegusa, T., Polym. Bull. *5*, 325 (1981)
29. Kobayashi, S., Saegusa, T.: ACS Polymer Preprints *20*, 815 (1979)
30. Saegusa, T., Yokoyama, T., Kimura, Y., Kobayashi, S.: Macromolecules *10*, 791 (1977)
31. Horn, H. G., Kolkmann, F.: Makromol. Chem. *183*, 1833 (1982)
32. Horn, H. G., Kolkmann, F.: Makromol. Chem. *183*, 1843 (1982)
33. Arbuzov, A. E.: J. Russ. Phys. Chem. Soc. *38*, 687 (1906)
34. Vogt, W.: Makromol. Chem. *178*, 3179 (1977)
35. Vandenberg, E. J.: U.S. Pat. 3,520,849 (1970)

15 Cationic Copolymerization of Heterocyclic Monomers

15.1 Statistical Copolymers

15.1.1 Scope and Limitation of The General Scheme Applied to Cationic Ring-opening Polymerization

The classical scheme of copolymerization, discussed in the first paragraph of this section, cannot be used for the analysis of cationic ring-opening polymerization without important changes.

The simplest general scheme of free-radical copolymerization (neglecting reversibility and influence of the penultimate unit) reads:

$$\begin{aligned} \sim m_1^* + M_1 &\xrightarrow{k_{11}} \sim m_1^* \\ \sim m_1^* + M_2 &\xrightarrow{k_{12}} \sim m_2^* \\ \sim m_2^* + M_1 &\xrightarrow{k_{21}} \sim m_1^* \\ \sim m_2^* + M_2 &\xrightarrow{k_{22}} \sim m_2^* \end{aligned} \qquad (15\text{-}1)$$

and is based on the following assumptions:

- there are only two kinds of active species, namely $\sim m_1^*$ and $\sim m_2^*$ and their reactivities are different,
- the reactivity of active centers does not depend on chain length;
- the stationary concentration of both active species are constant,
- the average kinetic chains are sufficiently long to neglect other processes involving monomer.

Solution of the set of differential equations, describing the general scheme of Eq. (15-1), leads to the well-known relationship between monomer feed ($[M_1]/[M_2]$) and copolymer composition ($d[M_1]/d[M_2]$) [1)]:

$$\frac{d[M_1]}{d[M_2]} = \frac{[M_1]}{[M_2]} \cdot \frac{r_1[M_1] + [M_2]}{r_2[M_2] + [M_1]} . \qquad (15\text{-}2)$$

This relationship allows $k_{11}/k_{12} = r_1$ and $k_{22}/k_{21} = r_2$ (the reactivity ratios) to be determined. Ideally random copolymerization (no specific interactions) gives $r_1 \cdot r_2 = 1$; the extent to which $r_1 \cdot r_2 < 1$ indicates the tendency of alternation.

If $r_1 \cdot r_2 > 1$ homoblock formation may expected, because homopropagations are faster than crosspropagations.

There are a few methods for determining r_1 and r_2 by using a modified versions of Eq. (15-2). Recently, the method of Kelen and Tüdös [2,3,4] has received increased attention and is more frequently used than the earlier methods of Fineman and Ross [5], Mayo and Lewis [1], and Ezrielev, Brokchina and Roskin [6]. Most of these methods are applicable only for low conversion data (for exceptions see Ref. [3]).

Application of these methods to cationic ring-opening copolymerization sometimes gives erroneous results. This is because some of these copolymerizations involve more than merely four propagation reactions.

Below we summarize the major reasons why Scheme (15-1) does not accurately describe at least some cationic ring-opening copolymerizations.

a. Reversibility

The heat of polymerization of a number of cyclic monomers is lower than that of unsaturated monomers, causing extensive reversibility of the propagation steps. Thus, Scheme (15-1) has to be complemented with reversible steps and this leads to a completely different functional dependence of $d[M_1]/d[M_2]$ ($= [m_1]/[m_2]$, i.e. copolymer composition) on $[M_1]/[M_2]$ (monomer feed). This is certainly the case with cyclic monomers with heats of copolymerization equal to or lower than $40\ kJ \cdot mole^{-1}$.

The effect of reversibility in Scheme (15-1) is shown by the influence of $[THF]_0$ and temperature on reactivity ratio values for the copolymerization of THF with PO. A quantitative description of a system with one monomer copolymerizing reversibly is given below.

b. Multiplicity of active centers

In cationic ring-opening polymerization the growing species may exist in various forms, i.e. ions, ion pairs, covalent species, etc. The reactivities of these species may differ and their proportions may change with concentrations, monomer feed and other variables, rendering the scheme inapplicable. However, the kinetic data described for THF in Vol. I of this monograph (Adv. Polymer Sci. *37*) indicate that the reactivities of ions and ion-pairs are the same, and that this element of multiplicity can be neglected at least for this monomer. On the other hand, copolymerization of THF with DXL (Sect. 15.1.3.2) in the presence of counterions providing covalent active centers yields random copolymers unavailable with initiators excluding formation of covalent species. Thus, multiplicity of active species have to be taken into account in the kinetic scheme. In the copolymerization of βPL with cyclic ethers (Sect. 15.1.3.3) the reactivity ratios, determined from the dependence of the copolymer composition on the monomer feed, differ from those calculated from microstructure. The simultaneous presence of "normal" and coordinate cationic species was proposed as an explanation. We wish to stress that the assumption of the multiplicity of active species rules out the application of a simple copolymerization scheme; thus, r_1 and r_2 become meaningless.

Multiplicity of active centers may also be due to the influence of penultimate (antepenultimate, etc.) units. This influence has been studied for free radical polymerization and the required equations are available [7]. In ring-opening copolymerizations, due to the onium structure of the growing species, this influence should be more pronounced than in free radical copolymerizations. Comparison of the idealized structures of the growing species:

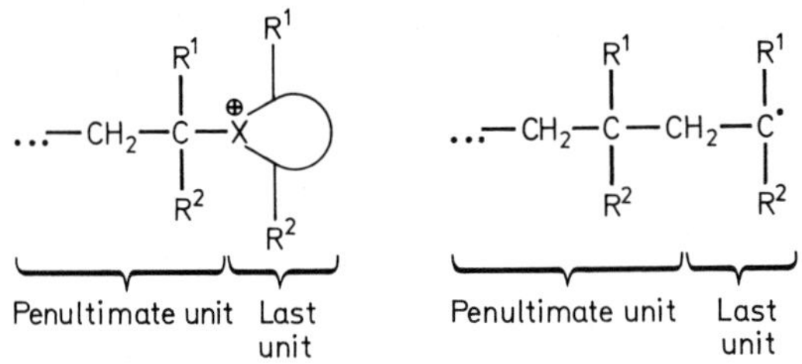

(15-3)

shows that the penultimate unit in onium-type ring-opening polymerization is to a large extent a part of the last unit, whereas in free radical polymerization it is much further apart from the growing center (cf. also Sect. 15.2.2.1).

c. Interconversions of active species

In free radical copolymerization interconversion of the growing species proceeds exclusively by crosspropagation. In cationic ring-opening polymerization interconversion of the growing species may occur by two routes i.e., the incoming monomer may attack either the endo- or the exocyclic carbon atom:

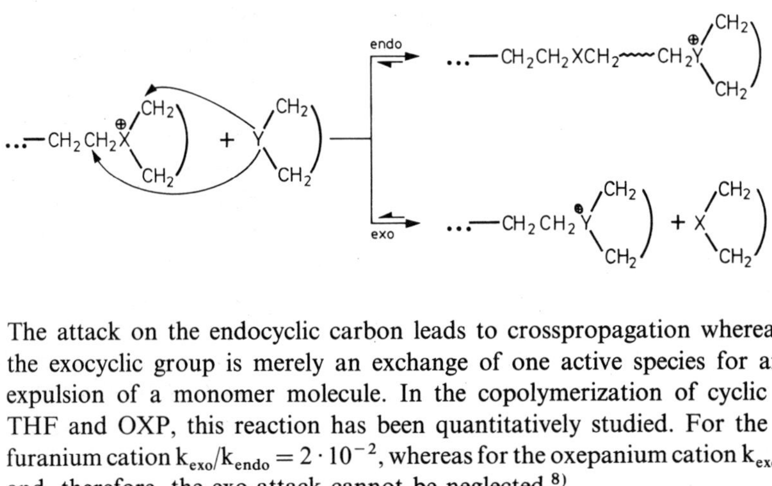

(15-4)

The attack on the endocyclic carbon leads to crosspropagation whereas attack on the exocyclic group is merely an exchange of one active species for another with expulsion of a monomer molecule. In the copolymerization of cyclic ethers, like THF and OXP, this reaction has been quantitatively studied. For the tetrahydrofuranium cation $k_{exo}/k_{endo} = 2 \cdot 10^{-2}$, whereas for the oxepanium cation $k_{exo}/k_{endo} = 0.5$ and, therefore, the exo-attack cannot be neglected [8].

d. Side reactions

In the polymerization of heterocyclic monomers, the polymer backbone possesses heteroatoms able to participate in chain transfer. The result of this reaction is the formation of inactive branched onium ions (termination). In other instances scrambling is observed leading to changes in the microstructure of the chain and interconversion of active centers. Intramolecular chain transfer to polymer is accompanied by the formation of macrocycles. Low-molecular-weight cyclic oligomers may escape detection, particularly when the polymer is separated by precipitation. Cyclic oligomers are better soluble and may remain in solution which leads to an erroneous material balance. This side reaction is important when the copolymerization is carried out close to the equilibrium concentration, and with easily cyclizing monomers.

Several of these features will be covered in more detail in connection with the kinetic solutions of schemes, including deviations from the classical scheme.

Recently, Tüdős et al. have reexamined the reactivity ratios reported for cationic ring-opening copolymerization proceeding by oxonium ions [9]. These authors concluded that except for five systems all others have been incorrectly interpreted. However, even these five systems include four that we would consider to be incompatible with the general scheme of copolymerization. Two concern the copolymerization of BCMO with THF and 2-methyl THF, which requires the incorporation of reversible steps into the scheme. The other two cases are the copolymerization of cyclic acetals with vinyl compounds (see Sect. 15.1.3.4), known to give mixtures of homopolymers and/or copolymers with long homoblocks and macrocyclics. This illustrates how little reliable information is available on onium ion ring-opening copolymerization.

Thus, in this chapter we will present a critical review of cationic ring-opening copolymerization on the basis of a restricted number of papers.

The reliable synthesis of a copolymer with a desired composition requires the knowledge of relative reactivities of both comonomers, i.e. the values of reactivity ratios r_1, and r_2, and the system has to conform to the simple Scheme (15-1). The reactivity ratios can then be obtained from the copolymer composition (Equation (15-2)) or from the microstructure of the chains. Agreement between the values calculated in these two ways indicates that the simple four parameter Scheme (15-1) can be applied.

Thus, we first briefly describe the theory relative to copolymer microstructure and then discuss systems obeying Scheme (15-1), to show the influence of monomer structure on reactivity. Subsequently we describe other systems that do not conform to Scheme (15-1) because of reversibility of propagation or multiplicity of active centers.

15.1.2 Microstructure and Reactivity Ratios

In several sections of this volume the microstructure of homo- and copolymers is discussed. The microstructure of the products can be used as a diagnostic tool to get information as to the uniformity of polymers formed, reactivities of comonomers, etc. In the absence of penultimate unit effect and subsequent segmental redistribution the microstructure of the copolymer chain is defined by the reactivity ratios and comonomer feed. Thus, the dyad and triad content may be a test of the applicability of the simple four-parameter scheme.

The probability of "11" addition (P_{11}) is proportional to the concentrations of the active center [$\sim\!\sim\!\sim m_1^*$], monomer [M_1], and the rate constant k_{11}. Similar relationship is valid for P_{12}. Therefore, their ratio is given by Eq. (15-5):

$$\frac{P_{11}}{P_{12}} = \frac{k_{11} \cdot [\sim\!\sim\!\sim m_1^*] \cdot [M_1]}{k_{12} \cdot [\sim\!\sim\!\sim m_1^*] \cdot [M_2]} = r_1 \cdot \frac{[M_1]}{[M_2]} = r_1 \cdot x \qquad (15\text{-}5)$$

(where $[M_1]/[M]_2 = x$

Since $P_{11} + P_{12} = 1$:

$$P_{11} = r_1x/(r_1x + 1)\,, \qquad P_{12} = 1/(r_1x + 1) \tag{15-6}$$

and by analogy

$$P_{22} = r_2/(x + r_2)\,, \qquad P_{21} = x/(x + r_2)\,. \tag{15-7}$$

Among these sequences there are dyads or triads. Let us analyse the proportion of dyads P(11), P(12) = P(21) and P(22) in the system. If P(12) is the probability of finding the sequence "12" among any sequences consisting of two units, we have:

$$P(11) + P(12) + P(21) + P(22) = 1 \tag{15-8}$$

$$P(1) + P(2) = 1 \tag{15-9}$$

obviously P(12) = P(21).

The probability of the formation of the dyad P(11) equals the product of probability of the P(1) unit and the conditional probability of addition P_{11}. Thus

$$P(11) = P(1) \cdot P_{11}\,. \tag{15-10}$$

Similarly for the triad P(112):

$$P(112) = P(11) \cdot P_{12} = P(1) \cdot P_{11} \cdot P_{12}\,. \tag{15-11}$$

Since P(12) = P(21), the equality similar to that for the steady-state conditions holds

$$P(1) \cdot P_{12} = P(2) \cdot P_{21}\,. \tag{15-12}$$

Remembering that P(1) = 1 — P(2) and knowing the probabilities P_{12} and P_{21}, proportions of the segments "1" and "2" in the copolymer can be calculated:

$$\begin{aligned} P(1) &= (1 + r_1x)/(r_1x + 2 + r_2/x) \\ P(2) &= (1 + r_2/x)/(r_1x + 2 + r_2/x) \end{aligned} \tag{15-13}$$

In a similar way the proportion of all dyads can be calculated:

$$\begin{aligned} P(11) &= P(1)\,P_{11} = r_1x/(r_1x + 2 + r_2/x) \\ P(12) &= 1/(r_1x + 2 + r_2/x) \\ P(22) &= (r_2/x)/(r_1x + 2 + r_2/x) \end{aligned} \tag{15-14}$$

If the penultimate unit affects the rate constants of propagation ($k_{111} \neq k_{211}$), the proportion of e.g. dyad (11) could be found from the expression:

$$P(11) = \frac{r_1'x \cdot \dfrac{1 + r_1x}{1 + r_1'x}}{r_1'x \cdot \dfrac{1 + r_1x}{1 + r_1'x} + 2 + r_2'/x \cdot \dfrac{1 + r_2/x}{1 + r_2'/x}} \qquad (15\text{-}15)$$

where $r_1 = k_{111}/k_{112}$, $r_1' = k_{211}/k_{212}$, $r_2 = k_{222}/k_{221}$, $k_2' = k_{122}/k_{121}$. The proportion of the triad e.g. P(112) was described by Eq. (15-11). Very often, however, this value is related only to the segments m_1. Therefore:

$$T(112) = P(112)/P(1) = P_{11} \cdot P_{12} = r_1x/(r_1x + 1)^2 \qquad (15\text{-}16)$$

Similarly:

$$\begin{aligned} T(111) &= (r_1x)^2/(1 + r_1x)^2 \\ T(212) &= 1/(1 + r_1x)^2 \\ T(211) &= r_1x/(1 + r_1x)^2 \end{aligned} \qquad (15\text{-}16a)$$

Thus, the total proportion of all triads having unit "1" in the center is $\Sigma T = 1$.

The proportions of triads when a penultimate effect operates are expressed by much more complicated formulas. The interested reader can find the solution of such schemes in the paper of Ito and Yamashita [10].

15.1.3 Simple Systems [Conforming to the Four Reactions Scheme]

By simple systems, we understand those that conform to the four-reaction Scheme (15-1). The contribution of the deviations described in Section 15.1.1 is small and may be neglected.

First we describe the copolymerization of oxetanes, as an example for microstructure and structure/reactivity studies. Then we discuss the copolymerization of thietanes, azetidines and oxazolines. Finally, we show the relationship between structure of heterocyclics (basicity, ring strain) and their nucleophilic reactivity.

15.1.3.1 Oxetanes

Microstructure

The cationic copolymerization of oxetanes provides a good example that shows agreement between two treatments, namely determination of the reactivity ratios from kinetic treatment and microstructure studies [11,12].

In the copolymerization of 3,3-dimethyloxetane (A) with oxetane (B) four possible triads may arise. Thus, taking the A unit in the central position we have:

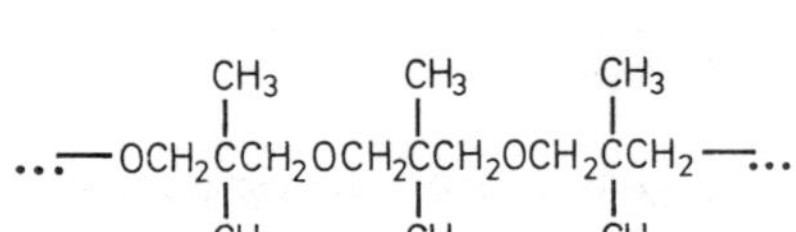 AAA (15-17a)

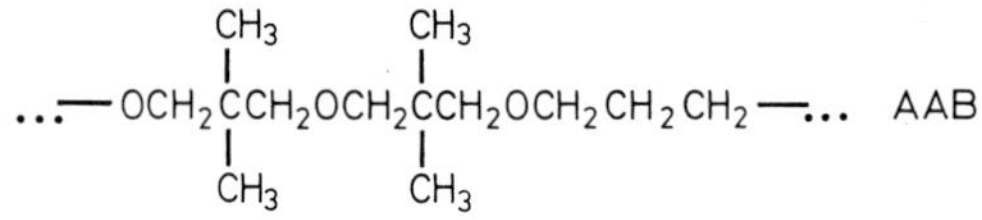

(15-17b)

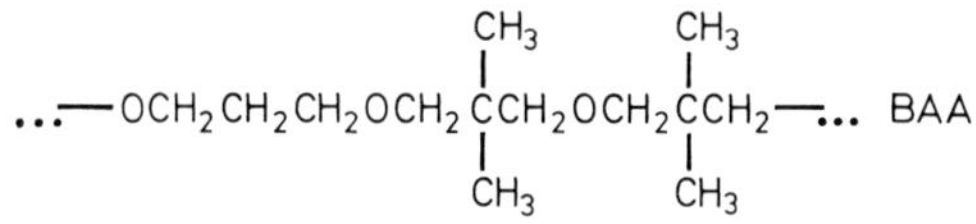

(15-17c)

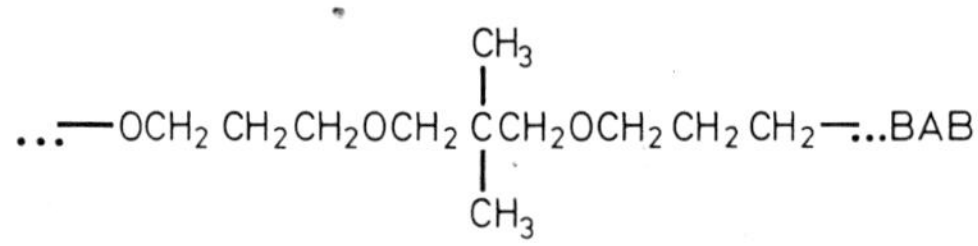

(15-17d)

The proportion of the triads (b) and (c) are equal by definition. ^{1}H-NMR spectroscopy was used to determine the triads content. Protons in the methyl groups give three distinctive resonances differing by 0.005 ppm [13]. The difference is larger for the methylene protons of A units reaching 0.03 ppm and four well resolved signals appear as shown in Fig. 15.1. Application of a curve resolver gave the areas under the corresponding peaks. These are proportional to the abundances of the different triads. The α-methylene groups in oxetanes absorb as triplets at a slightly lower field. A

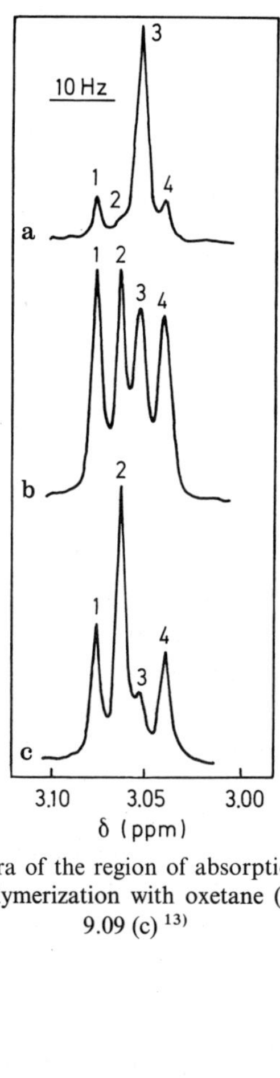

Fig. 15.1. 300 MHz ^{1}H-NMR spectra of the region of absorption of methylene CH_2 atoms in 3,3-dimethyloxetane (M_2) unit in copolymerization with oxetane (M_1) $[M_1]/[M_2] = 0.12$ (a), 0.96 (b), 9.09 (c) [13]

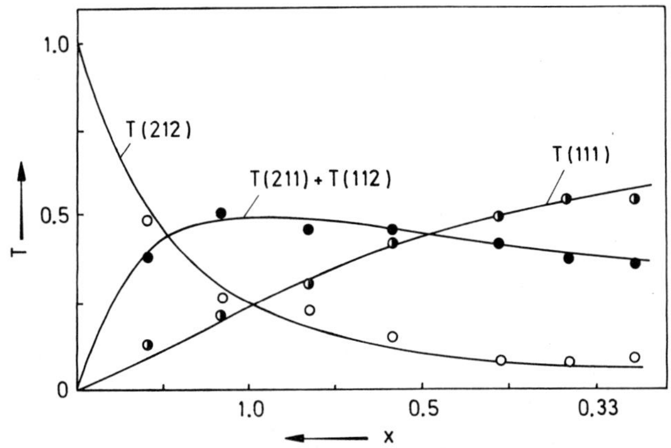

Fig. 15.2. Dependence of the relative proportions of different triads in copolymerisation of oxetane (M_1) with 3,3-dimethyloxetane (M_1) on the monomer composition $x = [M_1]/[M_2]$ [13]

multiplet in the region δ 3.43 was resolved in four triplets of the corresponding triads: BBB, BBA, ABB, ABA. Signals of the methylene protons give four triads, because sequences BBA and ABB or BAA and AAB have the same chemical environments but are not symmetrical.

The proportions of the triads vary with the monomer feed. As shown in Fig. 15.2, the proportions of the experimentally found triads agree well with the values calculated from Eq. (15-16a) using r obtained from the content of comonomers in the copolymer (method from Ref. [6]). This agreement means that in this system the penultimate effect, reversibility of propagation, interconversion of active species without propagation [Eq. (15-4)] and redistribution of the initially formed copolymer can be neglected and the process can be described by two reactivity ratios.

Structure-reactivity relationship

3-Methyl-2-chloromethyloxetane was copolymerized with a number of cyclic ethers, in order to establish the structure-reactivity correlation. Results are summarized in Table 15.1.

Copolymerization was carried out in C_2H_5Cl solvent at 0 °C and at 40 mol% total comonomer concentration with a $Al(C_2H_5)_3$-H_2O catalyst system [13]. Under these conditions, depropagation of THF and tetrahydropyran (THP) should be taken into account, and thus r_1 and r_2 derived from Scheme (15-1) cannot be equal for these monomers to the corresponding rate constant ratios (reactivity ratios). Therefore, only results for oxiranes and oxetanes can be discussed. It should be noted that in spite of the higher strain of the three-membered rings, their incorporation into the copolymer is slower than that of the more basic oxetanes.

It follows from comparison of r_2 and $1/r_1$ values with pK_b, given in Table 15.2, that the oxetanes react with the growing cation derived from BCMO, according to their basicities [15]. Some deviation is observed for monomers with small substituents, e.g. methyloxetanes. The lesser steric hindrance permits the approach of these molecules, increasing their nucleophilic reactivities. The product $r_1 \cdot r_2 \simeq 1$ indicates

Table 15.1. Copolymerization of 3-methyl-3-chloromethyloxetane (M_1) with cyclic ethers[a] [13]

M_2	r_1	$1/r_1$	r_2	r_1r_2
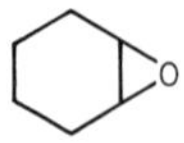	0.35	2.86	0.95	0.33
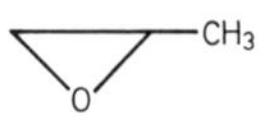	4.5[b]	0.2[b]	0.33[b]	1.49[b]
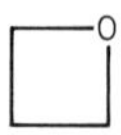	0.64	1.56	3.05	1.95
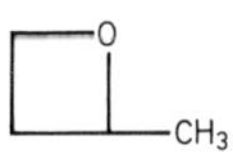	0.25	4.0	3.75	0.94
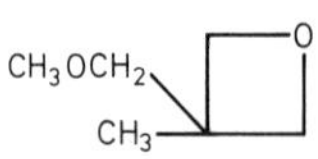	0.075	13.3	5.4	0.40
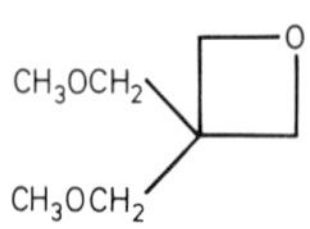	0.14	7.14	8.5	1.19
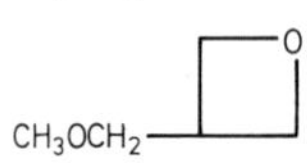	0.21	4.76	19.0	3.99
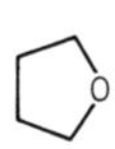	1.7	0.588 (0.23)[b]	0.17	0.28
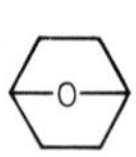	1.22	0.82	0.12	0.15
	8.5	0.12	0.02	0.17

[a] C_2H_5Cl solvent, 60 mol %, 0 °C, $AlEt_3—H_2O \approx 0.5$ mol %;
[b] Ref. [14]

that the copolymers formed are random and that the ratio of the rate constants is contant, i.e. $k_{11}/k_{12} = k_{21}/k_{22}$. Thus, specific interactions are absent. Similar results were observed in the copolymerization of oxetane (M_1), 3,3-dimethyloxetane (M_2) and 3,3-diethyloxetane (M_3) [12]. For the first two monomers $r_{12} = 1.19$ and $r_{21} = 0.95$ at 20 °C in CH_2Cl_2 solvent. For the latter two monomers $r_{23} = 1.1$ and $r_{32} = 0.9$ under the same conditions. The values of r_{13} and r_{31} can be predicted according to O'Driscoll [16] by the formula $r_{13} = r_{12} \cdot r_{23}$ provided that specific interactions are absent. The calculated values $r_{13} = 1.32$ and $r_{31} = 0.85$ and those determined experimentally $r_{13} = 1.76$ and $r_{31} = 0.56$ are in reasonable agreement.

Table 15.2. Copolymerization parameters of BCMO (M_1) ($pK_b = 5.65$) with oxetanes in CH_2Cl_2 at 20 °C [15]

M_2	r_1	$1/r_1$	r_2	r_1r_2	pK_b
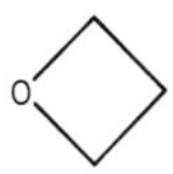	0.01	100	51	0.51	3.13
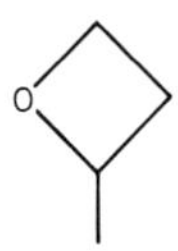	0.04	25	25	1.0	3.88
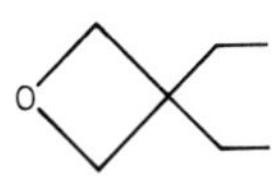	0.05	20	17	0.85	3.27
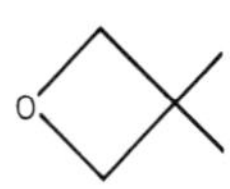	0.06	17	16	0.96	4.29
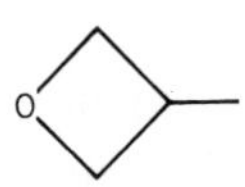	0.08	12.5	10	0.80	5.10
	0.20	5	47	0.94	4.59
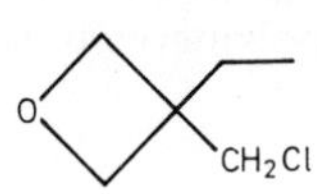	0.25	4	40	1.0	4.52
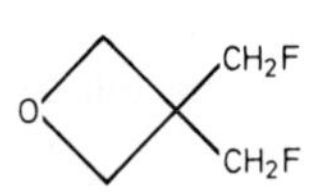	0.45	2.2	20	0.90	5.85

15.1.3.2 Thietanes and Azetidines

For this group of monomers both homo- and crosspropagations data are available.

Table 15.3 shows k_{homo} (k_{11} and k_{22}) and k_{cross} (k_{12} and k_{21}) values for three thietanes [17]. The monomers are of nearly the same basicity and differ only in substitution. Inspection of the data in Table 15.3 reveal that substitution does not much affect reactivity. The largest ratio of the rate constants of addition of various monomers to a given active species (inspect horizontal lines) is 4.5. On the contrary, substitution in the active species affects (up to $4 \cdot 10^3$ times) the rate constant of addition of the same monomer (see vertical lines). The activity ratios (r_1 and r_2) could thus be calculated from the corresponding ratios of the individual rate constants.

Table 15.3. Rate constants of homo- and crosspropagation of thietanes[a] [17)]

Monomer	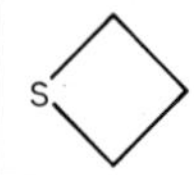	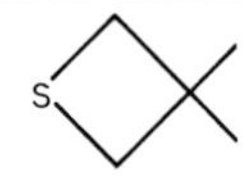	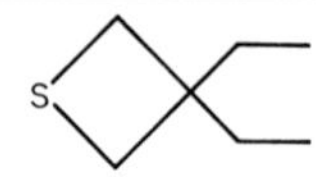
Basicity, $\Delta\nu$*) / Active centers	241	248	250
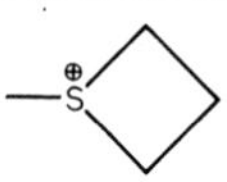	9000	5000	8000
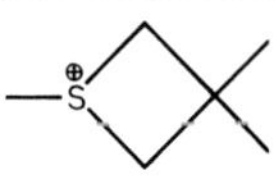	180	65	60
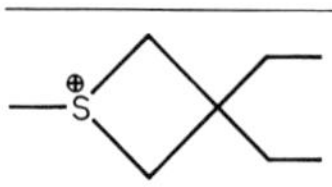	9	2	2

[a] Rate constants in $mole^{-1} \cdot l \cdot s^{-1} \cdot 10^4$ at 20 °C in CH_2Cl_2 solvent with $Et_3O^{\oplus}BF_4^{\ominus}$ as initiator; [b] Shift of the OH band in phenol, in cm^{-1}

Similar effects were observed in the copolymerization of azetidines [18)]. 1-Methylazetidine is three times more reactive than 1,3,3-trimethylazetidine towards azetidinium cations, and the 1-methyl-1-alkylazetidinium cation is 500 times more reactive than the 1-alkyl-1,3,3-trimethylazetidinium cation towards different azetidines at 80 °C in $C_6H_5NO_2$ solvent.

Thus, it appears that steric hindrance in the active species governing the rates of monomer additions is of decisive influence in cationic ring-opening copolymerization within a given group of monomers whose basicities are similar.

15.1.3.3 Oxazolines

Studies performed with these five-membered rings provide valuable information on the influence of basicity on reactivity ratios.

2-Oxazolines are known to polymerize with cationic initiators in polar solvents to nearly complete conversion. Although the formation of macrocycles cannot be avoided, this process is unimportant. Polymerization of these monomers proceeds with isomerization:

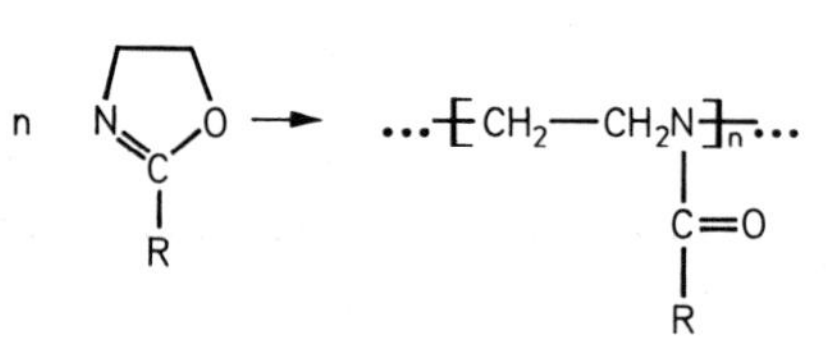

(15-18)

Kagiya et al. determined the reactivity ratios of 2-phenyl-2-oxazoline (M_1) in copolymerization with other substituted oxazolines in N,N-dimethylacetamide at 135 °C [19)]. The reported values of r_1 and r_2 are collected in Table 15.4: $1/r_1$ $(= k_{12}/k_{11})$ and

Table 15.4. Reactivity ratios in the copolymerization of 2-phenyl-2-oxazoline (M_1) [19]

Comonomer (M_2)	$\Delta\nu^a$ (cm^{-1})	r_1	$1/r_1$	r_2
2-isopropenyl-2-oxazoline	168	0.07 ± 0.01	14	7.9 ± 1.0
2-methyl-2-oxazoline	176	0.02 ± 0.02	50	10.2 ± 1.0
2-benzyl-2-oxazoline	191	0.01 ± 0.005	100	12.2 ± 1.0

[a] Shift of the O—D band of methanol-d; $\Delta\nu = 148\ cm^{-1}$ for M_1

r_2 ($= k_{22}/k_{21}$) are the measures of the reactivity of monomer M_2 toward active species m_1^*, and m_2^* relative to the reactivity of monomer M_1. According to these values monomer M_2 reacts many times faster with a given active species than monomer M_1. Thus, the reactivity of oxazolines increase with their basicity as shown by the data in Table 15.4.

15.1.3.4 Structure-reactivity Correlation in Cationic Ring-opening Copolymerization

A few papers are devoted to the correlation between monomer reactivity and structure. Monomers were characterized by basicity and ring strain. The reactivity was related to the $1/r_1$ parameter in copolymerization with a standard comonomer [most often to BCMO (M_1)].

Fukui et al. [20], using the simple molecular orbital theory, calculated the stabilization energy arising from π-conjugation in the transition state of radical and ionic polymerization of vinyl monomers. Further correlation of monomer structures with reactivities revealed that the rate constant of crosspropagation with one chosen active center increases with the basicity of the comonomer.

Aoki et al. [21] proposed the following formula relating $1/r_1$ to basicity and ring strain in the copolymerization of cyclic ethers with BCMO:

$$\log(1/r_1) = -0.086\Delta RS - 0.31(\Delta B) + 0.57 \tag{15-19}$$

where ΔRS is the free energy of polymerization of the corresponding models (cycloalkanes) and ΔB is the difference in pK_b of the comonomer and BCMO ($pK_b = 5.65$). According to Eq. (15-19) the relative reactivity, expressed by $1/r_1$, is much more sensitive to basicity than to ring strain (the corresponding parameters in Eq. (15-19) are 0.086 and 0.31).

Kagiya et al. [22] have correlated the reactivity ratio r_1 for BCMO, THF, 1,3-DXL, ε-CL with basicities, which were determined from the free energy of interaction with CH_3OD:

$$CH_3O{-}D + X\bigcirc \overset{K}{\rightleftharpoons} CH_3O{-}D\cdots X\bigcirc \tag{15-20}$$

and the homolytic bond dissociation energy ΔD in the involved monomer. The latter was estimated from activation energies, employing the Morse potential energy function. The empirical formula for cyclic ethers is:

$$\ln(1/r_1) = -\frac{1}{RT}[2.5\Delta\Delta G + 0.1\Delta\Delta D(C{-}O)] \tag{15-21}$$

A similar attempt was earlier made by Yamashita et al. [23].

These equations are the empirical relationships of copolymerization data with two independent parameters, however they neglect steric hindrance. Basicity is measured by protonation and it underestimates the influence of the steric hindrance, because of the small volume of the proton as compared with onium ions.

There are also other deficiences of these empirical approaches, and the linearity of the observed dependences should not be taken too seriously. The most important deficiency arises because of the unreliability of r_1 and r_2. As we stressed already, the most frequent error is due to the application of the four-reactions copolymerization scheme for systems requiring the introduction of reversibility. Thus, the reactivities of monomers having low ring strains, appear to be lower than they in fact are. This is particularly so for the copolymerization of THF, where the closer the system to the equilibrium homopolymerization conditions, the lower the determined reactivities of THF (cf. Sect. 15.1.3.1).

15.1.4 Copolymerization in Systems of Special Interest

In the preceding section we have disscussed the influence of monomer nucleophilicity on copolymerization behaviour. Only monomers of similar nucleophilicities can be randomly copolymerized. Attempted simultaneous copolymerization of e.g. amines with ethers, and sulfides or oxazolines with ethers gives homopolymers of the more nucleophilic monomer. However, stepwise addition of a more basic comonomer to a living polymer derived from the less nucleophilic heterocycle yields block copolymers (cf. Sect. 15.3.1.2).

15.1.4.1 Copolymerization of THF with Propylene Oxide (PO)

The copolymerization of THF with PO has partially been described in the section devoted to the reactive oligomers based on polyTHF (Chapter 6.5).

Reversibility of THF homopropagation leads to an additional difficulty in its copolymerization. This reversibility was neglected, however, by several groups and, therefore, erroneous values of r_1 and r_2 resulted (cf. Chapter 6.5).

The scheme including reversibility is shown by the following set of equations:

$$
\begin{aligned}
\ldots -m_1^* + M_1 &\underset{k_{-11}}{\overset{k_{11}}{\rightleftharpoons}} \ldots -m_1m_1^* \\
\ldots -m_1^* + M_2 &\underset{k_{-12}}{\overset{k_{12}}{\rightleftharpoons}} \ldots -m_1m_2^* \\
\ldots -m_2^* + M_1 &\underset{k_{-21}}{\overset{k_{21}}{\rightleftharpoons}} \ldots -m_2m_1^* \\
\ldots -m_2^* + M_2 &\underset{k_{-22}}{\overset{k_{22}}{\rightleftharpoons}} \ldots -m_2m_2^*
\end{aligned}
\qquad (15\text{-}22)
$$

where M_2 = THF. r_1 and r_2 values given in the literature have been calculated by the use of a much simpler four-reaction scheme (Table 15.5). Since r_1 and r_2 depend

Table 15.5. Reactivity ratios in copolymerization of PO with THF in $C_2H_4Cl_2$ solution ($[M_1]_0 + [M_2]_0 \simeq 7\ mol \cdot l^{-1}$)

No.	r_1 (PrOx)	r_2 (THF)	T, °C	Initiator	Ref.
1	0.05	0.5	+10	$BF_3 \cdot THF$	[28]
2	0.26	0.8	0	$BF_3 \cdot O(C_2H_5)_2H_2O$, $[PO]_0 > [THF]_0$	[27]
3	0.26	0.8	−10	$Ph_3C^{\oplus}SbF_6^{\ominus}$	[26]
4	0.34	0.63	0	$BF_3 \cdot O(C_2H_5)_2$ + 1,4-Butanediol	[24]
5	1.0	0.21	+20	$Ph_3C^{\oplus}SbF_6^{\ominus}$	[26]
6	1.0	0.09	0	$SbCl_5$ + 1,2-Propanediol	[26]
7	1.3	0.5	0	$BF_3 \cdot O(C_2H_5)_2$ + 1,4-Butanediol	[25]
8	1.8	0.6	0	$BF_3 \cdot O(C_2H_5)_2$ + 1,2-Propanediol	[26]
9	1.96	2.40	0	$BF_3 \cdot O(C_2H_5)_2H_2O$, $[PO]_0 < [THF]_0$	[27]

on the ratio of initial monomer concentrations, the simple scheme cannot correctly describe the copolymerization of THF/PO.

With all the reservation as to the meaning of r_1 and r_2, the following general comments can be made. The r_2 values in nearly all papers were reported to be less than unity, indicating that the reactivity of PO is higher than that of THF towards the tetrahydrofuranium cation. This result is surprising, because THF is much more nucleophilic than PO. In fact, THF rapidly enters the copolymer but it also leaves the chains, in contrast to the much more strained PO. Thus it appears as if PO were more reactive. The influence of temperature (No. 5 and 6 in Table 15.5) supports this explanation. At +20 °C, $r_1 > r_2$, but at −10 °C the reverse is observed under otherwise identical conditions. This is due to the lower $[THF]_e$ at −10 °C and the better fit to the four-equations scheme. At higher $[THF]_0$, r_2 values increase in line with the above given explanation (No. 7 and 8).

Entelis and Korovina introduced reversibility of THF homopropagation (k_{-22} but not k_{-21}) in their scheme [cf. Scheme (15-22) and (15-25)] and solved it numerically [28]. In this case $r_1 \ll r_2$, in agreement with the known order of nucleophilicities and $r_1 \cdot r_2$ is less than unity showing alternating tendency. This is due to the higher reactivity of the more nucleophilic THF and the low probability of formation of THF dyads because of reversible propagation.

In studies of THF/PO copolymerization, except in the cited-above paper [28], no attempts were made to consider reversibility of THF homopropagation.

Below we describe in more detail a reversible copolymerization with the THF/oxetane pair.

Copolymerization with one reversibly polymerizing monomer

The complete solution of Scheme (15-22), as shown by Wittmer [29] and Durgaryan [30], data becomes impossible without simplifying assumptions. Further derivations of various schemes applicable to particular systems are based on the Lowry's approach [31].
various schemes applicable to particular systems are based on Lowry's approach [31]. This involves elimination from the differential equations of growing species concentrations and the relation between rate constant ratios and copolymer composition on monomer feed, as in simple copolymerizations.

Cationic ring-opening polymerizations cannot be described by Lowry's treatment without major conceptual changes. Let us compare reversibility of chain growth in the homopolymerization of vinyl and heterocyclic monomers. In vinyl polymerization (independent of the mechanism of propagation) the last charge bearing unit is split off of the chain and the penultimate unit is becoming the new active species:

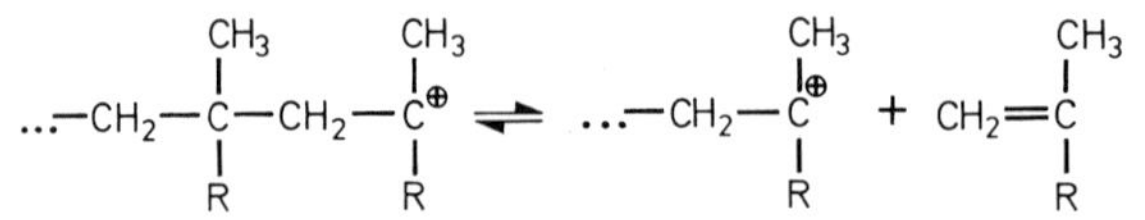

(15-23)

On the contrary, in cationic ring-opening polymerization we have:

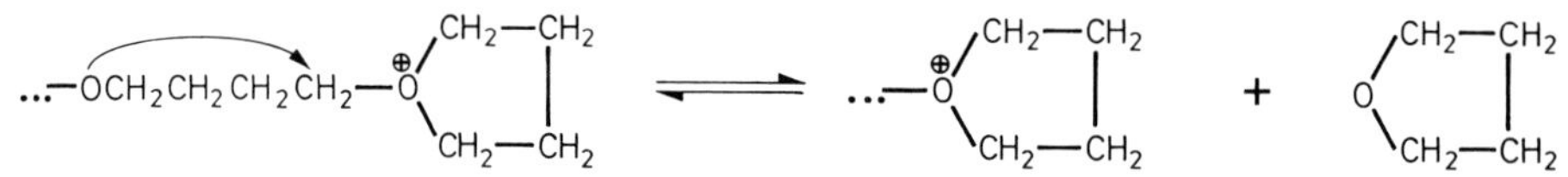

(15-24)

Thus, the splitting off of the last unit is accompanied by the formation of the active species by ring closure involving the penultimate unit. This closure is very unlikely for instance for 4- or 3-membered highly strained rings. Thus, depropagation is determined by the penultimate unit ring closure.

Yamashita et al. [32] have been first to recognize this difference between vinyl and ring-opening polymerizations and adapted Lowry's treatment to the cationic ring-opening polymerization of oxetanes with THF. Yamashita at al. described the copolymerization of BCMO (M_1) with THF (M_2) by the following scheme:

$$\begin{aligned}
\sim m_1^* + M_1 &\xrightarrow{k_{11}} \sim m_1 m_1^* \\
\sim m_1^* + M_2 &\xrightarrow{k_{12}} \sim m_1 m_2^* \\
\sim m_2^* + M_1 &\underset{k_{-21}}{\overset{k_{21}}{\rightleftharpoons}} \sim m_2 m_1^* \\
\sim m_2^* + M_2 &\underset{k_{-22}}{\overset{k_{22}}{\rightleftharpoons}} \sim m_2 m_2^*
\end{aligned} \qquad (15\text{-}25)$$

Two reversible reactions have been considered. It was rightly assumed that when m_1 (oxetane) is the penultimate unit there is no depropagation, because closing the four-membered ring is energetically unfavorable:

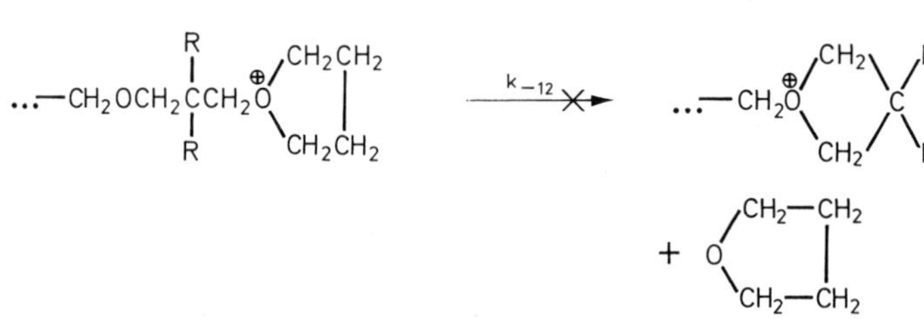

(15-26)

The solution of this scheme, assuming steady-state conditions of active centers leads to the following expressions:

$$\frac{[(m_2)^*_{n+1}]}{[(m_2)^*_n]} = \alpha = \frac{1}{2}\left\{\left(1 - \frac{K}{r_2}\cdot\sigma\cdot\beta + \frac{K}{r_2}[M_1] + K[M_2]\right) - \sqrt{\left(1 - \frac{K}{r_2}\cdot\sigma\cdot\beta + \frac{K}{r_2}[M_1] + K[M_2]\right)^2 - 4K[M_2]}\right\} \quad (15\text{-}27)$$

$$\frac{[(m_2)_n m_1^*]}{[(m_2)^*_n]} = \beta = \frac{r_1[M_1]}{r_1\sigma + r_1 r_2 \delta[M_1] + r_2\delta[M_2]} \quad (15\text{-}28)$$

and to the following copolymerization equation:

$$\frac{[m_1]}{[m_2]} = (1 - \alpha)\left(1 + r_1\frac{[M_1]}{[M_2]}\right) \quad (15\text{-}29)$$

where $K = k_{22}/k_{-22}$, $r_1 = k_{11}/k_{12}$, $r_2 = k_{22}/k_{21}$, $\delta = k_{11}/k_{22}$, $\sigma = k_{-21}/k_{21}$ and α is defined in Eq. (15-27).

The evaluation of the reactivity ratios is cumbersome and requires an arbitrary choice of the δ and σ parameters.

Recently, Scheme (15-25) has been modified by two of us assuming that no depropagation occurs in the THF-OX* unit ($m_2m_1^*$) because OX is highly strained and reacts before the THF penultimate unit has a chance to back-bite and depropagate [33].

Such a modified scheme (neglecting the reaction proceeding with the rate constant k_{-21}) has also been solved by Garrido et al. [34] leading to the same copolymerization equation as above but to a simpler expression for α:

$$\alpha = \frac{1}{2}\left\{\left(1 + \frac{K}{r_2}[M_1] + K[M_2]\right) - \sqrt{\left(1 + \frac{K}{r_2}[M_1] + K[M_2]\right)^2 - 4K[M_2]}\right\} \quad (15\text{-}30)$$

These authors have also studied by ^{1}H-NMR spectroscopy the microstructure of THF-3,3-dimethyloxetane (DMO) copolymers at 0 °C and calculated the dyad distribution (cf. Sect. 15.1.2).

$$\text{taking: } P(1) = P(12)\left(1 + \frac{r_1[M_1]}{[M_2]}\right) \quad (15\text{-}31)$$

where P(1) is the probability of finding unit m_1 in the copolymer, and P(12) is the probability for the m_1m_2 unit, r_1 was determined as 7.6 ± 0.3. Further, it was assumed that $K^{-1} = 2.0$ mole · l^{-1} and first from Eq. (15-29) α was determined from the slope of the plot of $[m_1]/[m_2]$ as a function of $(1 + r_1[M_1]/[M_2])$, and then r_2 from the rearranged Eq. (15-30).

$$\frac{[M_1]}{r_2} = \frac{\alpha + (1/\alpha - 1)\, K[M_2] - 1}{K} \tag{15-32}$$

This treatment gave $r_1 = 7.6$ and $r_2 = 0.13$. Neglecting depropagation (k_{-22}) and using the Fineman-Ross [5] or Kelen-Tüdős [2] treatment led to virtually the same results for r_1 and r_2. This is not surprising because the copolymerization conditions (high concentration of THF, relatively low temperature) did not favour depropagation. We think that the proper way to determine true r_1 and r_2 values for copolymerization with one reversibly polymerizing comonomer is by a combination of the usual approach of r_1 and r_2 calculation with microstructure analysis. This is particularly important when high temperatures and low concentrations of this comonomer are used, as shown by the data in Table 15.5.

We have given here an extensive treatment of the copolymerization with reversibility of propagation, because this phenomenon is still insufficiently recognized and there are still papers in which reversibility should have been considered; schemes that ignore reversibility lead to wrong conclusions. Thus the r values given in these papers do not have the meaning they pretend to have, and various authors have obtained different values for the same pair of monomers for the "supposed-to-be" reactivity ratios [98].

15.1.4.2 Copolymerization of THF with DXL

Copolymerization of THF with DXL proceeds in the unusual way. When these two monomers are mixed in solution and the initiator is added, some polymerization of THF is observed at the very beginning [35]. Then DXL polymerizes and throughout its polymerization the rate of THF polymerization is low. When DXL reaches its equilibrium concentration, THF polymerizes until its concentration becomes slightly lower than its homopolymerization $[THF]_e$ under similar conditions (cf. Fig. 15.3) [35, 36].

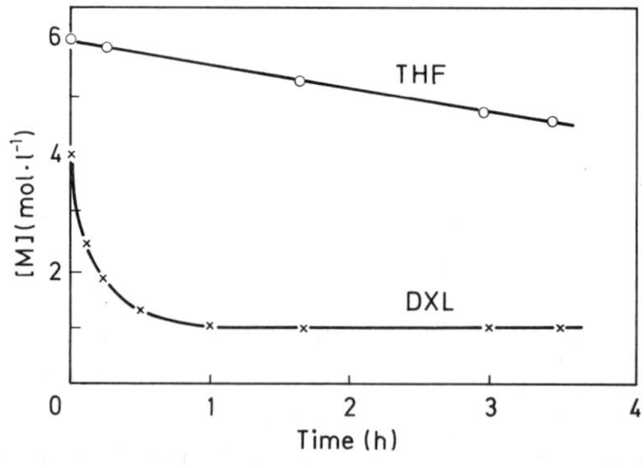

Fig. 15.3. Dependence of [THF] and [DXL] in the copolymerization in CH_2Cl_2 solution at 20° on time, $[(CH_3CH_2)_3O^+SbF_6^-]_0 = 2 \cdot 10^{-3}$ mol · l^{-1} [35]

The monomers polymerize slower than in homopolymerization [35, 36, 37, 38]. The final concentrations of the comonomers are close to those observed in homopolymerizations [37], which suggests a small extent of crosspropagation and the formation of a a mixture of homopolymers [39]. This conclusion was proven by the fractionation of the product [40].

This behaviour is unexpected. According to the correlations discussed in Sections 15.1.3.4 and 15.3.1.2, the more basic THF ($pK_a = -2.1$) should react faster than the less basic DXL ($pK_a = -4.9$).

The strange kinetic pattern of the THF-DXL "copolymerization" can be explained by the following scheme:

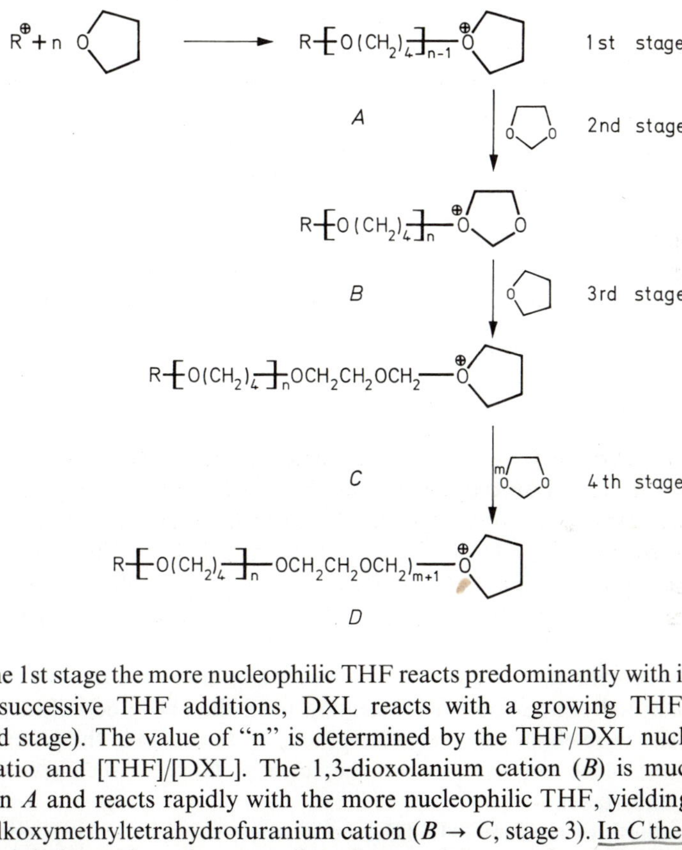

(15-33)

During the 1st stage the more nucleophilic THF reacts predominantly with initiator. After "n" successive THF additions, DXL reacts with a growing THF species (*A* → *B*, 2nd stage). The value of "n" is determined by the THF/DXL nucleophilic reactivity ratio and [THF]/[DXL]. The 1,3-dioxolanium cation (*B*) is much more reactive than *A* and reacts rapidly with the more nucleophilic THF, yielding predominatly 1-alkoxymethyltetrahydrofuranium cation (*B* → *C*, stage 3). In *C* the positive charge is mainly located on the exocyclic α-C-atom, because of stabilization due to the alkoxy substituent. Therefore, in *C* the exocyclic carbon atom is preferentially attacked by either monomers. Its reaction with THF is an exchange and does not result in chain growth. Attack of DXL on the exocyclic α-C-atom in *C* yields 1-alkoxymethyl-1,3-dioxolanium cations which rapidly react with THF reverting to 1-alkoxymethyltetrahydrofuranium cations (*D*) identical to (*C*), but with a longer chain. Thus, THF in the growing center cannot be incorporated at this stage into the polymer chain precluding copolymerization. On the contrary, DXL may insert in between the THF and the penultimate alkoxymethylene unit.

The endo-cyclic C-atoms in *C* or *D* can also be attacked by THF but with a much lower efficiency. In this case the *A* type species is formed and the whole process is repeated.

Model studies show that in the dimethylmethoxymethyloxonium ion the relative reactivity of methylene carbon (k_1) to methyl carbons (k_2) is $k_1/k_2 > 10^7$ [41].

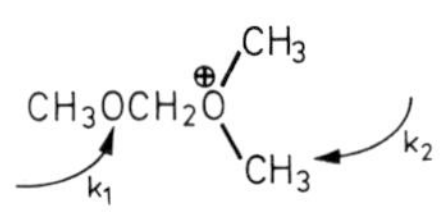

(15-34)

In line with this information the relative reactivity within the growing species in which the THF molecule is located at the end of the polyacetal chain (*C*) or (*D*) is most likely close to this value (in spite of the strain of the THF ring). This high ratio would eventually lead to nearly pure homopolyDXL with a few short THF sequences at the beginning of the chain.

When DXL comes to equilibrium, THF restarts to homopolymerize but slower than in homopolymerization, because some of the active species exist still in a *D*-type structure, hampering THF propagation.

The presence of homopolymers, instead of the block copolymer, is probably due to side reactions, e.g. hydride transfer. The alkoxycarbenium cations, present in polymerization in equilibrium with oxonium ions, are strong hydride acceptors:

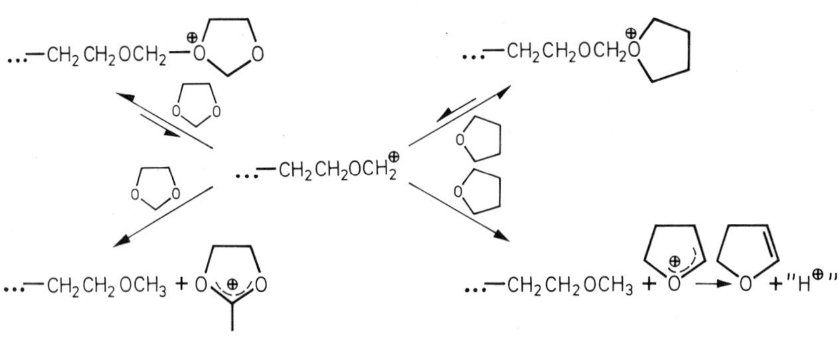

(15-35)

The low content of units due to the foreign comonomers could easily be overlooked in the (essentially) homopolymers resulting from the above described process.

If transfer were eliminated, i.e. at lower temperatures and in the presence of stable counterions, block copolymer could be formed; however such studies have not been yet undertaken.

The synthesis of random THF-DXL copolymers should also be possible either by changing the mechanism of polymerization (discussed below) or by transacetalization, i.e., by reshuffling the blocks. This reaction, although well-known for polyacetals and mixtures of polyacetals with polyethers [42], has not yet been applied for mixtures of polyDXL and polyTHF, probably because it is slow and is accompanied by side reactions.

Okada et al. [43] have claimed the successful synthesis of random THF-DXL copolymers using methyl fluorosulfonate initiator. With this type of initiator the ionic growth mechanism is accompanied by a covalent one, allowing the partial addition of one monomer to the growing species produced from the other.

This claim was based on the analysis of the ^{1}H- and ^{13}C-NMR spectra of the product formed in $CHCl_3$ solvent at 0 °C. Thus, the ^{1}H-NMR spectrum of the product obtained by Okada exhibits two signals of the oxymethylene groups at $\delta = 4.74$ and 4.77 ppm. According to model studies, the latter is due to the heterodyad $OCH_2CH_2O\underline{CH_2}O(CH_2)_4$, whereas the high-field signal is due to the homodyad (polyDXL). In the ^{13}C-NMR spectrum three signals appear around 96 ppm. The following assignments were given:

DDD	$OCH_2CH_2OCH_2OCH_2CH_2O\underline{CH_2}OCH_2CH_2OCH_2$	α	95.6 ppm
DDT	$OCH_2CH_2OCH_2OCH_2CH_2O\underline{CH_2}OCH_2CH_2CH_2CH_2$	β	96.1 ppm
TDD	$OCH_2CH_2CH_2CH_2OCH_2CH_2O\underline{CH_2}OCH_2CH_2OCH_2$	β	96.1 ppm
TDT	$OCH_2CH_2CH_2CH_2OCH_2CH_2O\underline{CH_2}OCH_2CH_2CH_2CH_2$	γ	96.4 ppm

(15-36)

Okada et al. correlated the proportion of dyads and triads using Markovian statistics which can be applied only to the simplest four-parameter system [e.g. Scheme (15-1)]. In such a system the proportion of β-triads (DDT + TDD) is equal to the geometrical average of α-(DDD) and γ-(TDT) triads ($\beta = 2\sqrt{\alpha \cdot \gamma}$). Analysis of Okada's data gives large deviations from this rule, e.g. the observed proportion of the β-dyad is half of that calculated by Markovian statistics ($\beta^{obsd} = 0.21$, $\beta^{calcd} = 0.47$). Evidently this statistics cannot be applied, probably because of reversibility and the diversity of active centers in the system.

We think, however, that in spite of the strong evidence produced by Okada in favour of the heterodyad presence, acid hydrolysis should be applied which is the most reliable proof for systems in which copolymers of a cyclic acetal with e.g. a cyclic ether is postulated.

Another point in Okada's work also requires comment. It has been proposed that covalent propagation proceeds via a four-center mechanism:

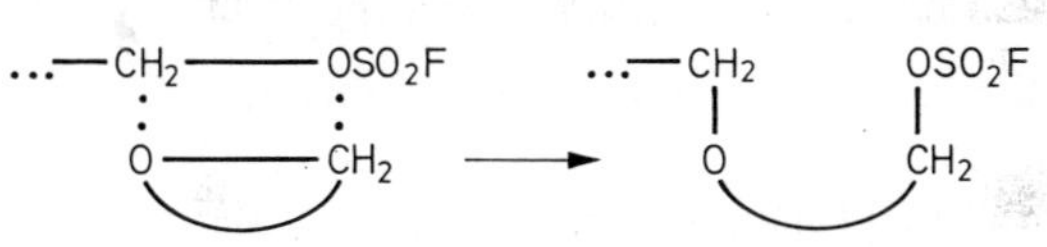

(15-37)

This reaction is quite unlikely because the simultaneous breaking and formation of two bonds should be accompanied by a very large negative activation entropy. Moreover, the reaction, in which two electron pairs participate, requires inversion of configuration according to the Woodward-Hoffman rules of the preservation of orbital symmetry [44]. The inversion would require an even number of electron pairs (parity rules [45]), and this is impossible for steric reasons:

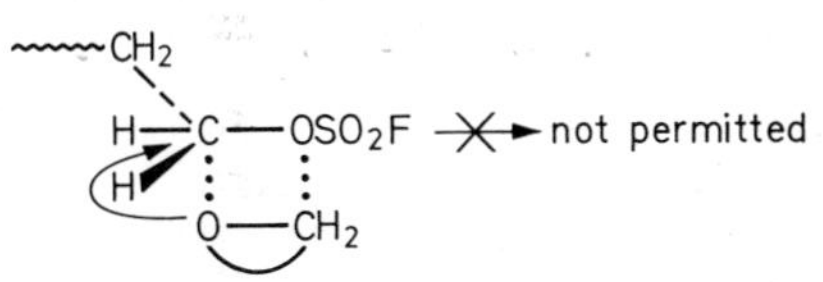

(15-38)

Retention of configuration at the α-methylene group is possible with an odd number of electron pairs. This could happen preferentially for the six-center and not the four-center mechanism described above. The latter, however, should be accompanied by a larger negative entropy of activation than the former. It has recently been found that in the polymerization of cyclic ethers covalent propagation proceeds by a conventional S_N2 mechanism, and we would prefer to propose that the same mechanism also operates in the THF-DXL copolymerization:

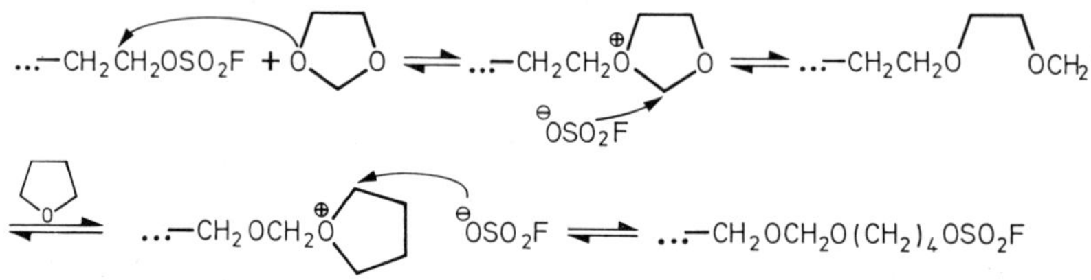

(15-39)

The formation of random copolymer becomes possible when the reaction between the anion and the endocyclic α-C-atom in the 1-alkoxymethyltetrahydrofuranium cation is faster than the reaction between DXL and the exocyclic α-C-atom:

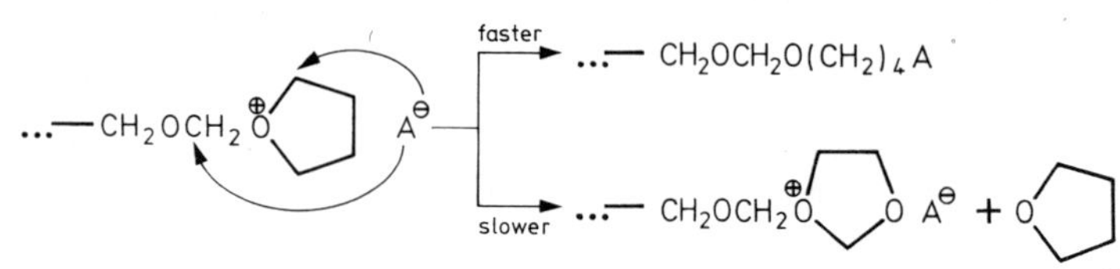

(15-40)

This requirement seems more likely because the $CF_3SO_3^{\ominus}$ anion is known to react with the tetrahydrofuranium cation with a rate constants comparable to that of THF addition, i.e., this reaction should be faster than that of DXL addition.

It seems that the "ester route" may provide practical conditions also for the formation of random copolymers from monomers giving, in purely cationic polymerization, cations differing in their chemical nature (e.g. carbenium and onium).

15.1.4.3 Copolymerization of Lactones with Cyclic Ethers

The basicity of esters is similar to that of related ethers, therefore, in copolymerization random structures could be expected. THF ($pK_b = 5.0$) and BCMO ($pK_b = 5.65$) were copolymerized with ε-caprolactone (εCL; $pK_b = 5.31$) and β-propiolactone (βPL, $pK_b = 10.06$), εCL and cyclic ethers enter the copolymer with nearly the same rates. The reactivity ratios determined by the Mayo-Lewis method are: r_1(BCMO) = 0.24, r_2(εCL) = 0.44 and r_1(THF) = 0.7, r_2(εCL) = 0.3 [46]. r_1 should be higher for THF(M_1) because the reversibility of THF polymerization has not been taken into account (cf. Sect. 15.1.3.1). r_1 and r_2 values suggest the formation of nearly random copolymers. The microstructure of BCMO-εCL copolymers was studied by NMR spectroscopy, and the new signals B and A due to heterodyad appeared [cf. Eq. (15-41)]. Simultaneously the ratio of signals C/D decreased from 2 to a lower value:

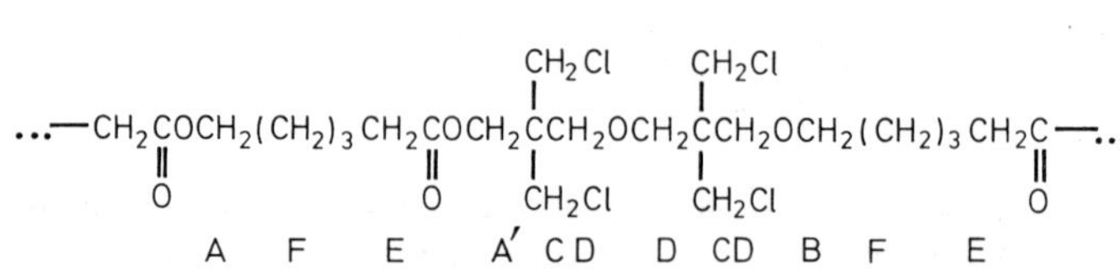

(15-41)

Analysis of the spectrum strongly indicates the formation of some random structures. The corresponding proportions of the different dyads were not calculated, however, and the agreement with the theoretically predicted values was not checked. Neither ^{1}H- nor the ^{13}C-NMR spectra of the THF-εCL copolymer indicate heterodyads in the product of alleged copolymerization. The MWD is bimodal and fractionation of benzene solutions with methanol as nonsolvent gives higher εCL contents for the lower-molecular-weight fractions. In the IR spectrum the crystalline band of polyTHF at 1000 cm^{-1} was absent. However, the X-ray diffraction pattern (11.2 mol% εCL) showed strong diffractions at $2\theta = 20°$ and 24° ascribed to the crystalline peak of polyTHF. No peak at $2\theta = 21.4°$ of poly εCL was observed however. The DSC thermogram of the copolymer showed one endotherm at the melting point of polyTHF (30–37 °C) but again no peak for polyεCL (55–60 °C). Thus, long crystalline sequences of polyTHF still existed in the copolymer, while polyεCL homoblocks, even if they existed, did not show any tendency to crystallize. With copolymers richer in εCL (27 mol%) the peaks of crystalline polyεCL appear in X-ray diffraction scans and DSC plots. Thus, the copolymer has long THF and εCL sequences, although this could not be expected from the reactivity ratios. The authors of the original paper proposed two types of active species: a "cationic" and a "coordinated" cationic, each being responsible for the predominating incorporation of the ether or ester [46].

βPL is much less basic than εCL, and its copolymerization with BCMO is more difficult. The following reactivity ratios were calculated: $r_1(\text{BCMO}) = 10$, $r_2(\beta\text{PL}) = 0.03$ with $BF_3 \cdot OEt_2$ initiator and $r_1 = 8$, $r_2 = 0.04$ with $SnCl_4$ [47]. The ^{1}H-NMR spectrum indicated the presence of heterodyads as well as both homodyads, but the content of the βPL-βPL homodyads measured in this way was found to be at least twice higher than that determined from reactivity ratios. Again, the authors suggest that enrichment in βPL is due to the coordinate cationic mechanism operating simultaneously with a conventional cationic process [47]. In the copolymerization of βPL, with THF, the presence of heterodyads was clearly observed [48].

$$\ldots -OCH_2(CH_2)_2CH_2- \; OCH_2CH_2\overset{\overset{\displaystyle O}{\|}}{C}-OCH_2CH_2\overset{\overset{\displaystyle O}{\|}}{C}O \; -OCH_2CH_2CH_2CH_2- \ldots \tag{15-42}$$

^{1}H	3.37	1.61	3.37	3.65	2.56	4.30	2.60	4.07	1.67	1.61	3.37
^{13}C	71.1	26.9	71.1	66.5	35.4	60.4	33.9	64.6	25.9	26.9	71.1

When THF was used above its equilibrium concentration ($[\text{THF}]_0 > [\text{THF}]_e$) mostly its homopolymer was formed. Only when $[\text{THF}]_0$ reaches $[\text{THF}]_e$ does βPL start to polymerize. Under these conditions one obtains two consecutive THF units separated by a number of βPL units, depending on the βPL content.

15.1.4.4 Attempted Copolymerization of Monomers that Belong to Different Classes of Chemical Compounds

Typical vinyl monomers, like styrene or its derivatives, are much less basic than heterocyclics. Thus attempts to prepare random vinyl/heterocyclic copolymers usually gave heterocyclic homopolymers or in a few systems mixtures of the two homopolymers, e.g., oxetane with N-vinyl carbazole [49], α-methylstyrene with βPL [50].

The less nucleophilic acetals copolymerize with vinyl compounds more readily. Perhaps, in these systems the alkoxycarbenium ions (... $-OCH_2^{\oplus}$) that coexist in equilibrium with oxonium ions facilitate copolymerization with vinyl compounds. Styrene copolymerizes with trioxane [51,52] and tetraoxane [53]. The latter system yields polytrioxane and trioxane-styrene copolymer together with 1,4-phenyl-1,3-dioxane. It was formed in 25% yield in ethylene dichloride at 30 °C after 1 hr using $[BF_3 \cdot OEt_2] = 10^{-2}$ mol · l^{-1}, $[\text{styrene}]_0 = [\text{tetraoxane}]_0 = 0.5$ mol · l^{-1}. The proposed mechanism of 4-phenyl-1,3-dioxane formation is shown below (cf. also Chap. 7):

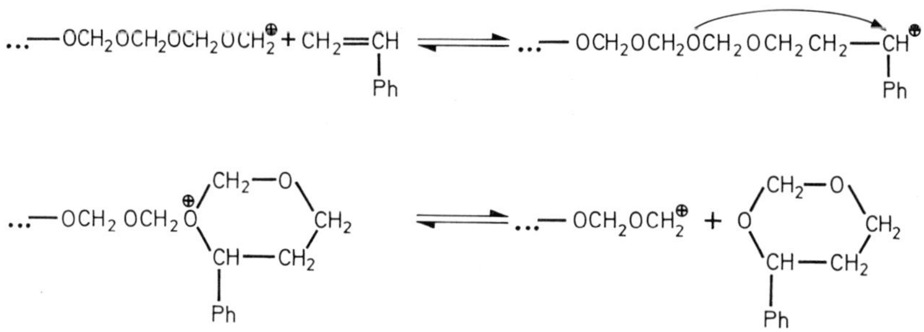

(15-43)

4-Phenyl-1,3-dioxane may further copolymerize with trioxane or tetraoxane.

Copolymerization of α-methylstyrene with tetraoxane produced, besides the linear products, 4-methyl-4-phenyl-1,3-dioxane [54].

In the copolymerization of tetraoxane with 1,1-diphenylethylene, 4,4-diphenyl-1,3-dioxane was also isolated [54].

Yamashita et al. copolymerized styrene with DXL and claimed copolymer formation [56]. GPC analysis indicated the presence of phenyl rings (UV) in the high polymer fraction and its absence in the low-molecular-weight oligomers. In the NMR spectra the intensity of the $-OCH_2O-$ protons decreases, indicating the presence of heterodyads. However, Plesch in his studies of reactions of olefins with DXL was able to obtain only low-molecular-weight cyclic oligomers [57].

DXL and isoprene form random copolymer, as it is evidenced from the microstructure analysis by NMR spectroscopy [58]. The copolymerization is very slow; several weeks at room temperature are required but polymers with high molecular weights ($\overline{M}_n \simeq 10^5$) are eventually formed. By increasing the DXL content, the proportion of 1,4-isoprene units decrease markedly.

In the copolymerization of BCMO with α-methylstyrene, *p*-methylstyrene, 2-chloroethyl vinyl ether and isobutylene, the latter three comonomers yield a benzene-soluble fraction that contains Cl [55]. Since PolyBCMO is insoluble in benzene, this result may indicate the formation of some products containing vinyl monomers and BCMO. Information about the products is rather scarce, and it is quite likely that copolymers did not form and that the Cl in the soluble fractions was due to secondary alkylation processes.

The copolymerization of an excess of isobutylene with DXP gave a main product with the following structure [59]:

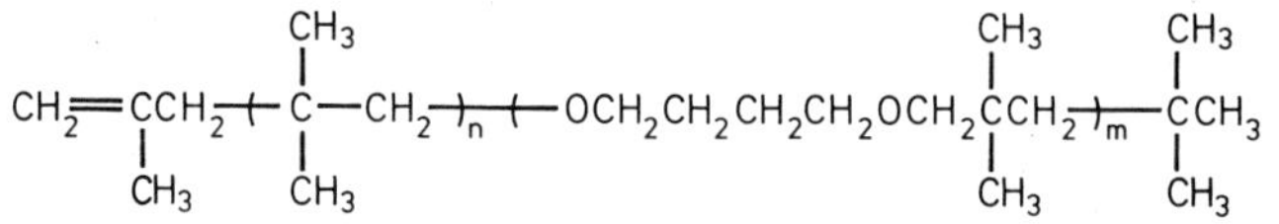

(15-44)

(where m $\simeq$ 3).

The presence of the ether bonds was established by ^{1}H-NMR spectroscopy, whereas, remarkably, acetal bonds were absent. The copolymer of the structure shown in Eq. (15-44) could have been formed by simultaneous transfer to polymer and monomer:

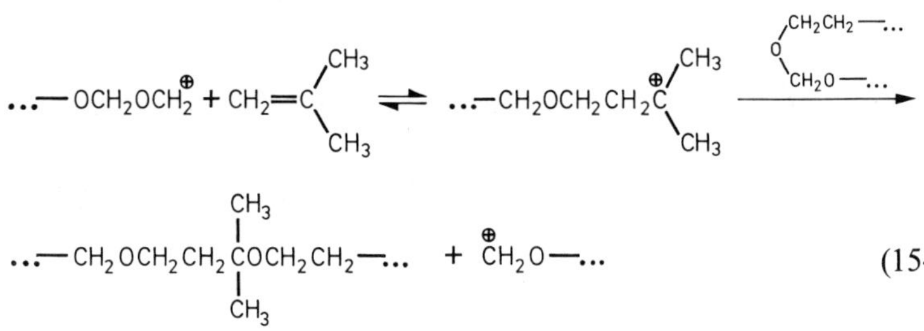

(15-45)

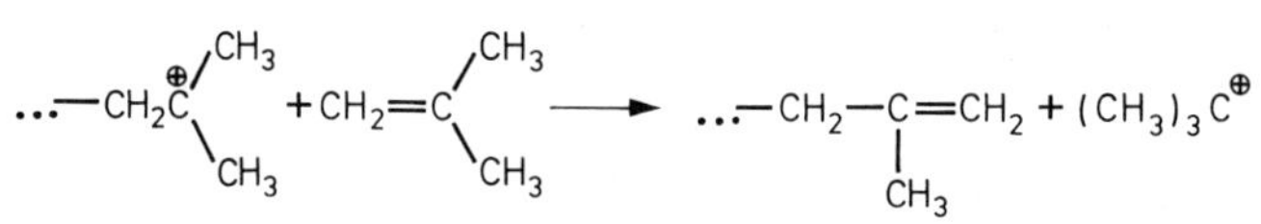

(15-46)

15.2 Alternating Copolymers

Alternating copolymers have regular structures $(AB)_n$ with the randomness parameter $\Psi = 2$ (cf. Chap. 2). They may also be considered to be homopolymers of the new structural unit AB. These copolymers can be formed in two ways:

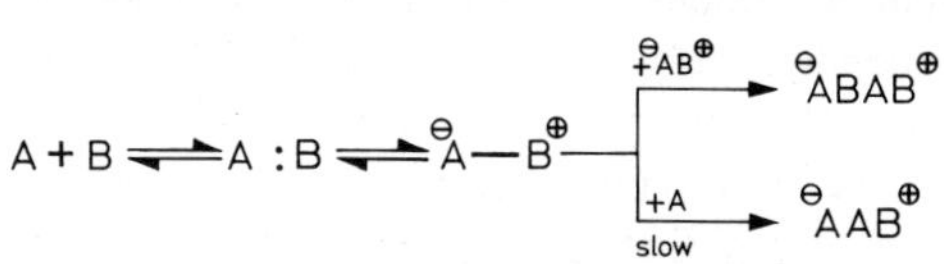

(15-47)

First, from e.g. strong donor — acceptor complex between comonomers or even a betaine structure. This intermediate undergoes faster ion-coupling than the corresponding ion-molecule reaction.

Second, by more rapid crosspropagation than homopropagation. This happens when the more reactive monomer cannot homopolymerize due to unfavorable thermodynamics; e.g. the system being above the ceiling or below the floor temperature.

15.2.1 Zwitterionic Copolymerization of Heterocyclic Monomers

Developments in this area have their roots in the alternating radical copolymerization [60,61,62], particularly when a donor-acceptor monomer pair forms a charge transfer (CT) complex. Depending on the position of the equilibrium in Eq. (15-48), copolymers of varying structures are formed:

$$D + A \overset{K}{\rightleftharpoons} DA \tag{15-48}$$

For small values of K ($K < 0.1$) random copolymer may be formed, whereas for higher values alternating copolymer is produced. With still higher K values ($K > 10$) spontaneous homopolymerization of both monomers can be observed [63].

The interaction of cationogenic and cationophilic comonomers may result in a betaine. Betaines of vinyl monomers often a collapse to cyclobutane derivatives [64, 65]:

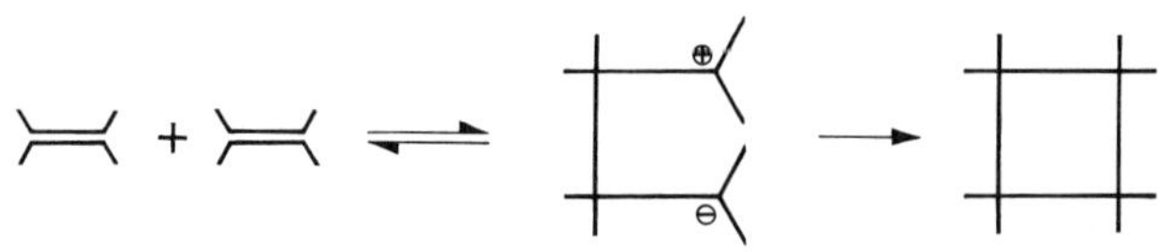

(15-49)

The self-neutralization of a heterocyclic betaine is much more difficult, and polymers are formed. To our knowledge the first example of such a process was described in 1963 [67, 68]:

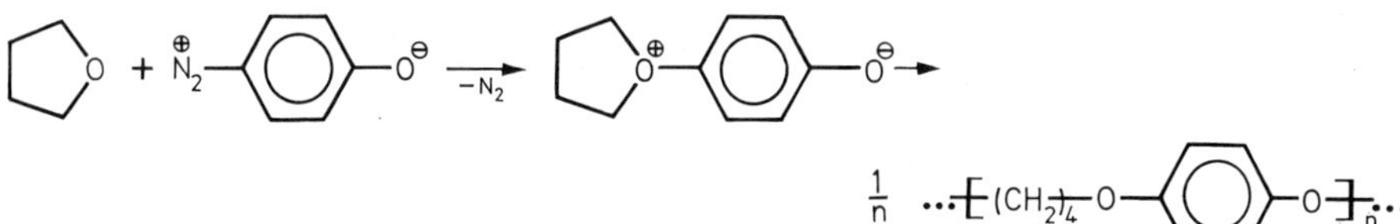

(15-50)

Later Kagiya et al. described the zwitterionic spontaneous copolymerization of β-propiolactone with different aziridines [69], and since 1972 Saegusa has started extensive studies of zwitterionic alternating copolymerization [70–79]. In most of these systems the molecular weights of the copolymers are low.

Recently, the synthesis of the first truly high-molecular-weight alternating copolymer, namely the copolymer of conidine (1-azabicyclo|4,2,0|octane) and β-propiolactone by a zwitterionic mechanism was reported [80]:

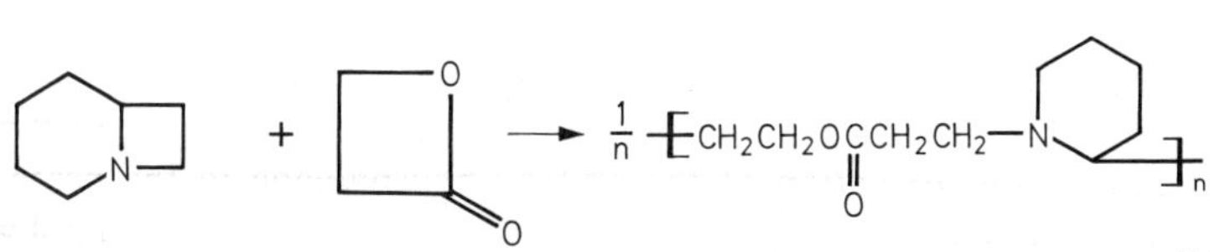

(15-51)

The structure was established by analysing the products of alkaline hydrolysis:

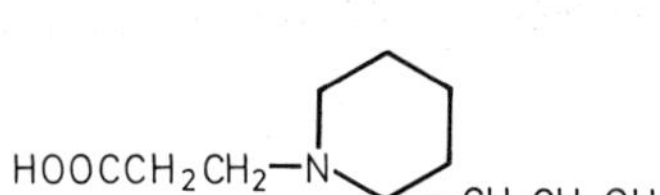

(15-52)

Molecular weights measured by VPO and by end-group titration were $\overline{M}_n = 30{,}000$ at 85% conversion in 1,4-dioxane after 3 hrs at 90 °C or as high as $\overline{M}_n = 70{,}000$ at 50% conversion in benzene after 107 hrs at 60 °C. In the more polar CH_2Cl_2, molecular weights were lower, although the overall rate of copolymerization was higher. According to the authors the copolymerization exhibits living character; fresh portions of commonomers copolymerized after complete conversion at the same rate as in the first stage. In solvents of low polarity the addition of separately prepared monomeric betaine accelerated the copolymerization. Thus betaine formation is the rate limiting step in these media. It seems that conidine and βPL are the only comonomer pair which homopolymerize by a living process with a sufficiently high rate. This may explain high-molecular-weight copolymer formation in contrast to the other systems discussed below.

Saegusa has tested a large number of nucleophilic and electrophilic monomer pairs (mostly heterocyclics) [70–79, 81, 82]. None of these pairs consists of monomers capable of giving living systems. Copolymerization of these monomers gave more than fifty different alternating copolymers. These investigation were authoritatively reviewed by Saegusa [70, 71, 72]. We confine ourselves to the description of the typical procedure and the basic features of the process, mostly following Saegusa's papers.

Usually, equimolar amounts of comonomers ($[M]_0 \gtrsim 1\ mol \cdot l^{-1}$) were dissolved in a polar solvent under N_2 in a glass tube which was then sealed off. The tube was heated, usually above 100 °C, for a few hours or even a few days. Then the copolymer was precipitated and dried in vacuum. Reprecipitation was carried out from chloroform solution into diethyl ether. The copolymer structure was established by NMR and IR spectroscopies.

Table 15.6 gives several examples of these systems showing the structures of comonomers and that of final product, reaction conditions and copolymer properties.

The molecular weights of copolymers described in Table 15.6 are usually low. The $\overline{M}_n$ may be increased by isolating and purifying the intermediate product (e.g. IP), as shown for No. 10 in Table 15.6.

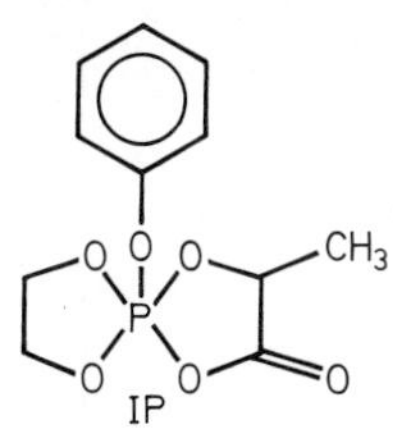

(15-53)

IP is formed at 0 °C in ether solution by mixing the ingredients. As Saegusa has pointed out, the formation of IP required reduction of the α-keto acid to α-hydroxy acid and simultaneous oxidation of the aryloxy phospholane to the penta-coordinated phosphorus compounds. IP polymerizes also by elimination of phenol when reacted with alcohols, amines or water, e.g.:

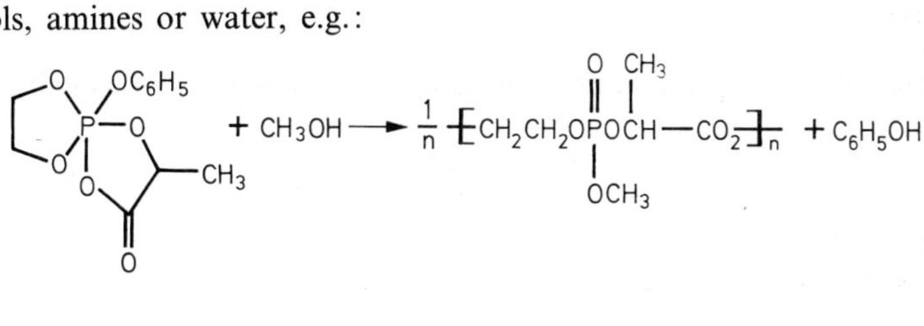

(15-54)

Table 15.6. Examples of zwitterionic, spontaneous, alternating copolymerization

No	Comonomer structures		Reaction conditions	Copolymer structures and properties		Ref.
1	2		3	4		5
1	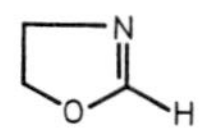	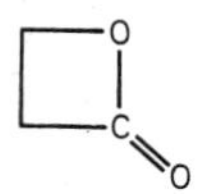	DMF, 30 min, 120 °C	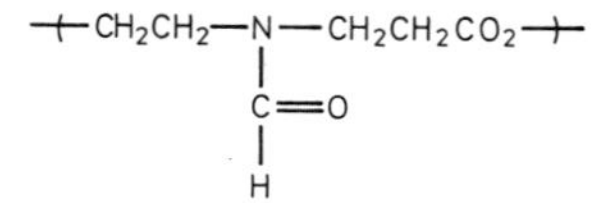	$\bar{M}_n = 2000$, 80% yield, 40% OXZ content	73,74)
2	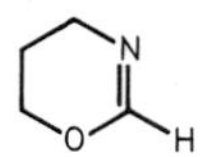	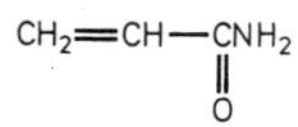	CH_3CN, 20 h, 70 °C	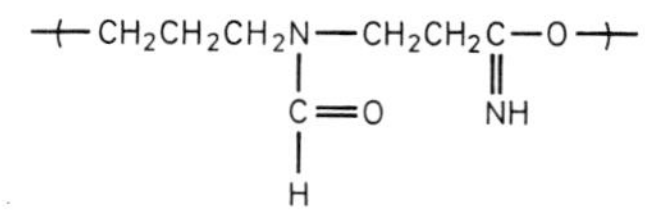	$\bar{M}_n = 1470$, 100% yield	75)
3	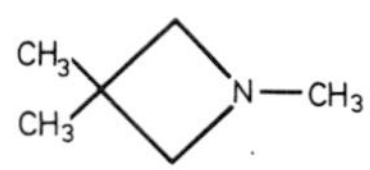	$CH_2{=}CH{-}CO_2H$	CH_3CN, 3.5 h, 80 °C	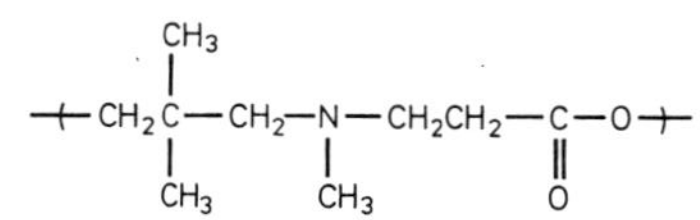	$\bar{M}_n = 1350$, 86% yield	76)
4	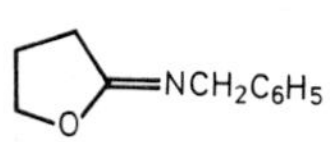	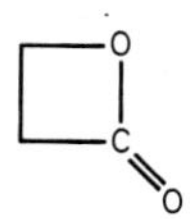	CH_3CN, 1.5 h, 80 °C	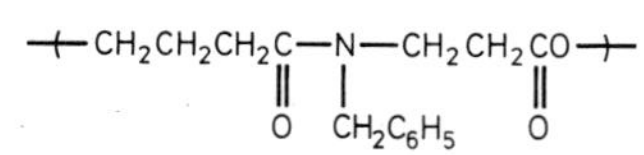	$\bar{M}_n = 5120$, 41% yield	77)
5	$C_6H_5CH{=}NC_6H_5$	$CH_2{=}CH{-}CO_2H$	CH_3CN, 110 h, 100 °C	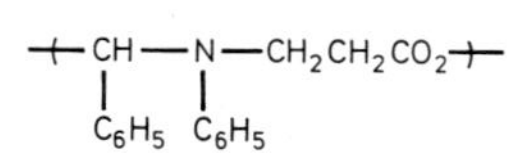	$\bar{M} = 800$, 67% yield, m.p. = 127 °C	78)

6	$C_6H_5CH{=}NC_6H_5$	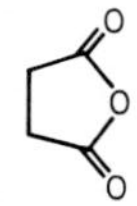	CH_3CN , 110 h , 100°C	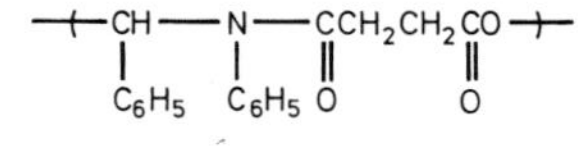	$\overline{M}_n = 614$, 90% yield m.p. = 155°C	78)
7	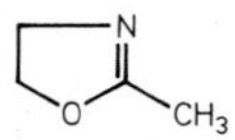	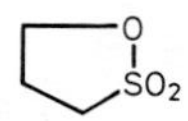	DMF , 10 min, 100°C	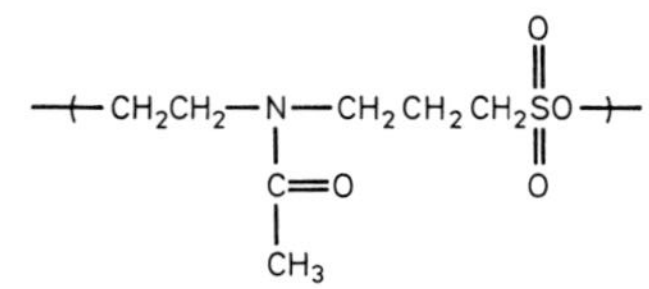	70% yield	79)
8	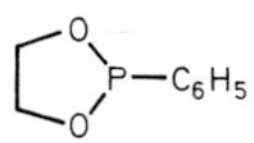	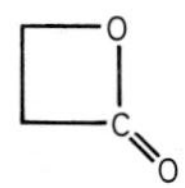	DMF , 20 h , 130°C	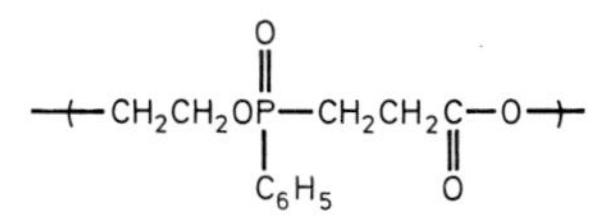	$\overline{M}_n = 1320$, 35% yield	81)
9	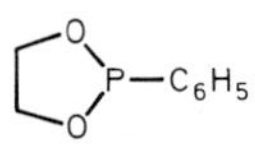	$CH_2{=}CH{-}C({=}O){-}CH_3$	DMF , 100 h , 130°C	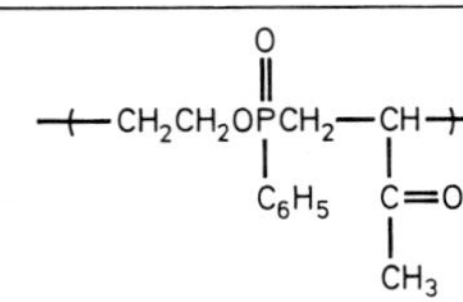	$\overline{M}_n = 1560$, 30% yield	72)
10	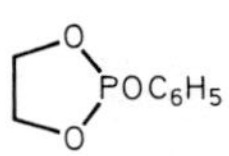	$RC({=}O)CO_2H$	DMF , 48 h , 120°C	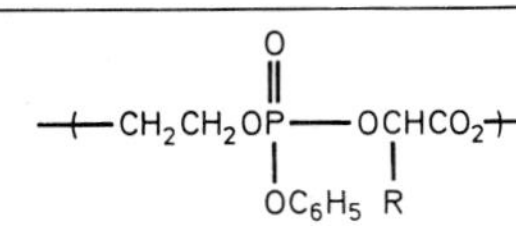	$\overline{M}_n = 7500$, 55% yield	82)
11	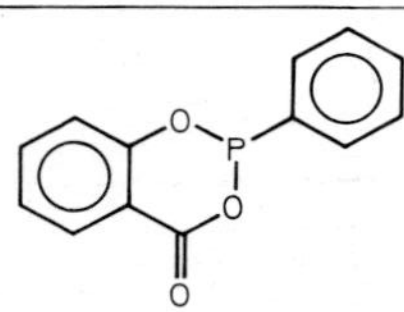	$CH_2{=}CH{-}C({=}O)CH_3$	DMF , 48 h , 110°C	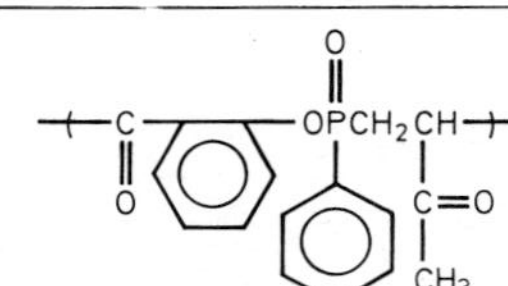	$\overline{M}_n = 1160$, 38% yield	72)

After 48 hrs at 120 °C a polymer with $\overline{M}_n = 4{,}180$ was formed in 35% yield. At room temperature after 69 days the corresponding polyphosphonic acid diester with $\overline{M}_n = 7{,}560$ was formed in 48% yield.

With amines two types of structural units were observed by ^{31}P-NMR spectroscopy:

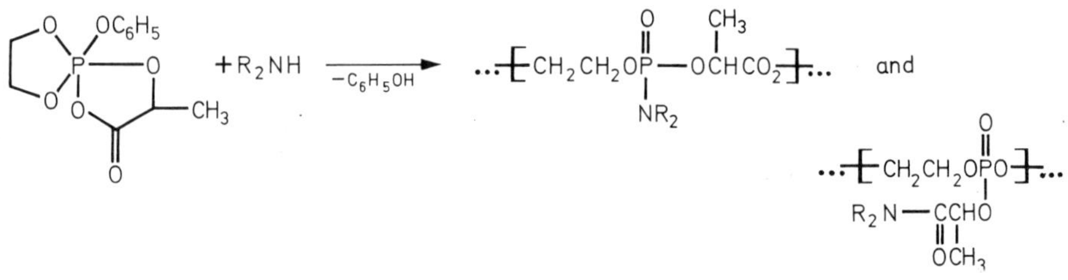

(15-55)

The yield was 32% and $\overline{M}_n = 3090$ with diethylamine at 100 °C after 70 hrs and both units appeared in equal amounts in the polymer.

Copolymerization with CO_2 or CS_2 gives rise to a 1:1:1 terpolymer. For example phospholane reacted with methyl acrylate at 120 °C under 50 kg/cm² CO_2 pressure gave a copolymer with $\overline{M}_n = 1{,}800$ in 61% yield after 300 hrs:

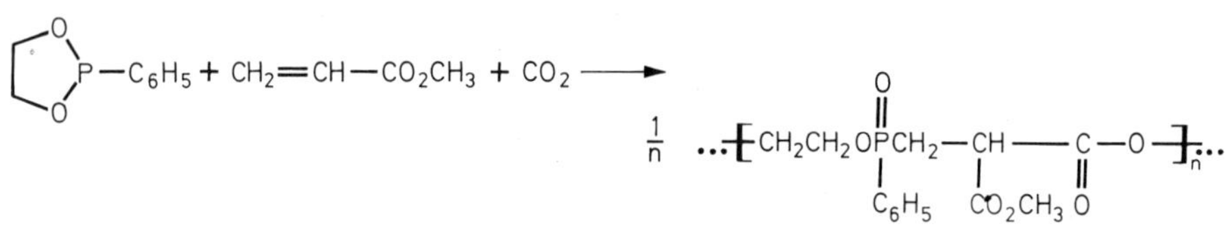

(15-56)

Recently phosphonite and 2-phenyl-1,3,2-dioxaphospholane were copolymerized with γ-butyrolactone and oxirane with yields not exceeding 50–70% to rather low-molecular-weight products ([η] < 0.06 dl/g) [83].

Side reactions

The majority of to date prepared copolymers, described in the preceding section, are of low molecular weights. The chain-breaking processes were not studied in any detail, but several routes may be proposed. The zwitterionic polymerization resembles a step-growth process. Therefore, high molecular weight products will be formed only at high conversions, although very often copolymerizations were stopped at relatively low conversions. In a few systems, in spite of complete conversion, molecular weights were still low (e.g. oxazine with acrylamide $\overline{M}_n = 1{,}470$).

Only a few termination processes have been detected; one of these is β-elimination [84] in aryl sulfonium zwitterion systems:

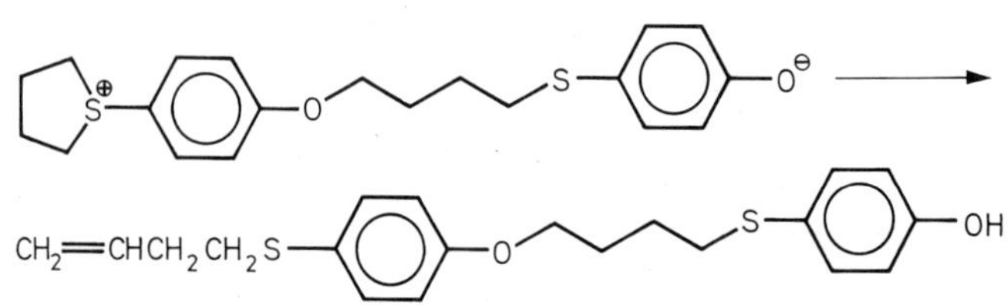

(15-57)

A possible transfer reaction is the rearrangement known to occur in the cationic homopolymerization of 1,3,2-dioxaphospholane [85]:

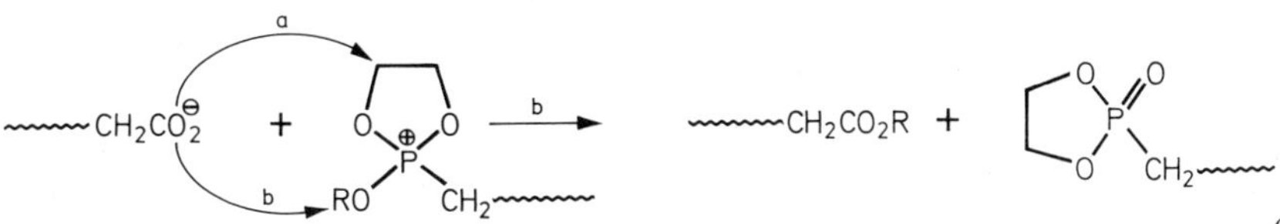

(15-58)

In the copolymerization with βPL, only route "a" has been considered by Saegusa, and transfer was not taken into account. Macrocyclization by end-to-end closure is another possibility, however, these macrocycles cannot easily be isolated and studied. The presence of macrocycles may be corroborated by comparison of $\overline{DP}_n$ determined from end-groups with those measured by other methods e.g. GPC, osmometry, etc., i.e. when cyclic and linear molecules are counted separately. Such macrocycles were indeed observed by Schmidt in the polymerization of aryl sulfonium zwitterion [86]:

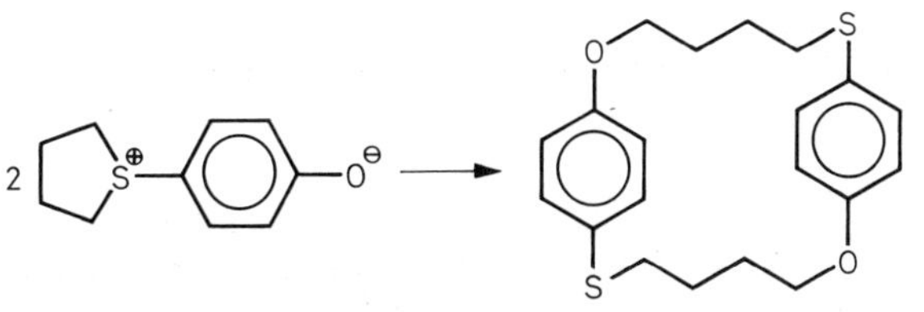

(15-59)

Recently Goethals et al. [87] observed formation of cyclic oligomers in copolymerization of 1-*tert*-butylaziridine (TBA = M_1) and βPL (M_2). Structures and concentrations have been analysed by GC-MS.

Concentrations and structures of the cyclic oligomers are given in Table 15.7. It is of interest to note, that only one of these cyclic oligomers has the alternate structure.

Table 15.7. Oligomers detected in the spontaneous copolymerization of 1-tert-butylaziridine (M_1) and β-propiolactone (M_2). $[M_1]_0 = [M_2]_0 = 0.8\ mol \cdot l^{-1}$, CH_3CN solvent, 50 °C [87]

No.	mol wt.	Structure	Relative intensity (GC)
1	342	$M_1M_2M_1M_2$ (cyclic)	74
2	342	$M_1M_1M_2M_2$ (cyclic)	10
3	369	$M_1M_1M_1M_2$ (cyclic)	100
4	441	$M_1M_1M_1M_2M_2$ (cyclic)	7
5	441	$M_1M_1M_1M_1M_2$ (cyclic)	98
6	468	$M_1M_1M_2M_1M_2$ (cyclic)	36

15.2.2 Alternating Copolymers from Monomers unable to Homopolymerize

The polymerization of the large majority of heterocyclics is accompanied by a simultaneous decrease of enthalpy and entropy. Thus, these monomers can be polymerized below their ceiling temperature T_c. Above T_c high homopolymer cannot be formed. Polymerizability of heterocyclics is reduced (decreasing in T_c), because of low ring strain (e.g. THP, tetrahydrothietane 1,4-dioxane) and increase in steric hindrance while passing from monomer to polymer. 2,2-Diphenyloxirane, which only dimerizes (similar to 1,1-diphenylethylene in vinyl polymerization), may serve as an example [88]:

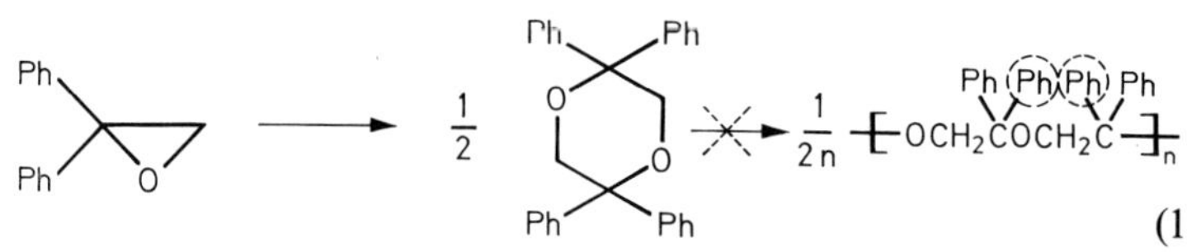

(15-60)

Insertion of a comonomer unit between two consecutive oxirane units would significantly decrease the repulsive interactions between the neighbouring phenyl substituents.

Alternating copolymerization is facilitated when the non-polymerizable monomer is more reactive. Unfavorable thermodynamics does not allow two consecutive additions of these monomers and the low reactivity of the second comonomer retards its propagation. Some examples are discussed below.

The copolymerization of BCMO with THF above its ceiling temperature will be examined. The more reactive THF forms alternating copolymer only above 60 mol% content of THF in the mixture. At lower THF concentrations BCMO-BCMO dyads are also formed [33].

Alternating copolymerization is possible in other systems because new, more stable chemical bonds are formed. Copolymerization of cyclic anhydride and THF (above T_c) to a polyester may serve as an example [89]:

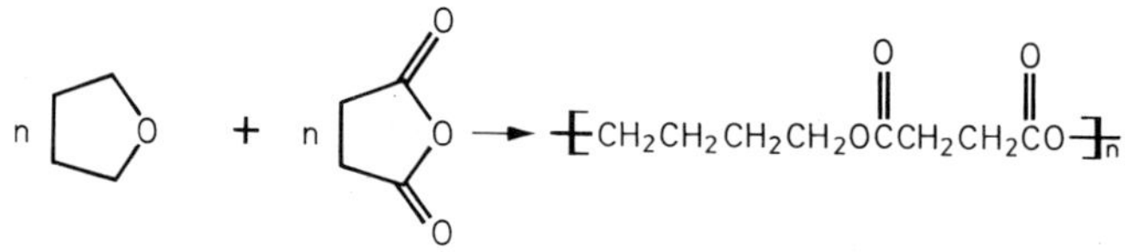

(15-61)

Sulfur dioxide and carbon dioxide, either of which is unable to homopolymerize, are known to easily give copolymers with cyclic compounds. In this way CO_2 copolymerizes cationically with aziridines [90]:

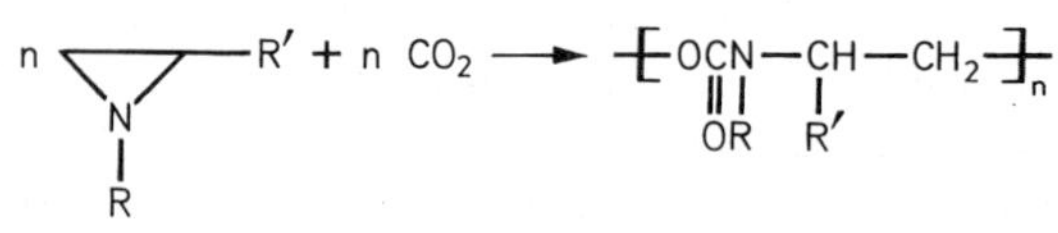

(15-62)

and SO_2 with propylene oxide to not-perfectly alternating poly(ether sulfides)[91, 92, 93]:

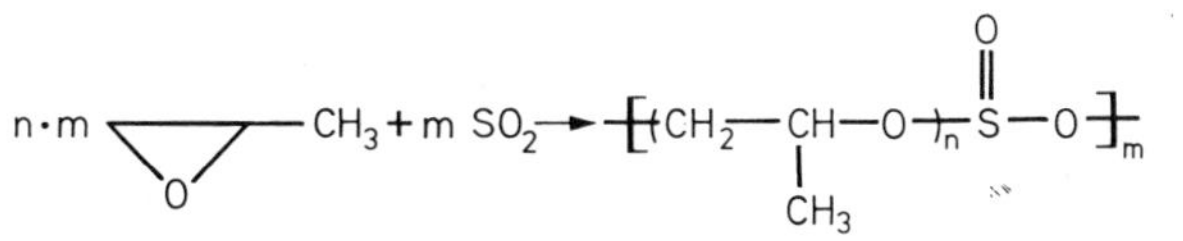

(15-63)

There are also examples of alternating terpolymers. In the presence of $Al(iBu)_3$ at 70 °C, THF (above T_c), ethylene oxide, and phthalic anhydride terpolymerize[94]:

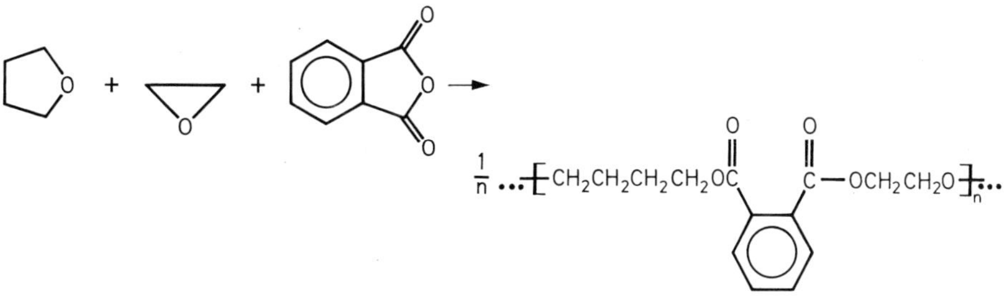

(15-64)

Formation of the alternating copolymer from the two non-homopolymerizable comonomers should involve the formation of more stable bonds either due to the new structure or because of the disappearance of steric hindrance prevailing in the homopolymer. Therefore, it is not sufficient to mix e.g. THF and THP above their ceiling temperature to obtain copolymer, because hindrance for depropagation does not exist. The small gain in the entropy of mixing in random copolymers disappears in alternating copolymers due to the regular structure. Thus only one example of this type of system exists, i.e., copolymerization of THF (above T_c) with cyclic anhydrides where new ester bonds are formed.

Copolymerization of monomers above their T_c

Similarly to other polymerizations, in cationic ring-opening polymerization also there are monomers that were found not to polymerize although they give copolymers with appropriately chosen comonomers. The conditions of copolymerization of such pairs and the properties of the resulting copolymers were given in Table 2.14 in Section 2.9. Closely related is another group of monomers unable to homopolymerize at a given concentration and temperature (i.e. below their ceiling temperature, T_c) but able to copolymerize under the same conditions.

For the first group of comonomers their T_c are unknown and may be at such a low temperature that homopolymerization is impossible for kinetic reasons, whereas for the second group T_c is known and under suitable conditions homopolymers are formed. It is possible that some monomers judged as unable to polymerize were studied only above their T_c.

Table 15.8, gives examples of copolymerization above T_c (depending on monomer concentration) and shows regular 1:1 or 1:2 copolymers under suitable conditions. When THF reacts with comonomer of similar or higher nucleophilic reactivity, an excess of THF is required to form the corresponding regular polymers (oxetanes,

Table 15.8. Examples of copolymerization of THF above T_c

No.	Comonomer	Conditions	T , °C	Product	Ref.
1	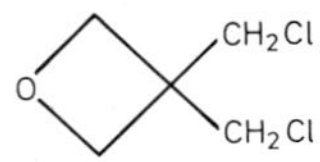	120 °C, bulk $Al(iBu)_3$, $[THF]_0 \approx 10\ mol \cdot l^{-1}$	70	2:1 copolymer	[33)]
2	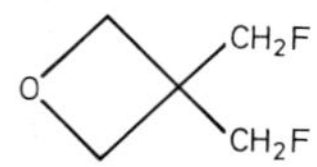	120 °C, bulk $Al(iBu)_3$, $[THF]_0 \approx 10\ mol \cdot l^{-1}$	70	1:1 copolymer	[33)]
3	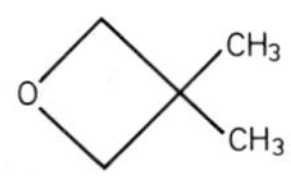	120 °C, bulk $Al(iBu)_3$, $[THF]_0 \approx 10\ mol \cdot l^{-1}$	70	1:1 copolymer	[33)]
4	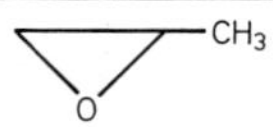	0 °C, BF_3, $[THF]_0 \approx 0.7\ mol \cdot l^{-1}$	—50	20% of THF	[27)]
5	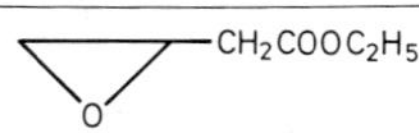	35 °C, CH_2Cl_2, $Et_3O^{\oplus}BF_4^{\ominus}$ $[THF]_0 = 2.95\ mol \cdot l^{-1}$	—20	2:1 copolymer	[95)]
6	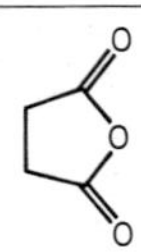	80 °C, $C_2H_4Cl_2$, PF_5 $[THF]_0 = 1.9\ mol \cdot l^{-1}$	—35	2:1 copolymer	[89)]
7	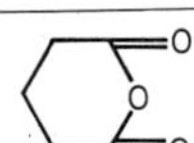	20 °€, $C_2H_4Cl_2$, PF_5 $[THF]_0 = 1.6\ mol \cdot l^{-1}$	—40	1:1 copolymer	[96)]
8[a]	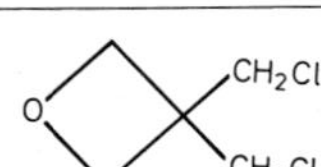	0 °C, BF_3, bulk $[2 \cdot MeTHF]_0 = 6\ mol \cdot l^{-1}$	—10	1:1 copolymer	[97)]

[a] 2-Me THF

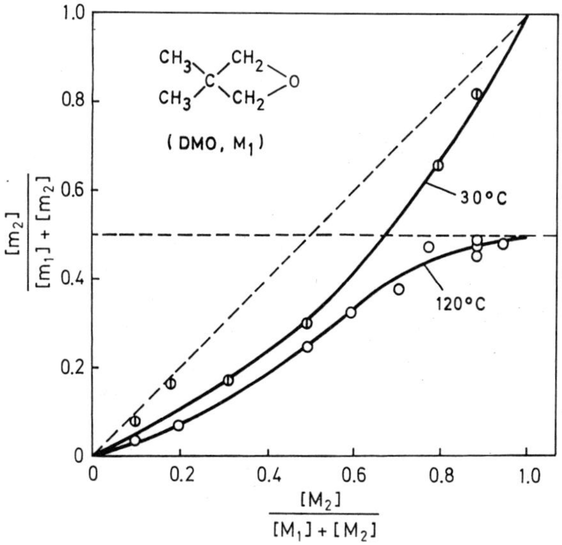

Fig. 15.4. Dependence of copolymer composition on comonomer feed in copolymerization of THF with DMO at 30° and 120° [33)]

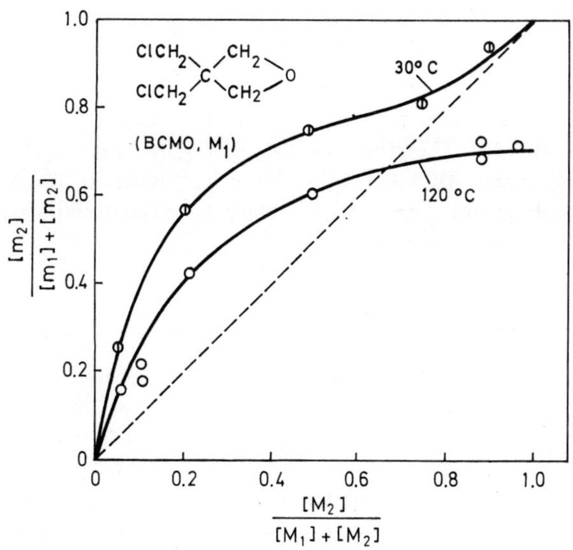

Fig. 15.5. Dependence of copolymer composition on comonomer feed in copolymerization of THF with BCMO at 30° and 120° [33)]

PO). At low THF concentrations, its content in the copolymer is low (PO). The dependence of the proportion of THF in the copolymer on comonomer feed is shown in Figs. 15.4 and 15.5 for two different oxetanes. When polymerization was carried out at 30 °C, i.e. below T_c for THF, the usual random copolymers were formed.

Polymerization above T_c clearly shows the limiting composition of THF at a high THF content in the monomer feed, because under these conditions THF cannot give longer homosequences. The limiting composition depends on the structure of comonomer. Thus, the DMO-THF pair gives a 1:1 composition (alternating copolymers) at $[THF]_0/[M_1]_0 \to \infty$. In the copolymerization of THF with BCMO the limiting composition is close to the structure $\{(BCMO)(THF)_2\}_n$ (although the final composition slightly exceeds 67% THF units indicating that a few BCMO–(THF–)$_3$ units are also present).

A rationale was put forward according to which the higher than 1:1 composition is due to the comonomer that stabilizes the THF units at the end of the chain against depropagation. The difference in ability to accommodate lower (DMO) and higher (BCMO) numbers of THF units above T_c has been explained by assuming that steric hindrance of THF depropagation is higher at the antepenultimate unit of BCMO, e.g.:

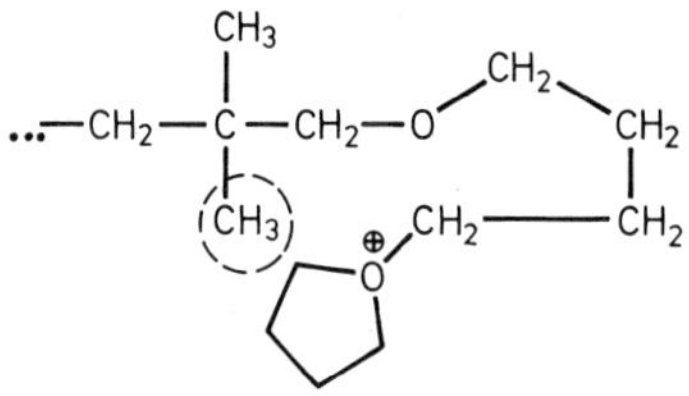 (15-80) 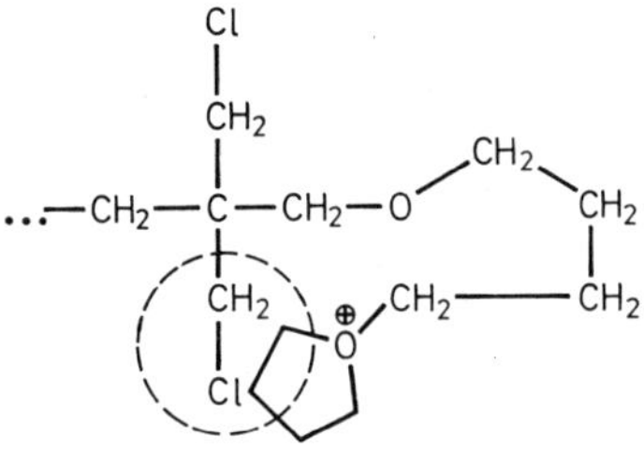 (15-64)

Thus, closing the ring from the antepenultimate unit, as required for depropagation from the $\sim\sim m_1 m_2 m_2^*$ unit, may be less perturbed during copolymerization with DMO than with BCMO. Therefore, the limited composition of BCMO copolymer is close to the repeating –BCMO–THF–THF– triad whereas for the other oxetane it is an alternating structure –DMO–THF–. The detailed analysis in terms of enthalpies and entropies of activation is given in Ref. [33].

15.3 Block Copolymers

The discovery of living anionic polymerization provided the most powerful and versatile method of preparing block copolymers [99, 100]. The living feature allows to prepare an active polymer from the first monomer, then the second monomer is added and a block copolymer results. Dianionically living polymer may give directly ABA copolymers. There are, however, a number of requirements that have to be met in order to obtain a block copolymer by using this scheme. The most important ones also apply to the cationic ring-opening polymerization:

1. In the first step strictly living conditions have to be obeyed.
2. The active species of the first block have to add the second monomer without side reactions.

The first requirement is obvious; the second one is of key importance in the preparation of block copolymers. It has been shown that the original anionic species of the first block had to be modified in some instances in order to successfully add the second monomer.

A number of papers describe the alleged formation of block copolymers by cationic ring-opening polymerization. Some of these claims are insufficiently substantiated, mostly because it was not shown without doubt that block copolymers have been obtained.

In our opinion, to make sure that block copolymer resulted, three products have to be obtained and their GPC traces compared, namely the first block, which has to be killed and isolated for this purpose, the alleged block copolymer, and the second block, prepared independently, and with the length ($\overline{DP}_n$) close to this, which has been assumed to be linked in the block copolymer.

Before the development and routine application of GPC, mostly the solubility was used to distinguish between a block copolymer and a mixture of two homopolymers. These data, in the absence of other evidence, have to be treated with great care.

In some attempted syntheses of block copolymers, random copolymers may result during the secondary processes of chain reshuffling. Thus, NMR methods, particularly ^{13}C-NMR spectroscopy, should be used to determine the presence of the heterodyads. In addition, DSC, light scattering, electron microscopy, small angle X-ray scattering, and thermomechanical analysis can also be used to distinguish between block and random structures.

15.3.1 Block Copolymers by Sequential Addition of two Monomers

No true living cationic vinyl polymers are known at present, although the papers of Skorokhodov on 1,2-dimethoxyethylene [101)] and of Higashimura on the I_2-initiated polymerization of vinyl ethers, p-methoxystyrene (Sect. 15.3.1.1.3) and of N-vinyl carbazole [102)] approach living features. It is possible, however, to prepare a vinyl polymer with an end-group that is inactive towards its own monomer but is able to convert a heterocyclic monomer into the onium ion, in this way starting the polymerization of a second monomer. The application of this approach for the synthesis of block copolymers is described below.

15.3.1.1 Polymerization of Heterocyclic Monomers by Living or Living-dormant Polymers Derived from Vinyl Monomers

15.3.1.1.1 Copolymers of Styrene with Aziridines

Polymerization of styrene in the presence of $HClO_4$ at ambient temperatures proceeds with considerable transfer i.e., several tens of macromolecules are formed from one initiator molecule [103)]. Therefore, there remains a small proportion of chains with reactive end-groups.

At a sufficiently low temperature (e.g. —78 °C) polymerization proceeds differently: after an initial fast period to limited conversions, the polymerization proceeds slowly (~2 hrs) to completion. This behaviour was explained assuming that polystyryl cations are the growing species during the initial fast period of polymerization. Combination with the perchlorate anions converts the macrocations to polystyryl esters that are unable to further react with styrene; in line with Szwarc we may call these species "living-dormant" [99]. Slow polymerization during the second stage may be attributed to minute amounts of reactive styryl carbenium ions in equilibrium with nonreactive macroester, or to the slow deprotonation of the macroester leading to initiation of new chains by proton transfer. Thus, the large majority of species is dormant (toward styrene).

Styrene (2–5 vol% in CH_2Cl_2) was polymerized with $HClO_4$ (0.01 mole · l^{-1}) and the system was kept at −78 °C a few minutes. Then, the second monomer 1-*t*-butyl aziridine (TBA) or 1(2-phenylethyl)-2-methyl aziridine (PEMA) was added to the solution [104]. The mixture was stored at −78 °C for several hours and then warmed up to room temperature. A sample of polystyrene solution was taken before the addition of aziridine and its $\overline{M}_n$ was determined. A block copolymer was isolated and its $\overline{M}_n$ was determined separately. GPC traces of both products showed unimodal molecular weight distributions. The GPC trace of the copolymer indicated a considerable molecular weight increase, as compared to the homopolymer. Two block copolymers were obtained with $\overline{M}_n$ = 4,100 (starting from polystyrene of $\overline{M}_n$ = 800) and $\overline{M}_n$ = 7,100 (from polystyrene of $\overline{M}_n$ = 1,800).

Copolymers with high (e.g., 88%) TBA content were water soluble and homopolystyrene could not be extracted by CCl_4 or CH_2Cl_2.

When the polymerization mixture was heated for a few minutes after the addition of TBA, a bimodal molecular weight distribution was obtained. Part of the polystyrene remained unreacted probably as a result of incomplete initiation of aziridine. The polystyryl ester may decompose e.g. by proton transfer before it can initiate TBA polymerization. Copolymer formation may proceed by the following route:

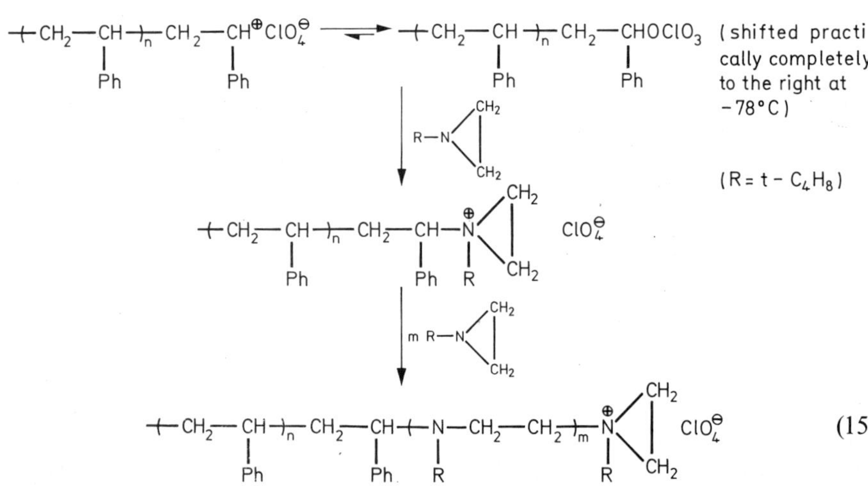

(1565)

15.3.1.1.2 Copolymers of Styrene with Oxazolines

Oxazoline is a relatively strong nucleophile and therefore it can be polymerized in the presence of such nucleophilic counterions as iodide or bromide anions. Thus, the synthesis of block copolymers of styrene (St) and 2-methyl-2-oxazoline (MOXL) was attempted by a method similar to that described in the previous section [105].

Polymerization of styrene was initiated by iodine and the resulting polySt contained about 10% iodine. This macromolecular iodide was used to initiate the polymerization of MOXL:

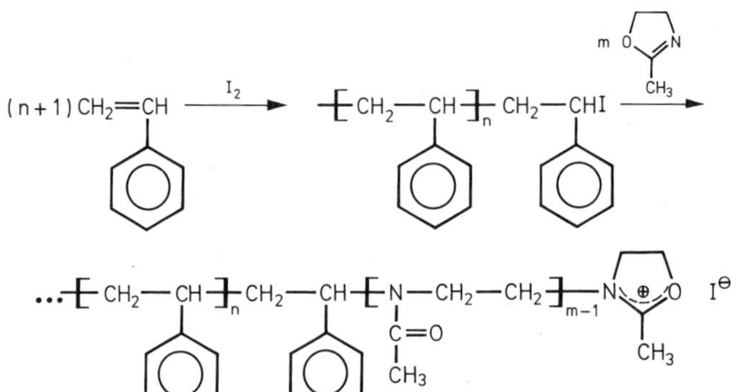

CHI
(15-66)

After the addition of MOXL the solution in DMF was refluxed for 36 hrs, and the resulting polymer was isolated and purified by precipitation from chloroform to diethyl ether. According to ^{1}H-NMR spectroscopy the product contained 66—83 mol % of units derived from MOXL. The claim that block copolymer was formed was based on solubility data. Recently, similar results have been reported for the styrene oxide/2-methyl-2-oxyzoline system [106].

15.3.1.1.3˙ Copolymers of p-Methoxystyrene with 1,3-Dioxolane

When *p*-methoxystyrene (*p*MOSt) is polymerized in the presence of iodine (I_2), iodoalkyl groups are formed and termination via proton transfer is not observed [107]:

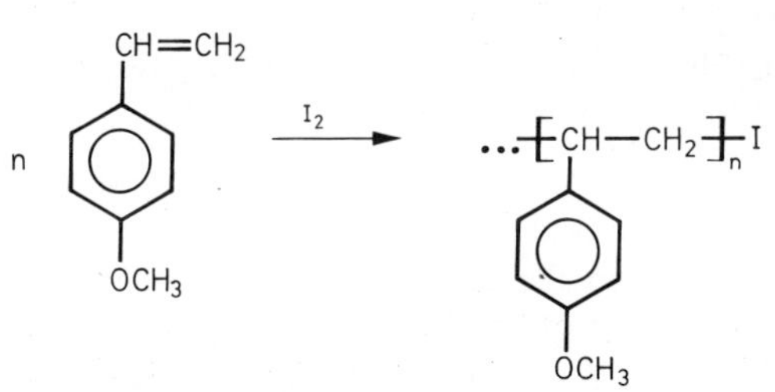

(15-67)

Poly*p*MOSt prepared in this way was used to initiate the polymerization of 1,3-dioxolane (DXL) [108]. A typical procedure was as follows: *p*MOSt (0.5 mol · l^{-1}) was polymerized with iodine (10 mmole · l^{-1}) initiator at 0 °C or —15 °C in CCl_4. After completion of *p*MOSt polymerization, DXL containing a dissolved neutral salt of low nucleophilic anion (such as $AgSbF_6$, $AgBF_4$ or Bu_4NClO_4) was added. After 64 hrs the polymerization was terminated and the polymer was precipitated in methanol. The conversion of DXL was unexpectedly high, 90%, in view of the equilibrium character of DXL polymerization ($[M]_e$ of DXL is $\approx$1.0 mole · l^{-1} at 0 °C).

The product was separated into methanol-soluble and methanol-insoluble fractions. The soluble fraction $\overline{M}_n$ = 5,000–9,000 by GPC with polystyrene standards was polyDXL (no absorption at 254 nm in UV). The insoluble part of $\overline{M}_n = 20{,}000$ exhibited unimodal MWD, however, it was broader and appeared at lower elution volumes than the original poly*p*MOSt. DSC analysis showed peaks at 30 °C and 60 °C which are also observed for polyDXL and are due to the complex morphology of the crystalline phase.

On the basis of these observations, block copolymer formation was claimed. It was reported, however, that some grafting also occurs when DXL is polymerized in the presence of preformed poly*p*MOSt. Thus, when dead poly*p*MOSt was added to the solution of living polyDXL, up to 70% of polyDXL was grafted by aromatic alkylation onto poly*p*MOSt:

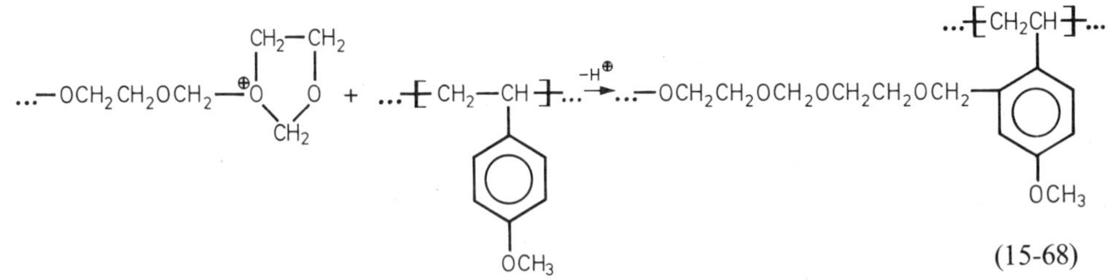

(15-68)

There is, however, still some doubt whether the alleged block copolymers are indeed copolymers and not predominantly a mixture of homopolymers containing some graft and block copolymers. The solubility criterion was mostly relied upon, and it was shown that 40% of DXL entered the copolymer (60% was isolated as methanol-soluble low-molecular-weight polyDXL). This criterion may be misleading as we already stressed, because, except for the very low-molecular-weight fraction, polyDXL is insoluble in methanol. Thus, the question whether or not block copolymers are formed in this system and if so, what is the efficiency of this reaction is still unresolved.

15.3.1.2 Polymerization of Heterocyclic Monomers by Living Polymers Derived from Heterocyclic Monomers

The synthesis of block copolymers by sequential polymerization requires a living polymer prepared from the less nucleophilic monomer (first block) and the addition of a more nucleophilic monomer to the active species located on this first block. The general order of nucleophilicities of heterocyclic monomers is as follows:

Siloxanes ≈ orthoesters < acetals < ethers < sulfides < oxazolines < amines. Depending on substitution and ring strain some changes may occur in these positions.

The choice of monomers is limited not only by the order of nucleophilicity. The first block should carry a living end capable to initiate the polymerization of the second monomer by addition. Up to now, only a few systems are known to meet these requirements. Thus, in the presence of stable counterions, living polymerization of DXL [109], DXP [109], THF [110,111,112], N-t-butylaziridine [113], 1,3,3-trimethylazetidine [114], conidine [115] and 1-(2-phenylethyl)-2-methylaziridine [116] may proceed. The last four monomers are highly nucleophilic cyclic amines and there is not too much chance that the active species derived from this class of monomers would initiate polymerization of other, less nucleophilic monomers. Thus, attempts directed toward the preparation of block copolymers of cyclic ethers or acetals were employed for the synthesis of the first block (living polymer) from which the second monomer was polymerized.

15.3.1.2.1 Copolymerization of Tetrahydrofuran with 3,3-Bis(Chloromethyl)Oxetane

The polymerization of 3,3-bis(chloromethyl)oxetane (BCMO) initiated with living polytetrahydrofuran (polyTHF) was studied by Saegusa [117]. In this system, due to the high equilibrium concentration of THF, some monomer remains in equilibrium with the polymer after the first stage is completed (i.e. at the polyTHF $\rightleftharpoons$ THF equilibrium). After addition of the second monomer, the remaining THF may participate in copolymerization with added BCMO (cf. Sect. 15.2.2.1., copolymerization above T_c), leading to random copolymer. Only after the complete consumption of THF the second monomer may form the required homoblock. Thus, the two homoblocks are separated by a third random BCMO/THF copolymer block:

$$[\mathrm{THF}]_m[(\mathrm{THF})_x(\mathrm{BCMO})_y]_n[\mathrm{BCMO}]_p \qquad (15\text{-}69)$$

Solubility data, and IR and DSC analysis, indicated triblock copolymer formation.

When the second stage leading to statistical copolymer was suppressed by lowering the temperature (and thus decreasing $[\mathrm{THF}]_e$), a copolymer whose structure was close to AB (diblock) was obtained.

15.3.1.2.2 Copolymers Based on Living PolyDXL

Yamashita et al. claimed the preparation of the following block copolymers, with living polyDXL as the first block [118]:

[homopolyDXL] [DXL-St random copolymer]

[homopolyDXL] [DXL-BCMO random copolymer]

[DXL-St random copolymer] [DXL-St-BCMO random copolymer]

Copolymerizations were carried out at ambient temperatures in CH_2Cl_2 solvent using $(C_2H_5)_3O^{\oplus}BF_4^{\ominus}$ initiator.

In the first copolymer the styrene content was very low (2.7%).

Addition of BCMO to the living polyDXL led to polymer which was separated into two fractions by boiling chloroform. The soluble part had a low chlorine content (2.2%) and an its molecular weight ($\overline{M}_n$ = 6100 vs. 6300) was that of the original polyDXL). Thus, this fraction consisted mainly of unreacted polyDXL. The insoluble fraction (36% of the total yield) contained 45% BCMO units.

A benzene soluble terpolymer was prepared in 36% yield with $\overline{M}_n$ = 5,800; it contained 86% DXL, 3% St and 11% BCMO units.

Only very rough characteristics of the polymers were given and the structures were not confirmed by extended solubility studies or GPC analysis.

15.3.1.2.3 Block Copolymers of Siloxanes

Kučera et al. reported the synthesis of the dicationically growing polyTHF from the "disilicenium cation" [119]. The exact structure of this initiator prepared from dimethylchlorosilane and $HClO_4$ is not yet known. The authors proposed two different ionic species, namely silicenium or oxonium cations in two consecutive papers [119, 143]. Thus:

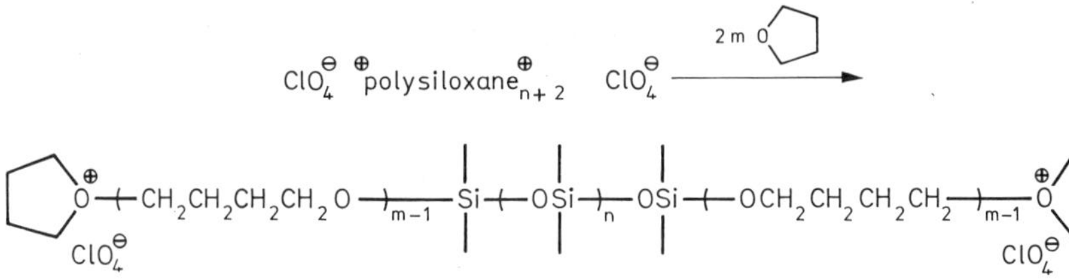

(15-70)

Remembering the order of basicity and reactivity given at the beginning of this section this experiment indicates that the synthesis of block copolymers from siloxanes and a variety of heterocyclic monomers may be possible. Recently, however, it was found that the efficiency of coupling between copolymer [Scheme (15-70)] and dianionically living polydimethylsiloxane is low. This was explained by the presence of large proportions of cycles in the alleged copolymer [120].

15.3.1.2.4 Other Systems

Preliminary results indicate the possibility of preparing ABA block copolymers containing polyacetal (as a middle block) and polyamine blocks [121]. When N-*t*-butylaziridine or 2-phenyl-2-oxazoline are added to a solution of living polyDXP, further polymerization ensues and the products have considerably higher molecular weights than the original polyDXP. NMR analysis confirmed the block character of the product.

Litt claimed formation of block copolymers of 2-lauroyl-2-oxazoline and 2-methyl-2-oxazine by sequential polymerization [122]. Methyl tosylate was the initiator and the polymerization was terminated by water. The proposed structure of the copolymer is:

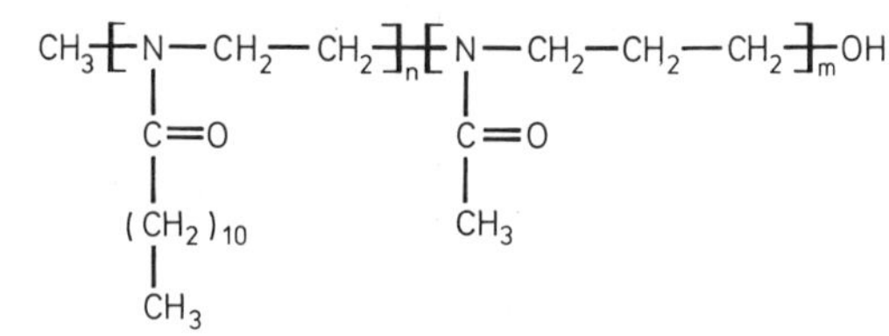

(15-71)

Lambert and Goethals attempted to prepare block copolymers of tetrahydrofuran (THF) with propylene sulfide, thiethane, 3,3-dimethylthiethane, aziridine and 1-(2-phenylethyl)-2-methyl-aziridine [123]. None of these systems gave pure block copolymers. Except with aziridine, mixtures of block copolymers and polyTHF was found from solubility data. With unsubstituted aziridine transfer led to a mixture of homopolymers:

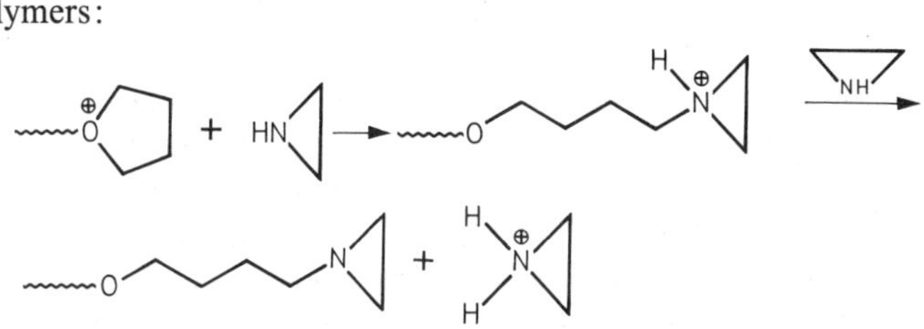

(15-72)

THF polymerization was initiated with triethyloxonium tetrafluoroborate. $BF_4^{\ominus}$ anions are known to decompose during THF polymerization, producing "dead" polyTHF carrying an alkyl fluoride end-group. Another reason for the formation of polyTHF may be chain transfer to polymer during THF polymerization, yielding dead blocks and dicationically living chains:

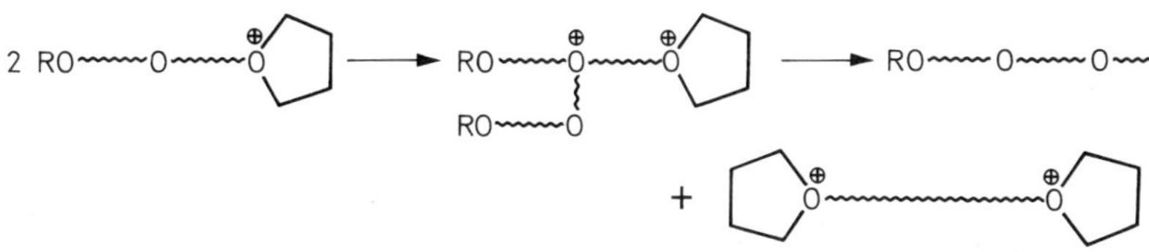

(15-73)

To prevent the formation of dead chains, initiators giving bidirectional growth have to be used [e.g. $(CF_3SO_2)_2O$]; then the chains will possess cations on both ends. This method has recently been used by Goethals and Tezuka for preparing dicationically living poly(*t*-butyl aziridine) [124].

15.3.1.2.5 Attempted Synthesis of Polyacetal-polyacetal Block Copolymers

After having established the living character of cyclic acetal polymerization (cf. Adv. Polymer Sci. *37* Sect. 4.1), the synthesis of block copolymers with two polyacetals blocks was attempted [109]. The polymers are living in the sense that the concentration of active species remains constant when equilibrium between the living polymer and monomer is established; however, at this stage transacetalization takes place, which, changes $\overline{M}_w$ but not $\overline{M}_n$. DXL was polymerized from living polyDXP and vice versa with $(CF_3SO_2)_2O$ initiator in CH_2Cl_2 solution at low temperature (—78 °C). Both monomers under these conditions give polymers whose $\overline{DP}_n$s are close to theoretical values calculated from $([M]_0 - [M]_e)/[I]_0$. The molecular weight of the final product has also been found to be equal to the sum of calculated molecular weights of the two blocks (Table 15.9).

The microstructure of the copolymers was studied by ^{13}C-NMR, spectroscopy. Instead of two signals for the OCH_2O group in two different homodyads:

DXL-DXL
$-OCH_2CH_2O\underline{\underline{CH_2}}OCH_2CH_2OCH_2-$ δ 95.36

DXP-DXP
$-OCH_2CH_2CH_2CH_2O\underline{\underline{CH_2}}OCH_2CH_2CH_2CH_2OCH_2$ δ 95.76

(15-74)

three signals were observed. The additional signal had a chemical-shift average between the homodyad signals and was attributed to DXL-DXP heterodyads. As judged by the intensity of the signals, the concentration of heterodyads was comparable to that of homodyads. The proportion of heterodyads increased with increasing polymerization time in the second stage, i.e. the stage in which the second block was supposed to form (in the presence of the preformed first block). For longer reaction

Table 15.9. Calculated and measured number average molecular weight of polymers isolated after first and second stage of sequential polymerization of 1,3-dioxolane and 1,3-dioxepane initiated by triflic anhydride. $[M]_0$ (A and B): from 2 to 3 mol · l^{-1}, $[I]_0$: from 10^{-3} to 10^{-2} mol · l^{-1} [109]

I stage B = 1,3-dioxolane				II stage A = 1,3-dioxepane			
T °C	Time in h	$\overline{M}_n$[a] (calcd)	$\overline{M}_n$ (osm.)	T °C	Time in min	$\overline{M}_n$[b] (calcd)	$\overline{M}_n$ (osm.)
−60	3	19350	18250	−60	45	38500	40000
−60	5.5	21500	24800	−60	15	48600	46500
−50	18	21300	21100	−50	15	44200	43300
−60	24	35000	44500	−60	48.60	85200	73000
B = 1,3-dioxepane				A = 1,3-dioxolane			
−70	2	19900	—	−70	15	33900	36700
−70	2	38400	—	−70	60	77800	74400
−70	2	40000	—	−70	2.60	78600	58600

[a] $\overline{DP}_n = ([M]_0 - [M]_e)/[I]_0(B)$;
[b] $\overline{DP}_n = ([M]_0 - [M]_e)/[I]_0(B) + ([M]_0 - [M]_e)/[I]'_0(A)$; $[I]'_0$-correlated for the volume change due to the addition of monomer A solution

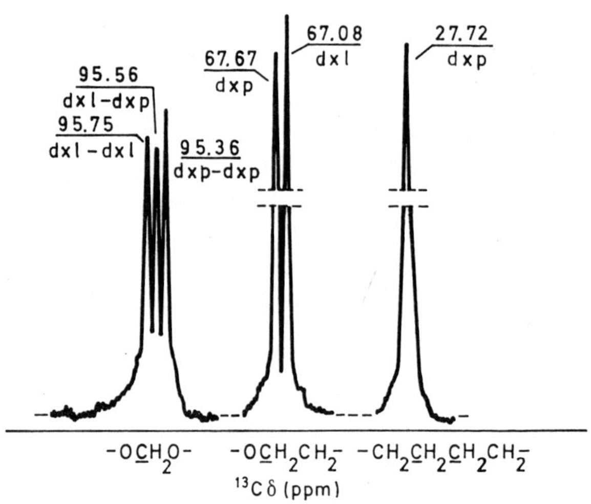

Fig. 15.6. ^{13}C-NMR spectrum of copolymer of 1,3-dioxepane with 1,3-dioxolane recorded in $CHCl_3$ solvent at 25° (1st experiment from Table 15.9) [109]

times (e.g. 24 hrs at −78 °C) the ratio of hetero- to homodyads is 2:1:1, which indicates that the copolymer is completely random. Apparently, when both blocks are polyacetals, transacetalization proceeds paralel to polymerization and prevents block copolymer formation. Reshuffling, transforms the original block copolymer first into multiblocks and later to completely random structures. This process proceeds simultaneously and at a rate comparable to that of propagation of the second monomer forming the second block [15, 125]. Therefore, pure block copolymer cannot be isolated at any stage of polymerization.

15.3.1.3 Block Copolymerization of Vinyl Monomers from Living Polyacetals

15.3.1.3.1 Copolymers of 1,3-Dioxepane and 1,2-Dimethoxyethylene

As described in the previous section, under suitable reaction conditions one can prepare living polyacetals (poly-1,3-dioxolane and poly-1,3-dioxepane) in almost quantitative yields having well defined molecular weights ($\overline{DP}_n = ([M]_0 - [M]_e)/[I]_0$), and a low content of cyclic fraction (a few percent).

Living bifunctional poly-1,3-dioxepane (polyDXP) was used as initiator for the polymerization of 1,2-dimethoxyethylene (DME) to prepare ABA-type block copolymers [121]:

$$^{\oplus}\text{polyDXP}^{\oplus} + \text{n DME} \rightarrow \text{(polyDME)-(polyDXP)-(polyDME)} \quad (15\text{-}75)$$

In the first stage, DXP was polymerized with $(CF_3SO_2)_2O$ initiator in CH_2Cl_2 solvent at −78 °C. After the completion of DXP polymerization, a sample was taken for $\overline{DP}_n$ determination of polyDXP and then a solution of DME was added. The system was kept at −78 °C for several hours and then the polymerization was terminated by sodium methanolate. The precipitated polymer was isolated, dried and characterized (yield, $\overline{DP}_n$, NMR). Results of typical experiments are given in Table 15.10. The molecular weights were close to the values calculated by assuming that each dicationic living polyDXP initiated DME polymerization.

GPC analysis showed that the precipitated copolymer had a unimodal molecular weight distribution, and the corresponding peak has shifted towards higher molecular weights as compared to the starting polyDXP. GPC analysis also showed that prior to precipitation with methanol, the copolymer contained some low-molecular-weight

Table 15.10. Polymerization of 1,2-dimethoxyethylene (DME) initiated with living dicationic poly-1,3-dioxepane (DXP) −78 °C, CH_2Cl_2

$[DXP]_0$ mol · l^{-1}	$[TFAn]_0$ mol · l^{-1}	Time hrs	Conversion (% of teor.)	$\overline{M}_n$ calcd	$\overline{M}_n$ measd
1.1	$2.55 \cdot 10^{-3}$	3.0	75.5 (110)	30200	37700
1.03	$4.40 \cdot 10^{-3}$	4.5	73 (110)	17480	21
0.85	$2.2 \cdot 10^{-3}$	1.7	60 (100)	21600	20900
1.14	$1.2 \cdot 10^{-3}$	2.5	63 (91)	65700	59600

$[DME]_0$ mol · l^{-1}		Time hrs	Conversion %	$\overline{M}_n$ calcd	$\overline{M}_n$ measd
1.82		1	50.5	73100	77500
1.02		2.25	56	41500	40400
0.76		5	70	43300	43100
1.74		3.0	63	160800	154200

fraction ($\overline{M}_n \simeq 1{,}000$ based on the universal calibration curve). According to ^{1}H-NMR spectroscopy this oligomer is rich in DME units (>90%).

Thus, it seems that polymerization of DME during the second stage is accompanied by some transfer, and in addition to copolymer a certain amount of low-molecular-weight products (probably DME homopolymer) are also formed.

In order to gain additional proof that ABA block copolymer has formed, degradation of the middle polyacetal block was attempted. This was done by heating a toluene solution of the polymer for 24 hrs at 60–80 °C in the presence of a catalytic amount (1 mol %) of triflic acid and subsequent evaporation of volatiles under vacuum:

$$[\text{DME}]_n - [\text{DXP}]_m - [\text{DME}]_n \xrightarrow[60-80^\circ\text{C}]{\text{HOSO}_2\text{CF}_3} 2[\text{DME}]_n + m\text{DXP} \tag{15-76}$$

According to its ^{1}H-NMR spectrum the polymer was practically pure polyDME (less than 5% of DXP units). Results of typical decomposition experiments are given in Table 15.11.

Table 15.11. Decomposition of polyDXP blocks [121]

Weight of sample g	Weight of recovered polymer g	wt % of recovered polymer	wt % of DME units in original sample by ^{1}H-NMR	$\overline{M}_n$ of original polymer	$\overline{M}_n$ of recovered polymer (polyDME)	$\overline{M}_n$ of polyDME block calcd for ideal ABA copolymer
0.338	0.200	59	60	65500	17050	13750
0.604	0.378	63	51	40400	10600	9300

Thus, it may be concluded that under the above conditions, polymerization of DME, initiated with living, bifunctional polyDXP, leads to the formation of ABA block copolymers with relatively high yields (50–60%). In addition to copolymer, low-molecular-weight polyDME is also formed, but it can be easily removed from copolymer by precipitation with methanol.

15.3.2 Star-shaped and Triblock Copolymers

Star-shaped polymers may be formed by coupling of living chains with multifunctional terminating agents or by using multifunctional initiators. The first approach is limited by the usually high viscosity of polymer solutions which lowers the efficiency of coupling of several chains by one multifunctional coupling agent. Hence, this method is mainly used for coupling of low-molecular-weight polymers.

A better approach is the use of multifunctional initiators. Application of some bifunctional initiators such as anhydrides of strong protonic acids, e.g. $(FSO_2)_2O$, $(ClSO_2)_2O$ or $(CF_3SO_2)_2O$, already have been described in Chapter 6.

Other bifunctional initiators are the di-1,3-dioxolan-2-ylium cation [127] dicationic living polysiloxane [119] and ditetrahydrofuranium cations [128].

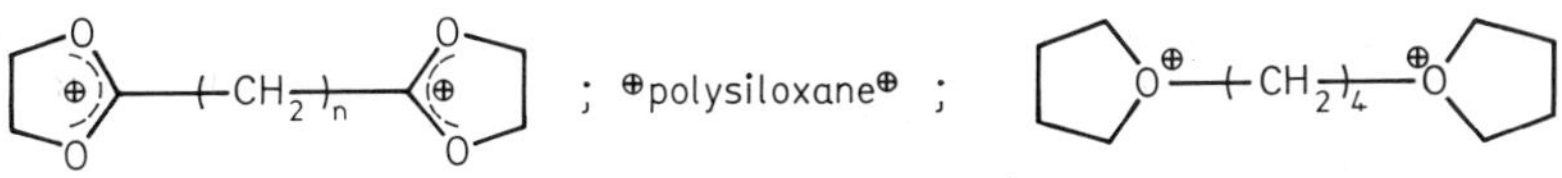

(15-77)

The easiest to prepare are acylium cations that may be obtained by the following methods [129, 130]:

a. acyl halides + Lewis acids

b. acyl halides + silver salts

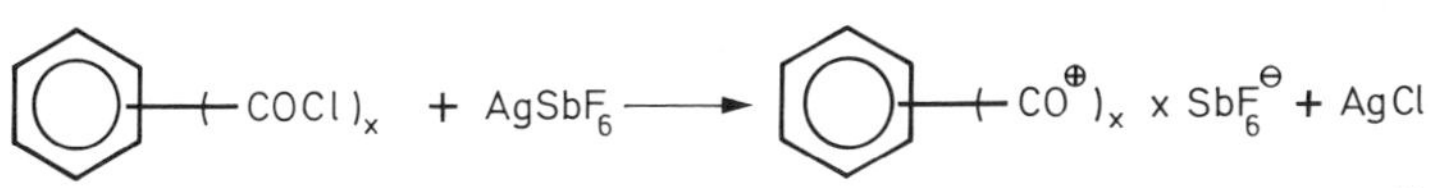

(15-78)

c. acyl chlorides + nitrosyl salts [131]:

$$R(COCl)_x + x\,NO^{\oplus}PF_6^{\ominus} \longrightarrow R(CO^{\oplus}PF_6^{\ominus})_x + x\,NOCl\uparrow$$

Method *a* is a direct synthesis and does not lead to by-products. The disadvantage is the necessity of using acyl fluorides instead or the more easily available chlorides. This, however, may be avoided, at least for some applications, by using the corresponding chloride and preparing the salt having mixed counterions.

$$RCOCl + SbF_5 \rightarrow RCO^{\oplus}SbF_5Cl^{\ominus} \qquad (15\text{-}79)$$

Method *b* employs acyl chloride, but the quantitative removal of AgCl is very difficult. Method *c* does not suffer this disadvantage, because the gaseous NOCl can easily be removed from the reaction medium; however, the starting material, i.e. the nitrosyl salt, may itself initiate the polymerization (e.g. polymerization of THF) [131]. Thus, this method cannot be used for the "in situ" formation of acyl cations, which is a frequent practice.

Aryl acylium salts with stable counterions are perfectly stable, when handled properly (vacuum). Aliphatic acylium salts (acetylium, propionylium, adipoylium) are less stable and are more reactive. In general, this class of compounds has advantages for a number of applications.

Trifunctional acylium salts were used as initiators of THF polymerization:

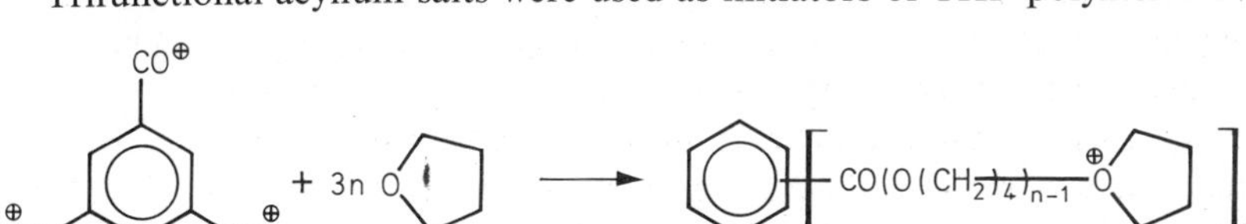

(15-80)

Table 15.12. Polymerization of THF initiated by different oxocarbenium salts: kinetics and number-average degrees of polymerization [130]

Initiator	Kinetics (25 °C) r^a		$\overline{DP}_n$ Calculated	Found, by: Head-group (initiation)		end-group (deactivation)		Osmometry
	bulk	CH_2Cl_2		UV	IR	chemical	UV	
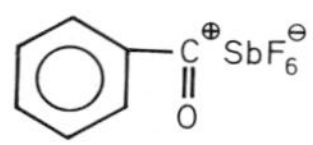	1.03		175	170	170	150		185
	1.00		370	360	355	330		365
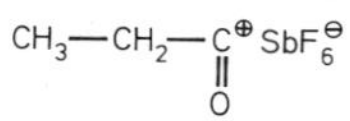	1.03		530			660		538
		1.00	360				375	370
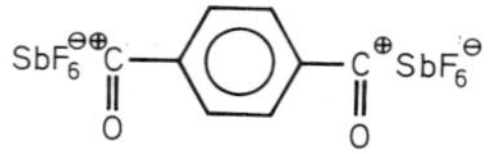	0.93		355	319		331		340
		0.96	255	230		266		245
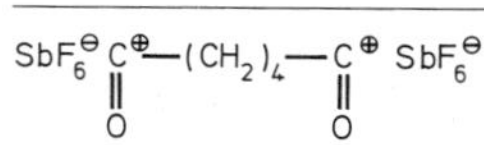	1.00		355				347	370
		1.00	230				215	240
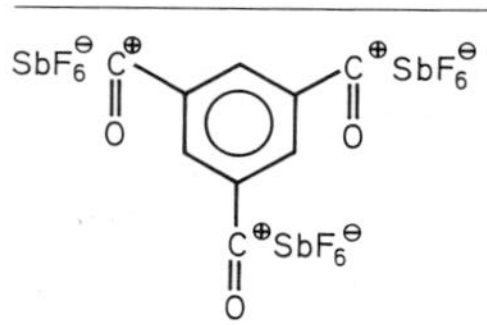		0.93	660			625		630
		0.93	366			348		340

[a] r is the efficiency of initiation, determined by comparing the measured rate constant of propagation (k_p) with k_p known for identical conditions

As shown by the data collected in Table 15.12, there is a very good agreement between the calculated ($\overline{DP}_n = ([M]_0 - [M]_e/[I]_0)$) and measured molecular weights. Also kinetic determinations of the propagation rate constants indicate that all three groups participate in the formation of growing chains with an efficiency close to 1. After hydrolysis, i.e., after breaking the link between initiator fragment and polymer chains, the molecular weights of polymers prepared with bifunctional terephthaloyl salts drop by 50%. For polymers prepared with trifunctional trimesoyl salts, the molecular weights decreased three fold ($\overline{M}_n$ dropped from 45,400 to 18,000).

Living poly(1-*t*-butylaziridine, polyTBA) with $\overline{M}_n = 3{,}000$ was coupled with terephthalic and pyromellitic acids [113] and polymers having unimodal molecular weight distributions and $\overline{M}_n = 6{,}000$ $(2 \times 3{,}000)$ and $\overline{M}_n = 11{,}600$ $(4 \times 3{,}000)$, respectively, were isolated. The following structure was confirmed by ^{1}H-NMR (300 MHz) spectroscopy for the latter case:

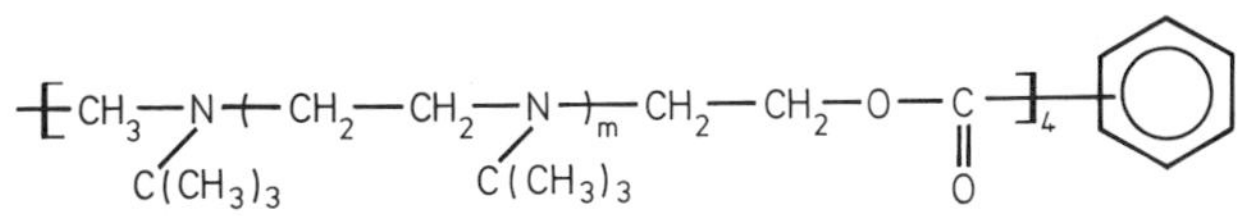

(15-81)

Living polyTBA$^\oplus$ with $\overline{M}_n = 5{,}500$ formed a star-shaped polymer containing up to five branches upon coupling with diethylenetriamine [113, 126]:

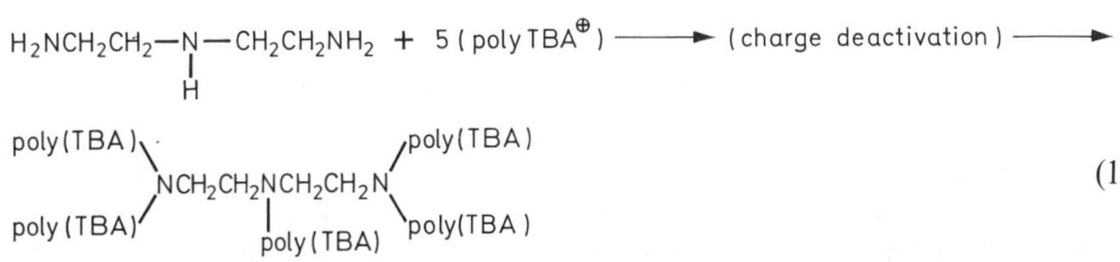

(15-82)

The product had a unimodal molecular weight distribution and $\overline{M}_n = 19{,}500$, indicating that some polymer with less than five branches has also formed. (The formation of a structure with five branches was confirmed, however, by ^{1}H-NMR spectroscopy).

15.3.3 Formation of Block Copolymers Starting from Living Anionic Polymers

Anionic polymerization frequently has been used to prepare well-defined living polymers such as polystyrene, poly(α-methylstyrene), polydienes, which may be transformed by two methods into block copolymers with cationically polymerizable monomers. When a living anionic polymer is mixed with a stoichiometric amount of a living cationic polymer the cationic and anionic species may couple. For example, anionic living polystyrene (St) or poly (α-methylstyrene) (MSt) were reacted with living cationic polytetrahydrofuran (THF). In the latter system the coupling efficiency was low, probably because of proton or hydride transfer [132]:

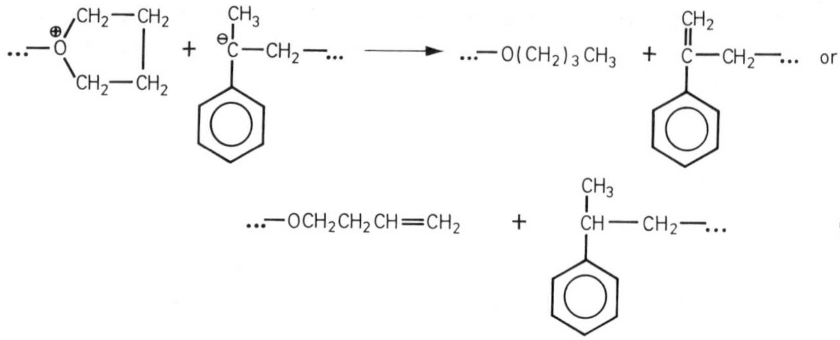

(15-83)

On the other hand, coupling efficiency between living polySt and living polyTHF was nearly quantitative [133]. This reaction was originally used by Vofsi and Levy [134] but the products obtained were insoluble. Later, Richards et al. under the same conditions prepared soluble multiblock copolymers, having the approximate structure [133]:

$$\left[(\mathrm{THF})_{170}\text{-}(\mathrm{St})_{10}\right]_{11} \tag{15-84}$$

Yamashita et al. modified the polystyryl end-groups by introducing carboxylate anions (by reaction with CO_2) [135]. After coupling this modified polySt with a living polyTHF and removing polyTHF by extraction with hot isopropanol, a solid copolymer, soluble in common organic solvents and having the structure:

$$\left[(\mathrm{THF})_{110}\text{-}(\mathrm{St})_{53}\right]_{30} \tag{15-85}$$

was obtained.

Modification of the polystyryl anion was also necessary for the preparation of block copolymers with living cationic poly(1-*t*-butylaziridine). In the absence of modification only a mixture of homopolymers was obtained. Thus, the carbanion was first converted with propylene sulfide to the thiolate anion [126] which was then coupled with the living aziridinium cation:

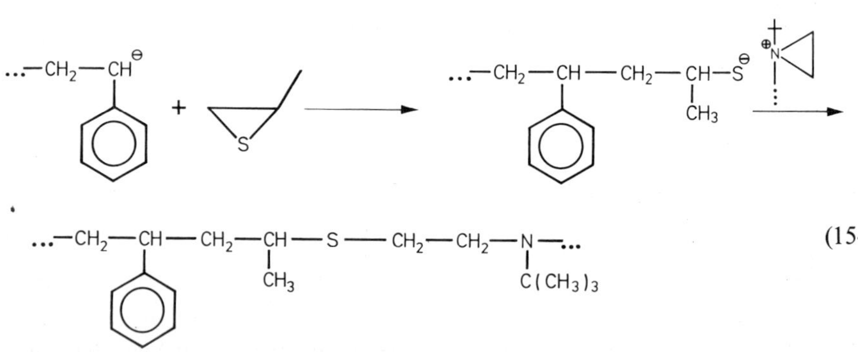

(15-86)

Using the same approach, ABA block copolymers (with TBA segments forming block A) were prepared using bifunctional telechelic amino- or carboxy-terminated polybutadiene or poly(butadiene-co-acrylonitrile). The latters are commercially available as oligomers (viscous liquids). The ABA block copolymers (whose block structure was proven by GPC and NMR analysis), were waxy or rubbery substances, depending on the length of the blocks [126].

Another method for preparing block copolymers is by the quantitative transformation of anionic species into groups able to initiate cationic polymerization. The possibility of forming cationic initiators of compounds having "active" halogen atoms by the use of silver salts (e.g. $AgSbF_6$) already has been described in Section 15.3.2. Direct bromination of living anionic polySt yields the desired starting material which is, however, contaminated by unreactive polymer with double molecular weights formed by Würtz condensation [136]:

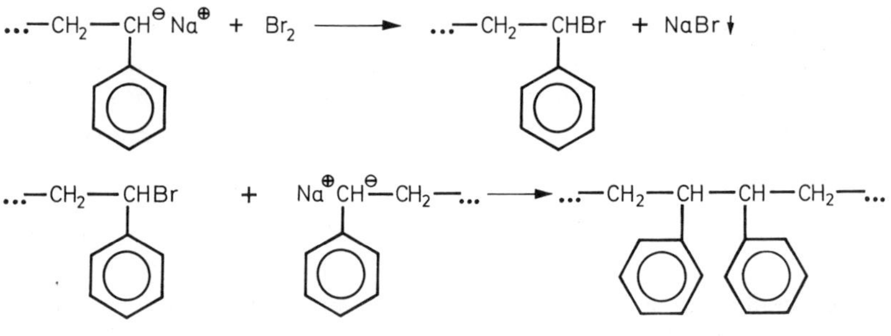

(15-87)

Coupling may be effectively reduced by first producing a Grignard reagent with anhydrous $MgBr_2$, and then treating it with a tenfold excess of Br_2 [137]:

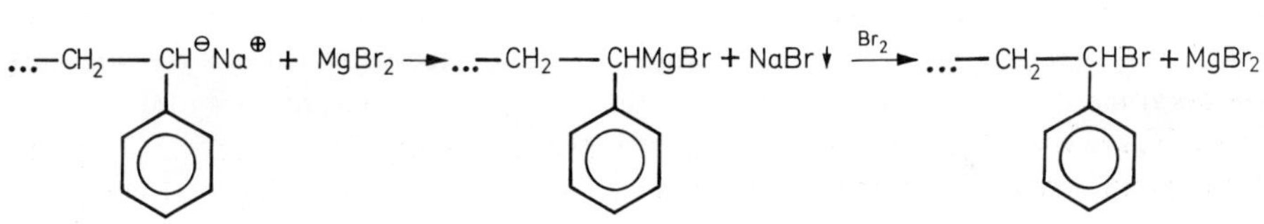

(15-88)

Reaction of brominated polySt with $AgPF_6$ or $AgClO_4$ in THF solution (*in situ* formation of the initiating cation) led to a mixture of block copolymer and both homopolymers. Homopolymers may have been formed by transfer i.e., protons elimination from the β-position [138].

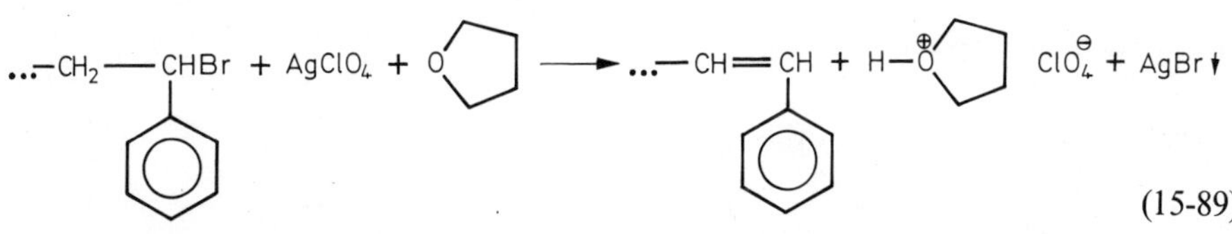

(15-89)

This undesirable reaction may be to some extent suppressed by lowering the temperature (<—10 °C). Another method calls for the use of xylylene dibromide as the terminating agent instead of bromine. The absence of hydrogen atoms in β-position eliminates the possibility of transfer [139]:

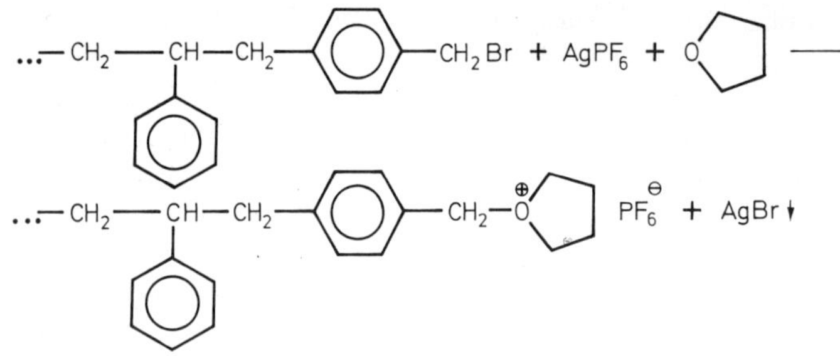

(15-90)

In the above discussed papers monofunctional polystyrenes with $\overline{M}_n = 2{,}000$ to 4,000 were used. The block copolymers had $\overline{M}_n = 12{,}000$ to 100,000. The molecular weight distribution was quite narrow for directly brominated polystyrenes ($\overline{M}_w/\overline{M}_n = 1.04$) and was slightly broader for xylylene dibromide modified polystyrenes ($\overline{M}_w/\overline{M}_n = 1.15$) (at -10 °C with $PF_6^{\ominus}$ counterion). The molecular weight distribution depended on both temperature and counterion [139].

Model experiments showed that initiation is quantitative only in some systems. Thus, benzyl bromide reacts with $AgSbF_6$ with 100% [140] yield, while $AgPF_6$ gives only 75% yield [136]. Mercury-, lead-, ferric halides and aluminium perchlorate were also used and generally gave lower initiation efficiencies and produced some homopolymer. This may be due to direct initiation by the added salts, in agreement with the recent data of Gandini et al., who observed initiation of vinyl and heterocyclic cationic polymerization by aluminium triflate [141].

Another method of preparing polystyrenes with active halogen end-groups (which may subsequently be converted into cationic species), is based on the reaction of living polystyrene with phosgene [142]. Thus, a 20-fold excess (over polystyryl sodium) of phosgene at -70 °C gave polymer with acyl chloride end-groups in 80% yield:

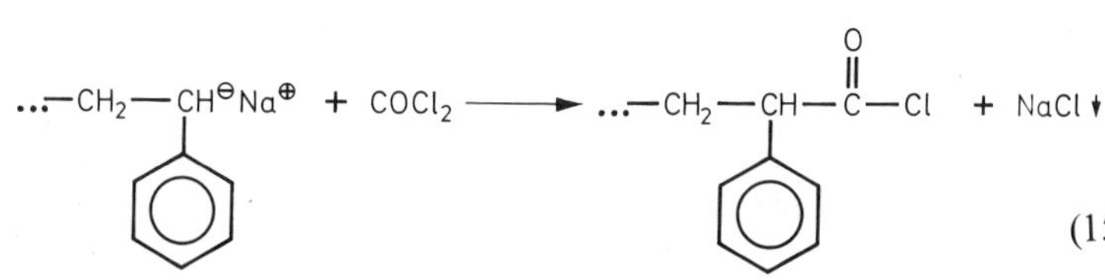

(15-91)

The remaining 20% consist mainly of dimerized polystyrene:

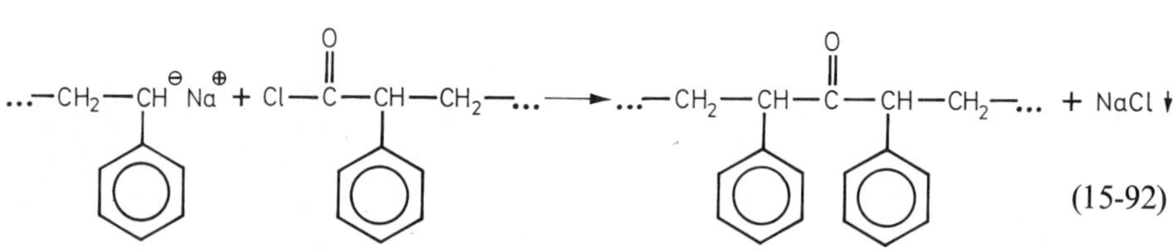

(15-92)

Dianionic living polystyrene reacts in 95% yield with dicationic silicenium initiator at -10 °C, transforming anionic to cationic growing centers [143]. The latter species were used as to initiate the polymerization of THF to form multiblock copolymers (cf. Sect. 15.3.1.2.3):

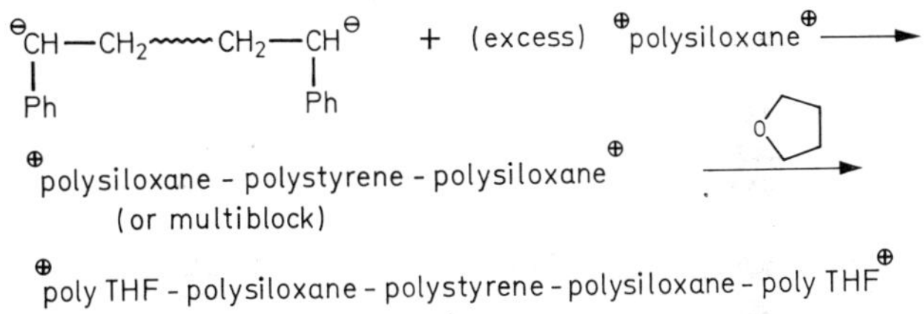

(15-93)

GPC analysis indicates, however, independent growth of homopolyTHF and a low efficiency of multiblock copolymer formation [143)].

The cationic polymerization of strongly nucleophilic monomers can be initiated by certain covalent end-groups (cf. Section 15.3.1.1.2). Thus, commercially available α,ω-polybutadiene glycol ([η] = 0.13 dl/g) was converted by tosyl chloride to polymer containing two tosylate end-groups per macromolecule (according to UV spectra the conversion was 100%). The addition of 2-oxazoline to the solution of this polymer in a DMF-toluene mixture at 100 °C gave an insoluble polymer with 61 mol% oxazoline units. Alkaline hydrolysis gave a chloroform soluble rubbery material with [η] = 0.33 dl/g and having ABA block copolymer structure (cf. Sect. 12.4) [144)].

15.4 Graft Copolymers

Synthesis of graft copolymers resembles that of block copolymers with an increased number of reactive sites per macromolecule. There are two techniques similar to those used in block copolymer preparation, namely grafting from and grafting onto preformed chains. Another method, grafting through, is similar to that used in conventional random copolymerization, where in-chain units function as comonomers.

15.4.1 Grafting from the Chains

Grafting from, exploits the active sites, which exist or can easily be generated on a polymer. Halogenated polymers are most frequently used for this purpose: poly(vinyl) chloride), polychloroprene, chlorinated EPDM, chlorobutyl rubber, bromobutyl rubber, chlorinated polybutadiene, chlorinated butadiene-styrene copolymers etc. [145)].

The grafting process usually requires the generation of active sites by the addition of a coinitiator, e.g. silver salts with anions of low nucleophilicity, such as $AgPF_6$ or $AgClO_4$. Among the grafted monomers are propylene oxide, 7-oxabicyclo-|2.2.1| heptane, styrene oxide, THF, 1,3-dioxolane, ε-caprolactone and thietane. Besides grafts homopolymers of oxiranes and lactones are formed using a PVC backbone [146)]. We have chosen one system from the literature, namely graft copolymerization of THF from polychloroprene to show the principle of this method [147, 148)].

Polychloroprene was reprecipitated three times from toluene, in methanol and dried in vacuum. Then a solution of $AgPF_6$ in CH_2Cl_2 was added to the solution of polychloroprene in THF. AgCl precipitated and the polymerization was terminated with triphenylphosphine or ammonium hydroxide under nitrogen. The copolymers were isolated by evaporating volatiles and drying in vacuum.

Polychloroprene (Neoprene W, $\overline{M}_n = 180{,}000$) was used as the backbone polymer. This polymer contains approximately 1.5% tertiary allylic chlorine atoms:

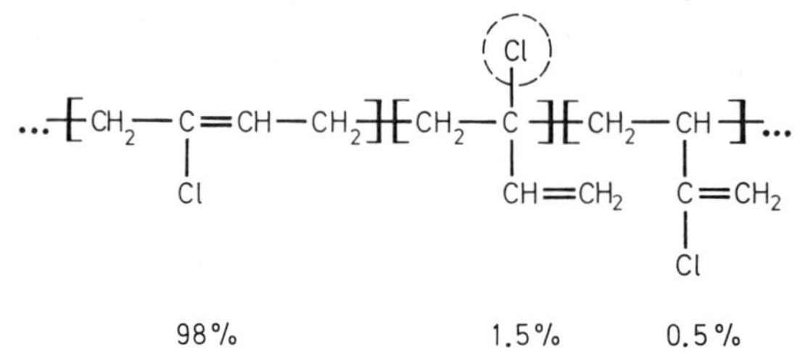

(15-94)

These chlorine atoms were activated by addition of $AgPF_6$ (in excess over the tertiary allylic chlorides) in the presence of THF; this monomer was quantitatively grafted from the neoprene backbone:

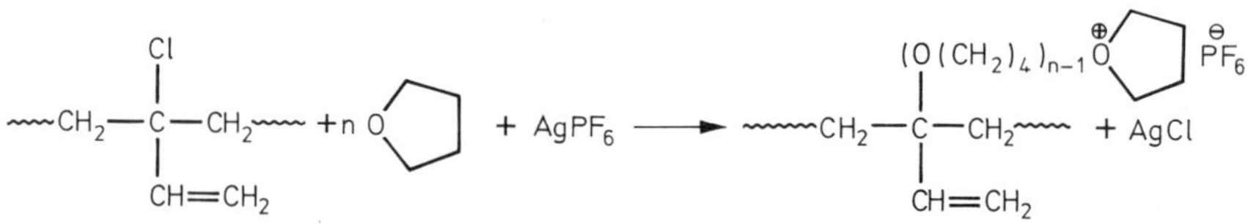

(15-95)

The product was soluble in ethyl acetate, a nonsolvent for neoprene. GPC analysis showed unimodal distribution, indicating the absence of low-molecular-weight polyTHF. The length of branches varied from $\overline{M}_n = 2{,}000$ to 50,000. The overall content of polyTHF in the graft copolymer was found to be from 14% to 90%. The copolymers exhibited one T_g which was between the T_gs of the corresponding homopolymers, and a crystalline melting point slightly lower than the T_m of polyTHF. Tests with uncured specimens indicated a decrease of elongation and an increase of tensile strength with an increase of the length of the polyTHF branches.

Polymerization of strongly nucleophiles heterocycles, e.g. oxazolines or cyclic amines, may proceed in the presence of chloride counterions and may be initiated by active chlorine-containing compounds without the addition of silver salts. In this way 2-methyl-2-oxazoline was grafted on copolymer of 1-chloro-1,3-butadiene with butadiene. Allylic chlorine sites in the starting polymer are reactive enough to initiate the polymerization of 2-methyl-2-oxazoline and thus generate poly(N-acetyl-ethylenimine) (polyoxazoline) grafts:

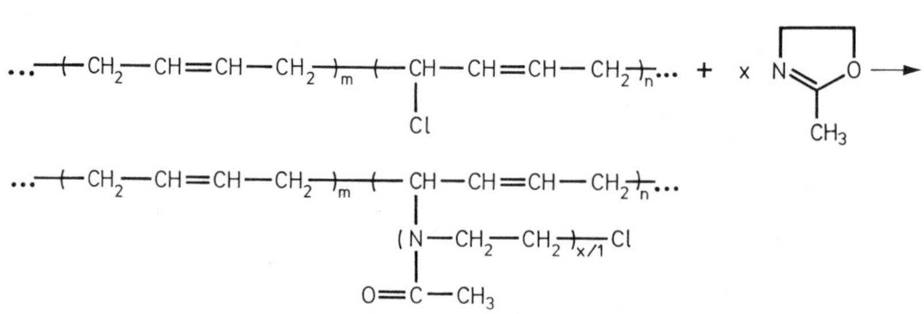

(15-96)

Grafting of 2-methyl-2-oxazoline from chloromethylated polystyrene beads in benzonitrile at 110 °C gave graft copolymers, which were hydrolysed to poly(styrene-*g*-ethyleneimine) and used as chelating agents.

Initiation of THF polymerization by the BF_3-oxirane system was also studied. Since BF_3 alone cannot initiate the polymerization of THF, the epoxide plays the role of a promoter (cf. Adv. Polymer Sci. *37*, Sect. 3.2.9):

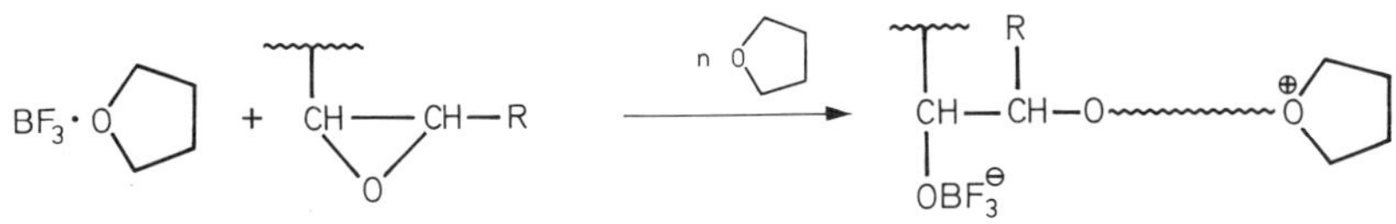

(15-97)

Kennedy and Dreyfuss have used as a grafting source Nordel 1440 epoxide [150]:

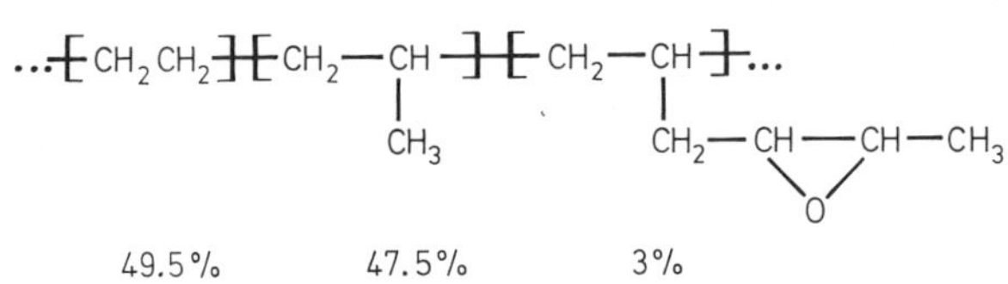

(15-98)

and Butyl LM 430 epoxide

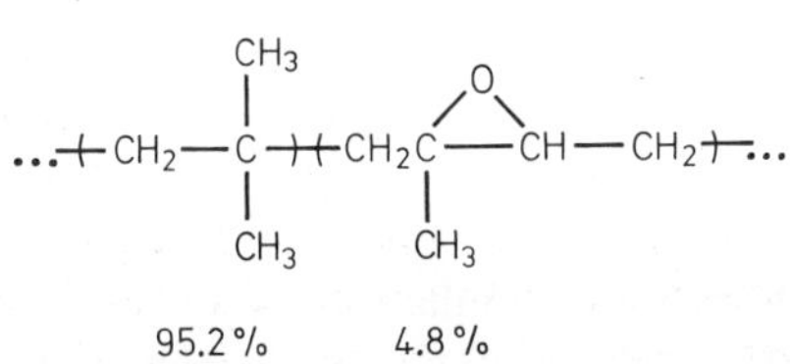

(15-99)

With Nordel epoxide and BF_3 the grafting efficiency exceeded 85% for THF, but the resulting polymer was insoluble due the copolymerization of epoxide rings (i.e., grafting 'through'). With stronger Lewis acids (PF_5), that are able to initiate the polymerization of THF on their own, grafting efficiency was lower and more than 70% of THF homopolymer was formed.

Grafting from butyl rubber epoxide did not occur, and it was suggested that the high steric hindrance in the trisubstituted epoxide prevented grafting.

Grafting with a well defined number of active sites has also been performed. It was established earlier that 2-phenyl-1,3-dioxolan-2-ylium salts initiate the polymerization of THF by direct addition and proceed with opening of the C-4/O-3 bond in the dioxolenium cation [151]. Thus, Okada et al. have prepared dioxolenium salts coupled to phenyl rings in polystyrene [152]:

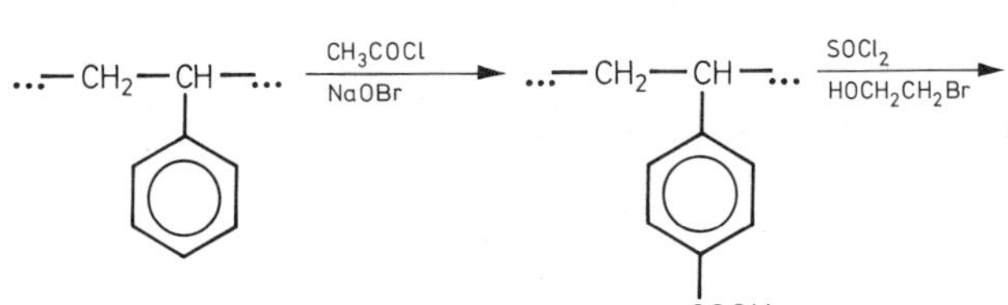

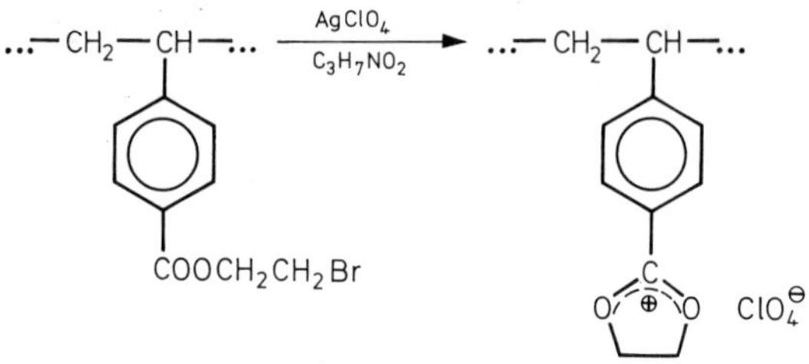
(15-100)

and used these salts, located on the macromolecules as initiating sites.

15.4.2 Grafting onto the Chains

This method of grafting requires the addition of a living cationic polymer to a pre-formed polymer having several nucleophilic centers in the chain. An example is epoxidized butyl rubber (see above). It was observed that the model of the reactive unit, namely the trisubstituted epoxide 2,3-epoxy-2,4,4-trimethylpentane, terminates THF polymerization [153]:

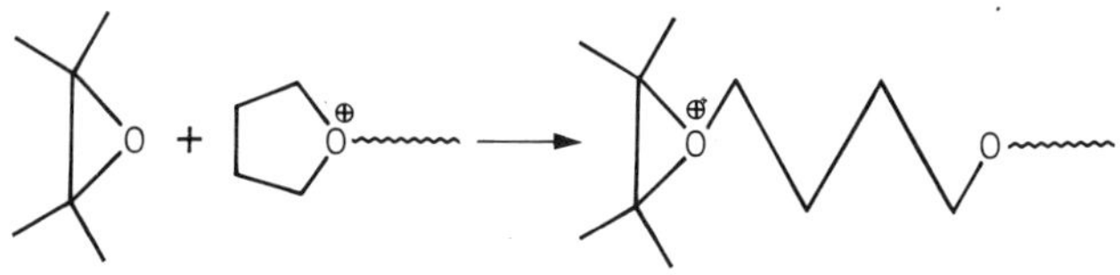
(15-101)

Formation of similar units in the chains occurs when butyl rubber containing some epoxy groups is mixed with living polyTHF. Some gel is also formed, indicating that grafting 'from', 'onto' and 'through' proceed simultaneously.

With more nucleophilic active sites only grafting 'onto' is possible. In this way Franta et al. grafted polyTHF onto poly(N,N-dimethylaminostyrene) and poly(2-vinylpyridine) [154]. The grafting process was in both cases quantitative, yielding polymeric quaternary ammonium salts:

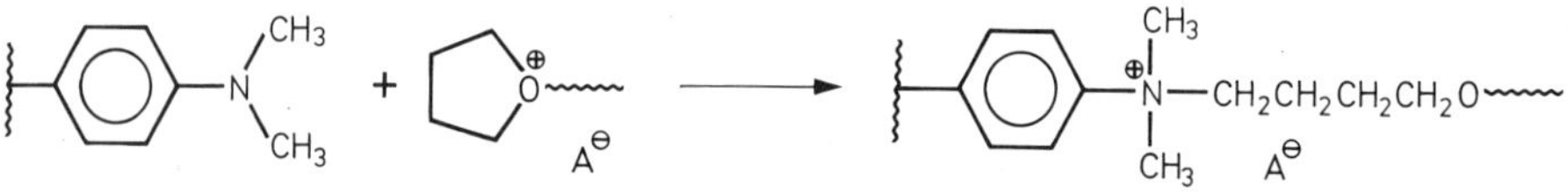
(15-102)

Random copolymerization of styrene and N,N-dimethylaminostyrene is less effective because some homopolymer of THF was also found among the products.

In a similar way Richards et al. have grafted polyTHF onto poly(4-vinyl pyridine) [155]:

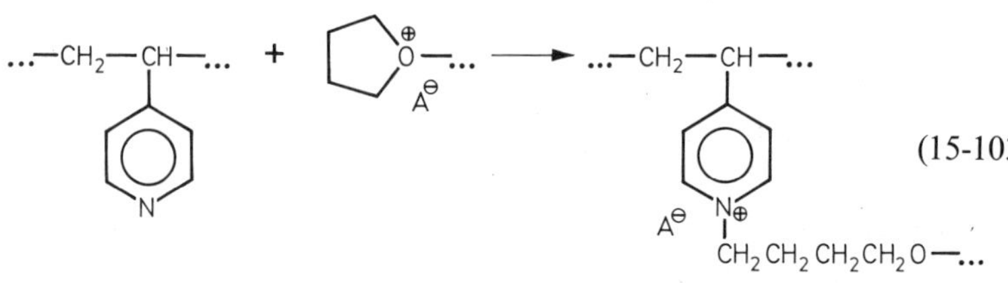
(15-103)

Due to the low solubility of poly(4-vinyl pyridine) in THF, a block copolymer of 4-vinyl pyridine with styrene has been used ($\overline{M}_n = 13{,}500$). Living polyTHF ($\overline{M}_n = 3{,}300$) has been grafted with efficiencies higher than 95%. About 20% of the pyridine units were consumed, yielding a graft copolymer with 80% polyTHF. No further data on this product were reported.

Using the grafting onto technique Munir and Goethals bound polyTBA to silica [156]. Two routes were used: the first, in which silica was reacted with 3-aminopropyl triethoxysilane, followed by reaction with living polyTBA, and the second one, in which polyTBA was reacted with 3-aminopropyl triethoxysilane, followed by reaction with silica:

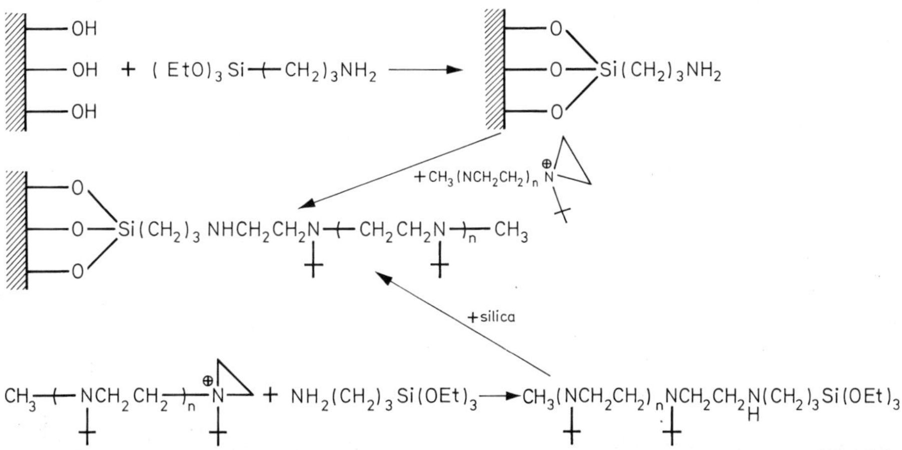

(15-104)

The second method gave higher grafting efficiencies than the first (15% versus 10%). In the second method the unreacted polymer can be reused for grafting, since it still contains siloxane end-groups. In the first method the polymer self-terminates and cannot further be applied. A higher grafting efficiency was found for polymer with $\overline{M}_n = 3{,}000$ than for polymers with $\overline{M}_n = 5{,}000$ or $\overline{M}_n = 10{,}000$. This is due to the small mean pore size of the silica used (80 Å).

Asami has reported grafting of living polyTHF on polymer with pendant alkali phenolates [157]. Poly(p-hydroxystyrene-*co*-styrene) gave 100% grafting efficiency at 0 °C with $Na^{\oplus}$ countercations; extraction with isopropanol gave 30%–78% copolymer the rest being insoluble gel.

$\overline{M}_n$ of the graft copolymer was estimated from $\overline{M}_n = 2 \cdot 10^5$ to $9 \cdot 10^5$ (universal calibration GPC). The number of branches was calculated as 10 to 40 per macromolecule. The branch lengths were measured independently for polyTHF ($\overline{M}_n = 2 \cdot 10^3$ to $4 \cdot 10^3$).

15.4.3 Grafting Through

This type of grafting has been observed for living polyTHF and polymers containing epoxide rings, e.g. epoxidized polymer. The living and growing polyTHF reacts at a certain stage with the oxirane ring within the polymer chain, and the active species thus formed adds further THF molecules [158]. Therefore, grafting through comprises two consecutive processes: first grafting onto and then grafting from.

When the next chain reacts by its epoxide group, with polyTHF growing from another chain, crosslinking becomes possible, leading to gel. This has been observed experimentally. Thus, one of the difficulties in the copolymerization of THF with epoxide-containing polymers is the stopping of the process before the second epoxide-incorporation takes place. This requires stopping the polymerization of THF at a low conversion. Otherwise, the soluble fraction content will be limited. Thus, with e.g. epoxidized Nordel the soluble fraction content was 15% of the total product. The n-heptane soluble graft copolymer contained 85 wt% polyTHF [158].

If each hydrocarbon chain contains only one epoxide ring, gelation would be depressed and grafts with larger numbers of side chains could form. Preparation of such well-defined reactive polymeric chains can be accomplished by introducing the active group via initiation or termination reactions. Such macromolecular monomeric species (macromonomers) are discussed in the following Section.

15.4.4 Macromonomers

An alternate route to graft copolymers is by synthesis of macromolecules possessing exactly one polymerizable group in the chain and their subsequent polymerization, i.e. grafting through. Such reactive polymers are macromolecular monomers, i.e. Macromers® as abbreviated by Milkovitch [159], who was the first to call the attention of the scientific community to the importance of this field although macromonomers have been prepared and copolymerized as early as 1962 [160].

Macromonomers lead to comb-like, graft and star-shaped polymers. The correct choice of elements constituting the backbone and branches allows the synthesis of products meeting the criteria of a thermoplastic elastomer [159]. This has been achieved by the use of methacrylic macromonomers with polystyrene chains and their copolymerization with butyl acrylate. The polystyrene side chains give the hard segments equivalent to crosslinking sites and the polyacrylate backbone is the soft segment joining the hard clusters.

Macromonomers were recently prepared by different techniques, including living cationic ring-opening and anionic polymerizations, transfer processes, polyaddition reactions and transformation of monofunctional polymers by introducing polymerizable groups. In nearly all macromonomers, the reactive groups have acrylate-type or *p*-substituted styrene-type structures. Therefore, predominantly radical and in some instances anionic processes were used for homo- and copolymerization of macromonomers. Cationic ring-opening polymerization has been used only once for 2-oxazoline end-groups [161].

There are three methods of synthesis of macromonomers. The reactive end-group can be introduced by initiation, by transfer and by a terminating agent. These methods are described below.

Initiation with oxocarbenium cations

One of us with Franta et al.[162] used methacryloyl chloride and silver hexafluoroantimonate to initiate the polymerization of THF with subsequent termination with sodium phenolate:

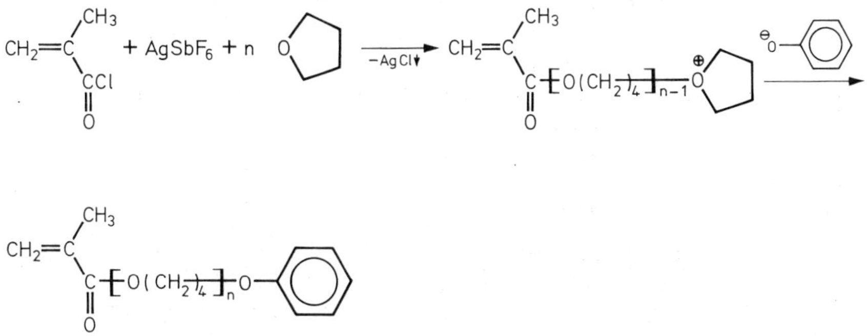

The molecular weights of macromonomers were in the range of $\overline{M}_n = 1{,}000$ to 5,000 with polydispersities: $\overline{M}_w/\overline{M}_n = 1.15$. The methacrylate end-groups were determined by double-bond analysis (IR and NMR).

Introduction of reactive end-groups by termination

This method requires not only strictly living conditions but also the quantitative conversion of the living growing centers into reactive end-groups. At present, living polymerization of THF and 1-t-butylaziridine have been used for this purpose, e.g.:

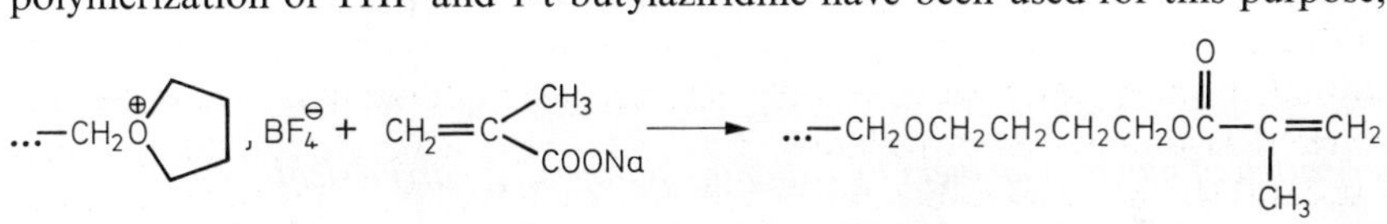

(15-107)

Table 15.13. Macromonomers by end-capping with compounds providing reactive groups

No.	Monomer	Initiator	End-capping agent	Conditions	Properties	Ref.
1	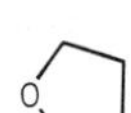	$Et_3O^{\oplus}BF_4^{\ominus}$	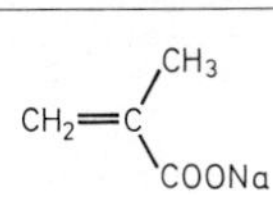	0 °C, 40 min	3% yield, $\overline{M}_n$ $\overline{M}_n = 9000$ F (functionality) = 1.0	[163]
2	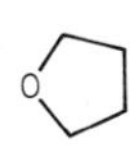	$Et_3O^{\oplus}BF_4^{\ominus}$	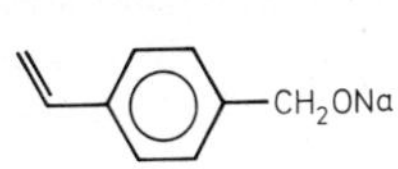	0 °C, 40 min	5% yield, $\overline{M}_n = 6000$ F = 1.0	[164]
3		$Et_3O^{\oplus}BF_4^{\ominus}$	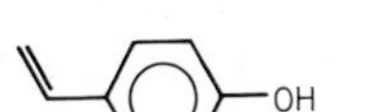 + NaOtBu	0 °C, 60 min	4% yield, $\overline{M}_n = 10000$ $\overline{M}_w/\overline{M}_n = 1.1$, F = 0.95	[165]
4		$CH_3OSO_2CF_3$	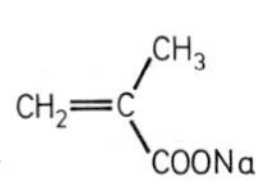	15 °C, 30 min, THF/HMPT	96% yield, $\overline{M}_n = 5000$ F = 1.0	[166]

The reaction conditions and product properties are given in Table 15.13 for different macromonomers.

Macromonomers by transfer reactions

The reactive end-groups can be introduced by intentionally added transfer agents or by spontaneous transfer. The latter was observed, e.g. in the polymerization of 2,2-dimethyloxetane initiated with $(C_2H_5)_3O^{\oplus}PF_6^{\ominus}$ [167]. A low-molecular-weight polymer ($\overline{M}_n = 1{,}000$) containing mainly macromolecules with terminal double bonds was formed in 30–90% yield. The reactivity of this macromonomer is unknown, however:

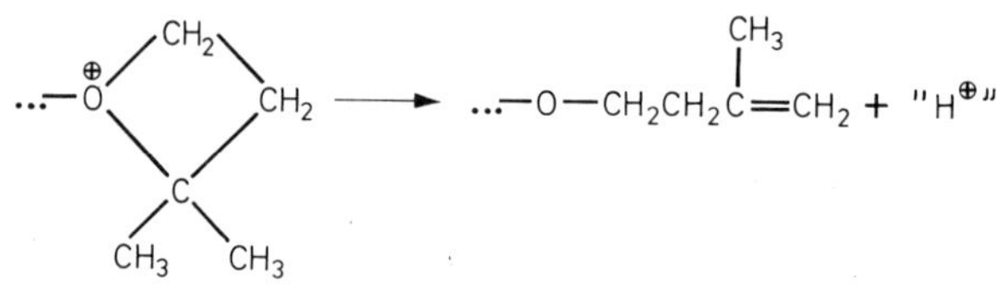
(15-108)

The cationic polymerization of THF in the presence of acrylic or methacrylic anhydrides yields polyTHF with two acrylic or methacrylic end-groups (cf. also Chap. 5) [168]:

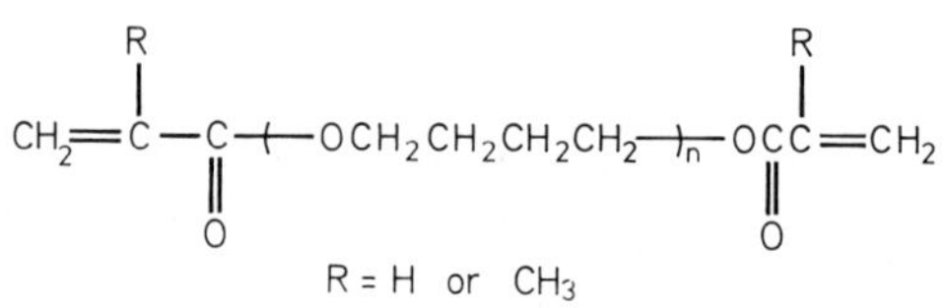
(15-109)

Polymerization of macromonomers by cationic ring-opening polymerization

Macromonomer has been prepared from 2-(p-hydroxyphenyl)-2-oxazoline and ethylene oxide using n-butyl lithium to initiate the anionic polymerization of EO [161]. The living chains were terminated with CH_3I or H_2O yielding macromonomers with $\overline{M}_n = 1{,}010$ and $\overline{M}_n = 1{,}930$, respectively:

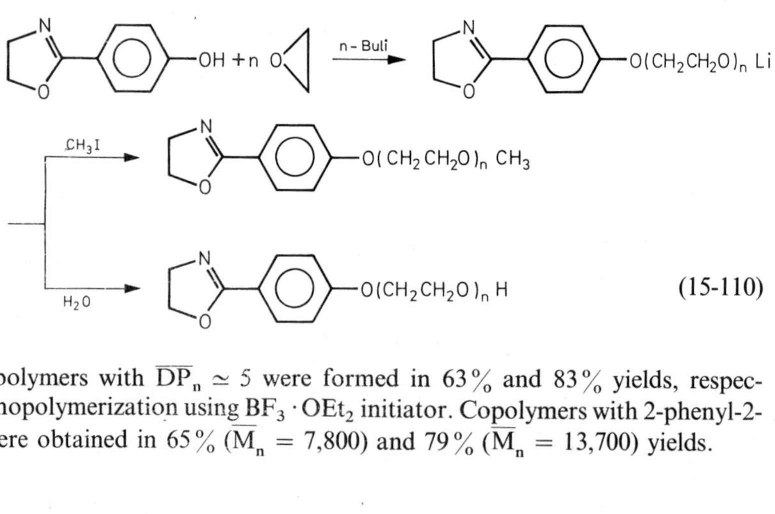
(15-110)

Comb-like polymers with $\overline{DP}_n \simeq 5$ were formed in 63% and 83% yields, respectively, inhomopolymerization using $BF_3 \cdot OEt_2$ initiator. Copolymers with 2-phenyl-2-oxazoline were obtained in 65% ($\overline{M}_n = 7{,}800$) and 79% ($\overline{M}_n = 13{,}700$) yields.

15.5 References

1. Mayo, F. R., Lewis, F. M.: J. Amer. Chem. Soc. *6* , 1862 (1941)
2. Kelen, T., Tüdős, F.: J. Macromol. Sci., Chem. *A9*, 1 (1975)
3. Tüdős, F. et al.: J. Macromol. Sci., Chem. *A10*, 1513 (1976)
4. Tüdős, F., Kelen, T.: J. Macromol. Sci., Chem. *A16*, 1283 (1981)
5. Fineman, M., Ross, S. D.: J. Polym. Sci. *5*, 259 (1950)
6. Ezrielev, A. J., Brokchina, E. L., Roskin, E. S.: Vysokomol. Soed. *A11*, 1970 (1969)
7. Merz, E., Alfrey T., Goldfinger, G.: J. Polym. Sci. *1*, 75 (1946)
8. Matyjaszewski, K.: Europ. Pol. J, *19*, 787 (1983)
9. Tüdős, F. et al.: 5th Intern. Symp. on Cationic Polymerization, Kyoto 1980, p. 171
10. Ito, K., Yamashita, Y.: J. Polym. Sci., *A3*, 2166 (1965)
11. Bucquoye, M., Goethals, E. J.: Europ. Polym. J. *14*, 323 (1978)
12. Bucquoye, M.: PhD Thesis, Ghent 1980
13. Geller, N. M., Kropachev, V. A., Dolgoplosk, B. A.: Vysokomol. Soed. *9*, 575 (1967)
14. Alferova, L. V., Kropachev, V. A., Dolgoplosk, B. A.: Vysokomol. Soed. *7*, 1065 (1965)
15. Okamura, S., Miyake, M., Shimazaki, N.: Enc. of Pol. Sci. and Techn., Mark. H., Bikales, N., Eds., Wiley-Interscience, New York 1969, Vol. 9. p. 668
16. O'Driscoll, K. F.: J. Polym. Sci. *A2*, 4201 (1964)
17. Goethals, E. J.: J. Polym. Sci., Polym. Symp. *56*, 271 (1976)
18. Goethals, E. J. et al.: Polymer J. *12*, 571 (1980)
19. Kagiya, T., et al.: J. Macromol. Sci., Chem. *A6*, 1631 (1972)
20. Fukui, K., Yonezawa, T., Nagata, C.: Bull. Chem. Soc. Jap. *27*, 423 (1954)
21. Aoki, S. et al.: J. Polym. Sci. A-1, *6*, 2585 (1968)
22. Kagiya, T., Sumida, Y., Inoue, T.: Polymer J. *1*, 312 (1970)
23. Yamashita, Y. et al.: J. Polym. Sci., A-1, *4*, 2121 (1966)
24. Hammond, J. H., Hooper, J. F., Robertson, W. G. P.: J. Polym. Sci., *A-1*, *9*, 265 (1971)
25. Blanchard, L. P., Singh, J., Baijal, M. D.: Can. J. Chem. *44*, 2679 (1966)
26. Malhotra, S. L., Blanchard, L. P.: J. Macromol. Sci.-Chem., *A9*, 1485 (1975)
27. Blanchard, L. P., Asselin, J. C., Malhotra, S. L.: Polymer J. *7*, 326 (1975)
28. Entelis, S. G., Korovina, G. V.: Makromol. Chem. *175*, 1253 (1974)
29. Wittmer, P.: Adv. Chem. Ser. *99*, 140 (1971)
30. Durgaryan, A. A.: Vysokomol. Soed. *8*, 790 (1966)
31. Lowry, G. G.: J. Polym. Sci. *45*, 169 (1960)
32. Yamashita, Y. et al.: Makromol. Chem. *117*, 242 (1968)
33. Kubisa, P., Penczek, S.: J. Macromol. Sci., Chem. *7*, 1509 (1973)
34. Garrido, L. et al.: J. Polym. Sci., Polym. Chem. Ed. *20*, 3377 (1982)
35. Enikolopyan, N. S. et al.: Vysokomol. Soed. *A19*, 1924 (1977)
36. Rozenberg, B. A., Efremova, A. I., Enikolopyan, N. S.: Vysokomol. Soed. *14*, 2576 (1972)
37. Efremova, A. I. et al.: Dokl. Akad. Nauk SSSR *190*, 872 (1970)
38. Ludvig, E. B. et al.: Dokl. Akad. Nauk SSSR *186*, 1351 (1969)
39. Binet, R., Harvey, P. E., Leonard, J.: Int. Symp. on Cationic Polymerization, Rouen 1973, 1973, paper C-1
40. Yamashita, Y. et al.: Makromol. Chem. *135*, 75 (1970)
41. Szymański, R.: PhD Thesis, Łódź, 1980
42. Fejgin, J., Penczek, S.: J. Polym. Sci. *84*, 615 (1966)
43. Yokoyama, Y., Okada, M., Sumitomo, H.: Polymer J. *11*, 629 (1979)
44. Woodward, R. B., Hoffman, R.: J. Amer. Chem. Soc. *87*, 395 (1965)
45. Mathieu, J., Rassat, A.: Tetrahedron *30*, 1753 (1974)
46. Yamashita, Y. et al.: Polymer J. *3*, 389 (1971)
47. Yamashita, Y. et al.: Macromolecules *2*, 613 (1969)
48. Khomyakov, A. K., Lyudvig, Ye. B., Shapet'ko, N. I.: Vysokomol. Soed. *A22*, 2649 (1980)
49. Gumbs, R. et al.: Macromolecules *2*, 77 (1969)
50. Ito, K., Umehara, K., Yamashita, Y.: Kogyo Kagaku Zasshi *70*, 2040 (1967)
51. Hohr, L., Cherdron, H., Kern, W.: Makromol. Chem. *52*, 59 (1962)
52. Okada, M., Yamashita, Y., Ishii, Y.: Makromol. Chem. *94*, 181 (1966)
53. Higashimura, T. et al.: J. Polym. Sci. A-1, *5*, 1937 (1967)

54. Miki, T., Higashimura, T., Okamura, S.: J. Polym. Sci. A-1, *8*, 505 (1970)
55. Higashimura, T., Masuda, T., Okamura, S.: J. Polym. Sci. A-1, *7*, 1115 (1969)
56. Yamashita, Y., Okada, M., Hirota, M.: Makromol. Chem. *122*, 284 (1969)
57. Plesch, P. H.: private communication
58. Okada, M. et al.: Polymer J. *1*, 1 (1970)
59. Matyjaszewski, K., Penczek, S.: unpublished
60. Bartlett, P. D., Nazaki, K.: J. Amer. Chem. Soc. *68*, 1495 (1946)
61. Mayo, F. R., Walling, C.: Chem. Rev. *46*, 191 (1950)
62. Gaylord, N.: Macromolecular Chemistry-8 (IUPAC, Helsinki 1972) Butterworths, London, p. 35
63. Iwatsuki, S., Yamashita, Y.: Progr. Pol. Sci. Jap. *2*, 1 (1971)
64. Stille, J. K. et al.: J. Macromol. Sci. *19*, 745 (1975)
65. Stille, J. K., Chung, D. C.: Macromolecules *8*, 114 (1975)
66. Hall, H. K., Jr. et al.: ACS Symp. Series *59*, 285 (1977)
67. Stille, J. K., Cassidy, P.: J. Polym. Sci., Polym. Letters *1*, 565 (1963)
68. Kunitake, T., Price, C.C.: J. Amer. Chem. Soc. *85*, 761 (1963)
69. Kagiya, T. et al.: J. Polym. Sci. *B*, *3*, 617 (1965)
70. Saegusa, T., Kobayashi, S., Kimura, Y.: Pure & Appl. Chem. *48*, 307 (1976)
71. Saegusa, T. et al.: ACS Symp. Series *59*, 345 (1977)
72. Saegusa, T.: Pure & Appl. Chem. *53*, 691 (1981)
73. Saegusa, T., Ikeda, H., Fujii, H.: Macromolecules *5*, 354 (1972)
74. Saegusa, T., Kobayashi, Kimura, Y.: Macromolecules *7*, 1 (1974)
75. Saegusa, T., Kimura, Y., Kobayashi, S.: 29th Annual Meeting of Chem. Soc. Jap. 1973
76. Saegusa, T. et al.: Macromolecules *7*, 956 (1974)
77. Saegusa, T., et al.: Macromolecules *7*, 546 (1974)
78. Saegusa, T., Kobyashi, S., Furukawa, J.: Macromolecules *8*, 703 (1975)
79. Saegusa, T. et al.: Macromolecules *8*, 259 (1975)
80. Beresfatskaya, T. Z., Nekrasov, A. V., Eltsefon, B. S.: Vysokomol. Soed. *A23*, 2581 (1981)
81. Saegusa, T. et al.: Macromolecules *9*, 724 (1976)
82. Saegusa, T. et al.: Macromolecules *10*, 791 (1977)
83. Rajendran, G. P., Mahadevan, V., Srinivasam, M.: Europ. Polym. J. *18*, 803 (1982)
84. Schmidt, D. L.: ACS Symp. Ser. *59*, 318 (1977)
85. Kawakami, Y., Miyata, K., Yamashita, Y.: Polym. J. *11*, 175 (1979)
86. Schmidt, D. L. et al.: J. Polym. Sci. *A-1*, *10*, 2951 (1972)
87. Goethals, E. J., Bouquet, G., Simonds, R.: IUPAC 6th International Symposium on Cationic Polymerization, Ghent 1983, p. 131
88. Ishida, S.: Bull. Chem. Soc. Jap. *33*, 924 (1960)
89. Berner, G., Reichert, K. H. W.: 1st International Symp. Ring-Opening Polymerization, Jablonna 1975, Preprints p. 98
90. Kagiya, T., Matsuda, T.: Polymer J. *3*, 398 (1971)
91. Soga, K., Ikeda, S.: J. Polym. Sci., Polym. Chem. Ed. *12*, 121 (1974)
92. Schaeffer, J.: Macromolecules *1*, 111 (1968)
93. Schaeffer, J., Kern, R., Katnik, R. J.: Macromolecules *1*, 107 (1981)
94. Hsieh, H. L.: J. Macromol. Sci., Chem. *A7*, 1525 (1973)
95. Saegusa, T. et al.: Polymer J. *11*, 463 (1979)
96. Ochsner, W., Reichert, K. H. W.: Makromol. Chem. *150*, 1 (1971)
97. Tsuda, T., Yamashita, Y.: Makromol. Chem. *99*, 297 (1966)
98. Malanga, M., Vogl, O.: J. Polym. Sci., Polym. Chem. Ed. *20*, 2033 (1982)
99. Szwarc, M.: Carbanions Living Polymers and Electron Transfer Processes, Wiley, New York 1968
100. Rempp, P., Franta, E., Herz, J.: ACS Symp. Ser. *166*, 59 (1981)
101. Stepanov, V. V. et al.: Vysokomol. Soed. *A13*, 821 (1976)
102. Higashimura, T., Teranishi, H., Sawamoto, M.: Polymer J. *12*, 393 (1980)
103. Pepper, D. C.: J. Polym. Sci., Polym. Symp. *50*, 51 (1975)
104. Bossaer, P. K. et al.: Europ. Polym. J. *13*, 489 (1977)
105. Seung, S. L. N., Young, R. N.: Polym. Bull. *1*, 481 (1979)
106. Seung, S. L. N., Young, R. N.: J. Polym. Sci., Polym. Letters *18*, 89 (1980)
107. Higashimura, T., Kishira, O.: Polymer J. *9*, 87 (1977)

108. Ikeda, I. et al.: Polymer J. *12*, 423 (1980)
109. Chwialkowska, W., Kubisa, P., Penczek, S.: Makromol. Chem. *183*, 753 (1982)
110. Rosenberg, B. A. et al.: Vysokomol. Soed. *6*, 2030 (1964)
111. Sims, D.: J. Chem. Soc. *1964*, 864
112. Vofsi, D., Tobolsky, A. V.: J. Polym. Sci. *A3*, 3261 (1965)
113. Munir, A., Goethals, E. J.: J. Polym. Sci., Polym. Chem. Ed. *19*, 1985 (1981)
114. Schacht, E. H., Goethals, E. J.: Makromol. Chem. *167*, 155 (1973)
115. Rasvodovskii, E. F. et al.: J. Macromol. Sci.-Chem. *A8*, 241 (1974)
116. Bruggeman, P., Goethals, E. J.: 1st Int. Symp. Ring-Opening Polym., Jablonna 1975, Preprints p. 111
117. Saegusa, T., Matsumoto, S., Hashimoto, Y.: Macromolecules *3*, 377 (1970)
118. Yamashita, Y., Okada, M., Hirota, M.: Angew. Makromol. Chem. *9*, 136 (1969)
119. Kucera, M., Bozek, F., Majerova, K.: Polymer *20*, 1013 (1979)
120. Kucera, M. et al.: Polymer *24*, 217 (1983)
121. Kubisa, P. et al.: 6th IUPAC Symposium on Cationic Polymerization, Ghent 1983
122. Litt, M., Herz, J., Turi, E.: Polymer Preprints *10*, (2), 905 (1969)
123. Lambert, J. L., Goethals, E. J.: Makromol. Chem. *133*, 289 (1970)
124. Goethals, E. J., Tezuka, Y.: Europ. Polym. J. *18*, 991 (1982)
125. Jaacks, V., Kern, W.: Makromol. Chem. *83*, 71 (1965)
126. Goethals, E. J., Munir, A., Bossaer, P.: Pure & Appl. Chem. *53*, 1753 (1981)
127. Yamashita, Y., Chiba, K.: Polymer J. *4*, 200 (1973)
128. Hoene, R., Reichert, K. H. W.: Makromol. Chem. *177*, 3545 (1976)
129. Olah, G. A. et al.: J. Amer. Chem. Soc. *84*, 2733 (1962)
130. Franta, E. et al.: J. Polym. Sci., Polym. Symp. *56*, 139 (1976)
131. Eckstein, Y., Dreyfuss, P.: J. Polym. Sci., Polym. Chem. Ed. *17*, 4115 (1979)
132. Richards, D. H., Kingston, S. B., Souel, T.: Polymer *19*, 806 (1978)
133. Richards, D. H., Kingston, S. B., Souel, T.: Polymer *19*, 68 (1978)
134. Berger, G., Levy, M., Vofsi, D.: J. Polym. Sci. *B4*, 183 (1966)
135. Yamashita, Y. et al.: Macromolecules *4*, 548 (1971)
136. Burgess, F. J. et al.: Polymer *18*, 719 (1977)
137. Burgess, F. J., Richards, D. H.: Polymer *17*, 1020 (1976)
138. Burgess, F. J. et al.: Polymer *18*, 726 (1977)
139. Burgess, F. J. et al.: Polymer *18*, 733 (1977)
140. Zilliox, J. G. et al.: IUPAC Meeting, Mainz 1979, Vol. 1 p. 56
141. Collomb, J., Gandini, A., Cheradame, H.: Makromol. Chem. Rapid Commun. *1980*, 489
142. Rempp, P., Franta, E.: Vysokomol. Soed. *21*, 2529 (1979)
143. Kucera, M., Bozek, F., Majerova, K.: Polymer *23*, 207 (1982)
144. Saegusa, T., Ikeda, M.: Macromolecules *6*, 805 (1973)
145. Dreyfuss, P., Kennedy, J. P.: J. Polym. Sci., Polym. Symp. *56*, 129 (1976)
146. Dreyfuss, P., Adaway, T., Kennedy, J. P.: J. Appl. Polym. Sci., Appl. Polym. Symp. *30*, 183 (1977)
147. Lehmann, J., Dreyfuss, P.: Adv. Chem. Ser. *176*, 587 (1979)
148. Eckstein, Y., Dreyfuss, P.: J. Polym. Sci., Polym. Chem. Ed. *18*, 1799 (1980)
149. Saegusa, T., Yamada, A., Kobayashi, S.: Polymer J. *11*, 53 (1979)
150. Dreyfuss, P., Kennedy, J. P.: J. Appl. Polym. Sci., Appl. Polym. Symp. *30*, 153 (1977)
151. Yamashita, Y. et al.: Polymer J. *2*, 43 (1971)
152. Okada, M., Sumitomo, H., Kokazawa, T.: Makromol. Chem. *162*, 285 (1981)
153. Dreyfuss, P., Kennedy, J. P.: J. Appl. Polym. Sci. Appl. Polym. Symp. *30*, 165 (1977)
154. Dondos, A. et al.: Makromol. Chem. *179*, 2549 (1978)
155. Cunliffe, A. V. et al.: Polymer *22*, 101 (1981)
156. Munir, A., Goethals, E. J.: IUPAC Symposium, Strasbourg 1981, Vol. I, p. 299
157. Asami, R. et al.: Polymer J. *13*, 141 (1981)
158. Dreyfuss, P., Kennedy, J. P.: J. Appl. Polym. Sci., Appl. Polym. Symp. *30*, 179 (1977)
159. Milkovitch, R.: ACS Symp. Ser. *166*, 41 (1981)
160. Greber, G., Reese, E.: Makromol. Chem. *55*, 96 (1962)
161. Kobayashi, S. et al.: Polym. Bull. *9*, 169 (1983)
162. Sierra-Vargas, J. et al.: Polym. Bull. *3*, 83 (1980)
163. Takaki, M., Asami, R., Kuwabara, T.: Polym. Bull. *7*, 521 (1982)

164. Asami, R. et al.: Polym. Bull. *2*, 713 (1980)
165. Asami, R. et al.: Polymer J. *15*, 139 (1983)
166. Goethals, E. J., Vlegels, M. A.: Polym. Bull. *4*, 521 (1981)
167. Kops, J., Spanggaard, H.: Macromolecules *15*, 1225 (1982)
168. Kress, H. J., Heitz, W.: Makromol. Chem., Rapid Commun. *2*, 427 (1981)

Acknowledgement: This work was accomplished mostly within a program of the Polish Academy of Sciences. The senior author of this monograph, S. Penczek, wishes to thank Professor R. C. Schulz for offering him a visiting professorship in 1980 at the Institute of Organic Chemistry, University of Mainz, where the work on this book was started and considerably advanced. Particular thanks are extended to Professors R. C. Schulz and H. Ringsdorf as well as to their coworkers for friendly help and numerous discussions throughout the entire stay in Mainz.

Editor: J. P. Kennedy
Received September 3, 1984

Author Index Volumes 1–68/69

Subject Index